Praise for *Mobile Game Development with Unity*

"If you want to build any kind of game for mobile platforms, you've got to take a look at Unity. This book is an excellent, thorough, and seriously fun guide to putting together gameplay in one of the best game engines out there for indie developers."

—*Adam Saltsman, Creator of* Canabalt *and* Overland *at Finji*

"The best way to learn how to use a game engine is by getting your hands dirty and building your own projects. In this book, Paris and Jon guide you through the creation of two radically different games, giving you invaluable hands-on experience with a wide range of Unity's features."

—*Alec Holowka, Lead Developer of* Night in the Woods *and* Aquaria *at Infinite Ammo*

"This book changed my life. I now feel inner peace, and I'm pretty sure I can see through time."

—*Liam Esler, Game Developers' Association of Australia*

Unity移动游戏开发（影印版）

Mobile Game Development with Unity

Jon Manning,
Paris Buttfield-Addison 著

Beijing • Boston • Farnham • Sebastopol • Tokyo

O'Reilly Media, Inc.授权东南大学出版社出版

南京 东南大学出版社

图书在版编目(CIP)数据

Unity移动游戏开发:英文/(美)乔恩·曼宁,(美)帕里斯·巴特菲尔德著.—影印本.—南京:东南大学出版社,2018.2

书名原文:Mobile Game Development with Unity

ISBN 978-7-5641-7527-6

Ⅰ.①U… Ⅱ.①乔… ②帕… Ⅲ.①游戏程序-程序设计-英文 Ⅳ.①TP317.6

中国版本图书馆CIP数据核字(2017)第296217号

图字:10-2015-258号

Unity移动游戏开发(影印版)

出版发行:东南大学出版社
地　　址:南京四牌楼2号　　邮编:210096
出 版 人:江建中
网　　址:http://www.seupress.com
电子邮件:press@seupress.com
印　　刷:常州市武进第三印刷有限公司
开　　本:787毫米×980毫米　16开本
印　　张:29
字　　数:501千字
版　　次:2018年2月第1版
印　　次:2018年2月第1次印刷
书　　号:ISBN 978-7-5641-7527-6
定　　价:99.00元

本社图书若有印装质量问题,请直接与营销部联系。电话(传真):025-83791830

Table of Contents

Part III. Building a 3D Game: Space Shooter

Part IV. Advanced Features

Preface

Welcome to *Mobile Game Development with Unity*! In this book, we'll take you from nothing all the way up to building two complete games, and teach you both beginning and advanced Unity concepts and techniques along the way.

The book is split into four parts.

Part I introduces the Unity game engine, and explores the basics, including how to structure games, graphics, scripting, sounds, physics, and particle systems. Part II then takes you through the construction of a full 2D game with Unity, involving a gnome on a rope trying to get treasure. Part III explores the construction of a full 3D game with Unity, including spaceships, asteroids, and more. Part IV explores some of the more advanced features of Unity, including lighting, the GUI system, extending the Unity editor itself, the Unity asset store, deploying games, and platform-specific features.

If you have any feedback, please let us know! You can email us at *unitybook@secretlab.com.au*.

Resources Used in This Book

Supplemental material (art, sound, code examples, exercises, errata, etc.) is available for download at *http://secretlab.com.au/books/unity*.

Audience and Approach

This book is designed for people who want to build games but don't have any previous game development experience.

Unity supports a few different programming languages. We'll be using C# in this book. We will assume that you know how to program in a relatively modern language, but it doesn't have to be recent programming experience as long as you're somewhat comfortable with the basics.

The Unity editor runs on both macOS and Windows. We use macOS, so the screenshots shown throughout the book are taken from there, but everything we cover is identical on Windows, with one small exception: building iOS games with Unity. We'll explain when we get to it, but you can't do it on Windows. Android works fine on Windows though, and macOS can build for both iOS and Android.

The book takes the approach that you need to understand the basics of game design, as well as Unity itself, before you build some games, so we teach you that in Part I. Once that's done, parts II and III explore the construction of a 2D game and a 3D game, respectively, and then in Part IV we follow up with all the other Unity features that you should know about.

We will assume that you're fairly confident and comfortable navigating your operating system, and using your mobile devices (whether they be iOS or Android).

We won't be covering the creation of art or sound assets for your games, although we do supply assets for the two games you build through this book.

Conventions Used in This Book

The following typographical conventions are used in this book:

Italic
: Indicates new terms, URLs, email addresses, filenames, and file extensions.

`Constant width`
: Used for program listings, as well as within paragraphs to refer to program elements such as variable or function names, databases, data types, environment variables, statements, and keywords.

`Constant width bold`
: Shows commands or other text that should be typed literally by the user.

`Constant width italic`
: Shows text that should be replaced with user-supplied values or by values determined by context.

This icon signifies a tip or suggestion.

This element signifies a general note.

This icon indicates a warning or caution.

Using Code Examples

Supplemental material (code examples, exercises, errata, etc.) is available for download at *http://secretlab.com.au/books/unity*.

This book is here to help you get your job done. In general, if example code is offered with this book, you may use it in your programs and documentation. You do not need to contact us for permission unless you're reproducing a significant portion of the code. For example, writing a program that uses several chunks of code from this book does not require permission. Selling or distributing a CD-ROM of examples from O'Reilly books does require permission. Answering a question by citing this book and quoting example code does not require permission. Incorporating a significant amount of example code from this book into your product's documentation does require permission.

We appreciate, but do not require, attribution. An attribution usually includes the title, author, publisher, and ISBN. For example:

"*Mobile Game Development with Unity* by Jonathon Manning and Paris Buttfield-Addison (O'Reilly). Copyright 2017 Jon Manning and Paris Buttfield-Addison, 978-1-491-94474-5."

If you feel your use of code examples falls outside fair use or the permission given above, feel free to contact us at *permissions@oreilly.com*.

O'Reilly Safari

Safari (formerly Safari Books Online) is a membership-based training and reference platform for enterprise, government, educators, and individuals.

Members have access to thousands of books, training videos, Learning Paths, interactive tutorials, and curated playlists from over 250 publishers, including O'Reilly Media, Harvard Business Review, Prentice Hall Professional, Addison-Wesley Professional, Microsoft Press, Sams, Que, Peachpit Press, Adobe, Focal Press, Cisco Press, John Wiley & Sons, Syngress, Morgan Kaufmann, IBM Redbooks, Packt, Adobe Press, FT Press, Apress, Manning, New Riders, McGraw-Hill, Jones & Bartlett, and Course Technology, among others.

For more information, please visit *http://oreilly.com/safari*.

How to Contact Us

Please address comments and questions concerning this book to the publisher:

O'Reilly Media, Inc.
1005 Gravenstein Highway North
Sebastopol, CA 95472
800-998-9938 (in the United States or Canada)
707-829-0515 (international or local)
707-829-0104 (fax)

We have a web page for this book, where we list errata, examples, and any additional information. You can access this page at *http://bit.ly/Mobile-Game-Dev-Unity*.

To comment or ask technical questions about this book, send email to *bookquestions@oreilly.com*.

For more information about our books, courses, conferences, and news, see our website at *http://www.oreilly.com*.

Find us on Facebook: *http://facebook.com/oreilly*

Follow us on Twitter: *http://twitter.com/oreillymedia*

Watch us on YouTube: *http://www.youtube.com/oreillymedia*

Acknowledgments

Jon and Paris wish to thank their fabulous editors, especially Brian MacDonald (@bmac_editor (*https://twitter.com/bmac_editor*)) and Rachel Roumeliotis (@rroumeliotis (*https://twitter.com/rroumelio tis*)) for their work in bringing this book to fruition. Thanks for all the enthusiasm! Thanks also to the fabulous staff at O'Reilly Media, for making writing books such a pleasure.

Thanks also to our families for encouraging our game development, as well as all of MacLab and OSCON (you know who you are) for encouragement and enthusiasm. Thanks particularly to our fabulous tech reviewer, Dr. Tim Nugent (@the_mcjones (*https://twitter.com/the_mcjones*)).

PART I

The Basics of Unity

This book covers much of what you need to know to effectively build mobile games using the Unity game engine. The three chapters in this first part of the book introduce Unity, take you on a tour of the application, and discuss how programming works in Unity, using the C# programming language.

CHAPTER 1

Introducing Unity

To kick off our exploration of the Unity game engine, we'll start with the basics: what Unity is, what it's useful for, and how to get it. At the same time, we'll set up some useful constraints for the subject material we're looking at in this book; after all, you're holding a book that claims to be about mobile development, not *all* development. Such a book would be much heavier, or would make your reading software crash. We aim to spare you this misfortune.

Hello, Book

Before we dive into Unity itself, let's take a closer look at what we're talking about here: the field of mobile games.

Mobile Games

So, what is a mobile game, and how is it different from any other sort of game? More practically, how do these differences factor into your decisions when you're both designing and later implementing a game?

Fifteen years ago, a mobile game was likely to be one of two things:

- An incredibly simple game, with minimal interactions, graphics, and complexity

- A much more complex affair, available only on specialized mobile gaming consoles, and created by companies with access to expensive dev kits for said mobile gaming consoles

This split was the result of both hardware complexity and distribution availability. If you wanted to make a game that was in any way complex (and by *complex* we mean featured the incredible ability to have more than one thing moving on the screen at a time), you needed the more advanced computing power available only on expensive portable consoles, like Nintendo's handheld devices. Because the console owners also owned the distribution channels for the games, and wanted to have a high degree of control, getting permission to make games for more capable hardware became a challenge.

However, as more powerful hardware became cheaper over time, more options opened up for developers. In 2008, Apple made its iPhone available to software developers, and in the same year Google's Android platform became available. Over the years, iOS and Android have become extremely capable platforms, and mobile games are the most popular video games in the world.

These days, a mobile game is typically one of three things:

- A simple game, with carefully chosen interactions, graphics, and controlled complexity, because the game design was best supported by these facets
- A much more complex affair, available for anything ranging from specialized mobile game consoles to smartphones
- A mobile port of a game that debuted on a console or PC

You can use Unity to do all three of these; in this book, we'll be concentrating on the first approach. After exploring Unity and how it's used, we'll step through the creation of two games that fit those facets.

Hello, Unity

Now that we've elaborated a bit on what we're trying to make, let's talk about what we're going to make it *with*: the Unity game engine.

What's Unity For?

Over the years, Unity's focus has been on *democratizing game development*—that is, allowing anyone to make a game, and to make it available in as many places as possible. However, no single software package is perfect for all situations, and it's worth knowing what Unity is most suitable for, and when you should consider a different software package.

Unity is particularly great in situations like these:

When you're building a game for multiple devices.
: Unity's cross-platform support may be the best in the industry, and if you want to build a game that runs on multiple platforms (or even just multiple *mobile* platforms), Unity can be the best way to go about it.

When speed of development is important.
: You *could* spend months developing a game engine that contains the features you need. Or, you could use a third-party engine, like Unity. To be fair, there are other engines that exist, like Unreal (*https://www.unrealengine.com*) or Cocos2D (*http://www.cocos2d.org*); however, this leads us into the next point.

When you need a complete feature set, and don't want to build your own tools.
: Unity happens to contain a blend of features that are ideal for mobile games, and provides ways of creating your content that are very easy to use.

That said, there are some situations in which Unity is less useful. These include:

When you're building something that shouldn't redraw very often.
: Some kinds of games that aren't terribly graphically intense are less suited for Unity, because Unity's engine redraws the screen every frame. This is necessary for real-time animation, but uses more energy.

When you need very precise control over what the engine is doing.
: Unless you've purchased a source code license to Unity (which is possible, but less common), you don't have any way to control the lowest level behavior of the engine. That's not to say you don't have fine-grained control over Unity (and in most cases,

you don't need it anyway), but that there are certain things that are out of your hands.

Getting Unity

Unity is available for Windows, macOS, and Linux. Unity comes in three main flavors: *Personal*, *Plus*, and *Pro*.

At the time of this book's release (mid-2017), Linux support was experimental.

- The Personal edition is designed for solo developers who want to use Unity to make a game on their own. The Personal edition is free.
- The Plus edition is designed for solo developers or small teams. At the time of writing, the Plus edition costs $35 per month.
- The Pro edition is designed for small to large teams. At the time of writing, the Pro edition costs $125 per month.

Unity is also available in an Enterprise license, which is designed for large teams, but is not something that the authors have used much.

The features of the Unity software are largely the same across each edition. The main difference between the free and paid editions is that the Personal edition imposes a splash screen on your game, which shows the Unity logo. The free edition is only available to individuals or organizations that have a revenue of $100,000 a year or less, while the limit for Plus is $200,000. Plus and Pro also include slightly better services, such as priority build queues in Unity's Cloud Build service (discussed in more detail in "Unity Cloud Build" on page 423).

To download Unity, head to *https://store.unity.com*. Once you've installed it, you're ready to get going, and we'll see you in the next chapter.

CHAPTER 2
A Tour of Unity

Once you've got Unity installed, it's helpful to spend a bit of time learning your way around it. Unity's user interface is reasonably straightforward, but there are enough individual pieces that it's worth taking some time to review it.

The Editor

When you fire up Unity for the first time, you'll be asked to provide your license key, and you'll be asked to sign in to your account. If you don't have one, or if you don't want to sign in, you can skip the login.

If you don't log in, Cloud Builder and other Unity services will not be available to you. We'll look at Unity's services later in Chapter 17; we won't use them much when we're first starting out, but it's nice to be signed in.

Once you're past that point, you'll be taken to Unity's start screen, where you can choose to either create a new project, or open an existing one (Figure 2-1).

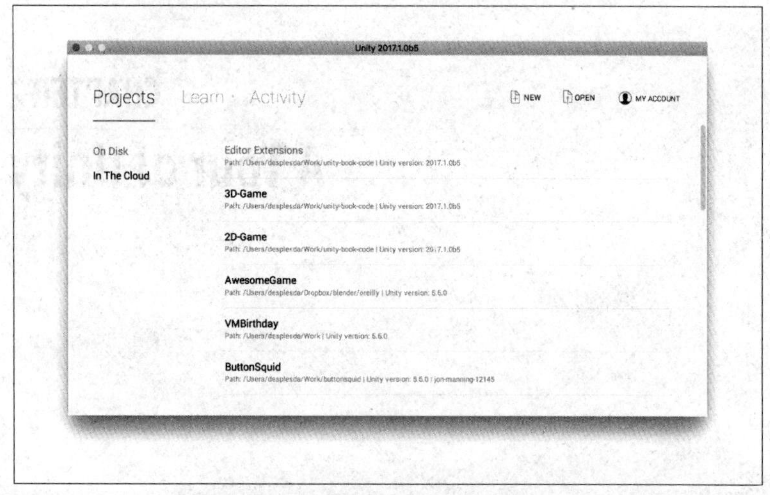

Figure 2-1. Unity's splash screen, when signed in

If you click on the New button at the top-right, Unity will ask you for some information for it to use while setting up the project (Figure 2-2), including the name of the project, where to save it, and whether you'd like Unity to create a 2D or 3D project.

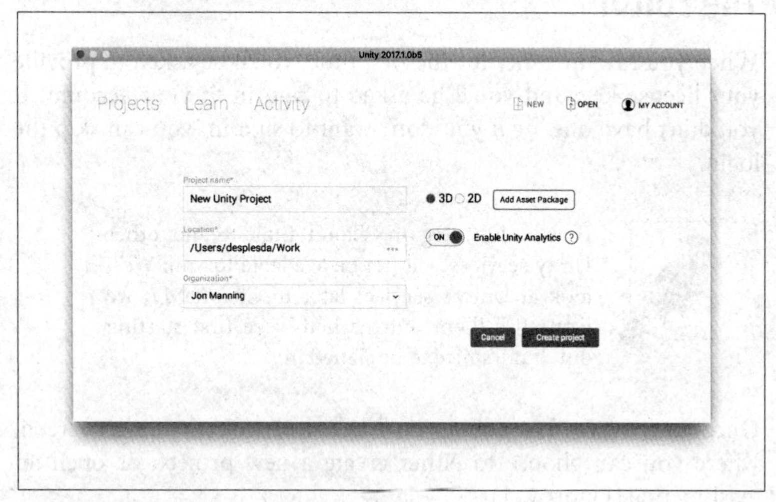

Figure 2-2. Creating a new project

The selection between 2D or 3D doesn't result in a huge degree of difference. 2D projects default to a side-on view, while 3D projects default to a 3D perspective. You can change the setting at any time, as well, in the Editor Settings inspector (see "The Inspector" on page 17 to learn how to access it).

When you click the "Create project" button, Unity will generate the project on disk for you and open it in the editor (Figure 2-3).

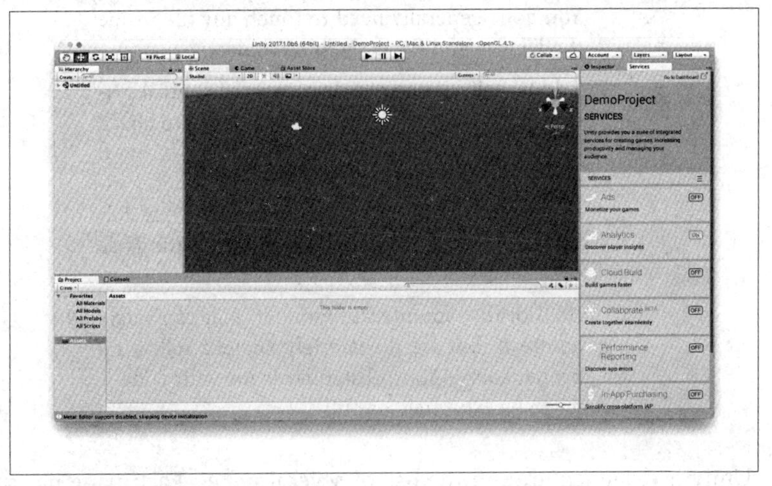

Figure 2-3. The editor

Project Structure

Unity projects are not single files; instead, they're folders, which contain three important subfolders: *Assets*, *ProjectSettings*, and *Library*. The *Assets* folder contains all of the files that your game uses: your levels, textures, sound effects, and scripts. The *Library* folder contains data that's internal to Unity, and the *ProjectSettings* folder contains files that contain your project's settings.

You don't generally need to touch any file inside the *Library* and *ProjectSettings* folders.

Additionally, if you're using a source control system like Git or Perforce, you don't need to check the *Library* folder into your repository, but you do need to check in the *Assets* and *ProjectSettings* folders in order to make sure that your collaborators have the same assets and settings as you.

If all of that sounded unfamiliar, you can safely ignore it, but we do strongly suggest following proper source control standards for your code—it can be extremely useful!

Unity is designed around the use of several *panes*. Each pane has a tab at its top left, which can be dragged around to change the layout of the application. You can also drag a tab out and make it a separate window. Not all of Unity's panes are visible by default, and as you build your game, you'll end up opening more of them via the Window menu.

If you ever get completely lost, you can always reset your layout by opening the Window menu and choosing Layouts → Default.

Play Mode and Edit Mode

The Unity editor exists in one of two modes: Edit Mode and Play Mode. In Edit Mode, which is the default, you create your scene, configure your game objects, and generally build your game. In Play Mode, you play your game and interact with your scene.

To enter Play Mode, click the Play button at the top of the Editor window (Figure 2-4). Unity will start the game; to leave Play Mode, click the Play button again.

You can also press Command-P (Ctrl-P on a PC) to enter and leave Play Mode.

Figure 2-4. The Play Mode controls

While in Play Mode, you can temporarily pause the game by pressing the Pause icon in the middle of the Play Mode controls. Press it again to resume playback. You can also ask Unity to advance a single frame and then pause again by clicking the Step button at the far right.

Any changes that are made to your scene are undone when you leave Play Mode. This includes both changes that happened as a result of gameplay, and changes that you made to your game objects without realizing you were in Play Mode. Double-check before making changes!

Let's now take a closer look at the tabs that appear by default. In this chapter, we'll refer to the location of the panes as they appear in the default layout. (If you can't see one of the panes, make sure you're using the default layout.)

The Scene View

The scene view is the pane in the middle of the window. The scene view is where you spend most of your time, since it's here that you're able to look at the contents of your game's *scenes*.

Unity projects are broken up into scenes. Each scene contains a collection of game objects; by creating and modifying game objects, you create your game's worlds.

You can think of a scene as a level, but scenes are also used to break up your game into manageable chunks. For example, the main menu of your game is usually its own scene, as well as each of its levels.

The Mode Selector

The scene view can be in one of five different modes. The mode selector, at the top-left of the window (seen in Figure 2-5), controls how you're interacting with the scene view.

Figure 2-5. The scene view's mode selector, shown here in Translate mode

The five modes, from left to right, are:

Grab mode
: When this mode is active, left-clicking and dragging the mouse will pan the view.

Translation mode
: When this mode is active, the currently selected objects can be moved around.

Rotation mode
: When this mode is active, the currently selected objects can be rotated.

Scale mode
: When this mode is active, the currently selected objects can be resized.

Rectangle mode
: When this mode is active, you can move and resize the currently selected objects using 2D handles. This is particularly useful when laying out a 2D scene, or working with a GUI.

You can't select any objects in Grab mode, but you can in the other modes.

You can switch the mode that the scene view is in using the mode selector; alternatively, you can press the Q, W, E, R, and T keys to quickly switch between them.

Getting Around

There are a few ways to get around in the scene view:

- Click the Hand icon at the top-left of the window to enter Grab mode, and left-click and drag to pan the view.
- Hold down the Option key (Alt on a PC) and left-click and drag to rotate the view.
- Select an object in the scene by left-clicking on it in the scene, or clicking on its entry in the Hierarchy (which we'll talk about in "The Hierarchy" on page 14), move the mouse over the scene view, and press F to focus the view on the selected object.
- Hold down the right mouse button, and move the mouse to look around; while you're holding the right mouse button, you can use the W, A, S, and D keys to fly forward, left, back, and right. You can also use the Q and E keys to fly up and down. Hold the Shift key to fly faster.

You can also press the Q key to switch to Grab mode, instead of clicking on the Hand icon.

Handle Controls

To the right of the mode selector, you'll find the handle controls (Figure 2-6). The handle controls determine where the handles—the movement, rotation, and scaling controls that appear when you select an object—should be positioned and oriented.

Figure 2-6. The handle controls; in this image, the handle's positions are set to Pivot, and the orientation is set to Local

There are two controls that you can configure: the position of the handles and their orientation.

The position of the handles can be set to either Pivot or Center.

- When set to Pivot, the handles appear at the pivot point of the object. For example, 3D models of people typically have their pivot point placed between their feet.
- When set to Center, the handles appear in the center of the object, and disregard the object's pivot point.

The orientation of the handles can be set to either Local or Global.

- When set to Local, the handles are oriented relative to the object you have selected. That is, if you rotate an object so that its *up* direction is now facing sideways, the *up* arrow will face sideways as well. This allows you to move the object in its "local" up direction.
- When set to Global, the handles are oriented relative to the world—that is, the *up* direction will always be straight up, ignoring the object's actual rotation. This can be useful when you need to move a rotated object.

The Hierarchy

The Hierarchy pane (Figure 2-7) appears at the left of your scene view, and displays the list of all objects in the currently open scene. If you have a complex scene, the hierarchy lets you quickly find an object by name.

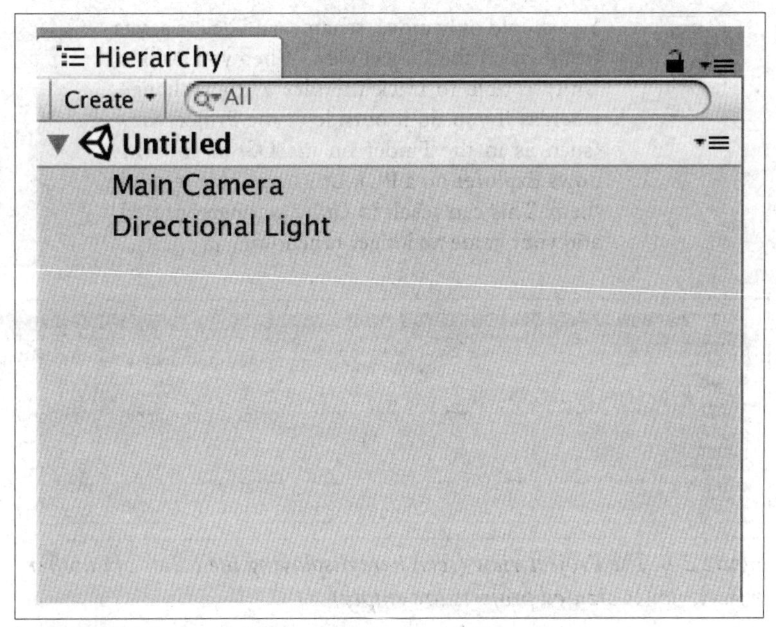

Figure 2-7. The Hierarchy pane

The hierarchy, as its name suggests, also lets you view the parent–child relationship of objects. In Unity, objects can contain other objects; in the hierarchy, you can explore this tree of objects. You can also drag and drop objects to rearrange them in the list.

At the top of the hierarchy, you'll find a search field, which you can use to type the name of the object you're looking for. This is particularly useful in complex scenes.

The Project View

The Project view (Figure 2-8), at the bottom of the Editor window, displays the contents of your project's Assets folder. From here, you can work with the assets in your game, and manage the folder layout.

You should only move, rename, and delete assets from within the Project view. When you do this, Unity is able to track the files as they change, whereas if you do it outside of the Project view (such as in the Finder on macOS, or in Windows Explorer on a PC), Unity isn't able to track them. This can result in Unity getting confused, and your game no longer functioning properly.

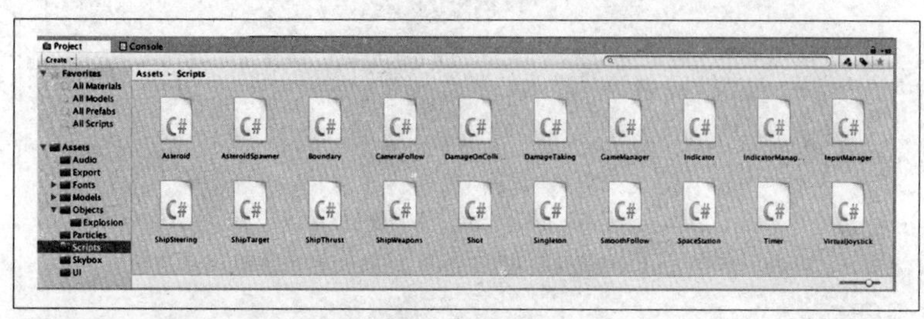

Figure 2-8. The Project view (seen here displaying the assets of another project; newly created projects are empty)

The Project view can be viewed in either a single-column layout, or a double-column layout. The double-column layout can be seen in Figure 2-8; on the left column, the list of folders appears, and on the right, the contents of the currently selected folder appear. The double-column view is best suited for wide layouts.

By contrast, the single-column view (Figure 2-9) lists all folders and their contents in a single list. This makes it ideal for narrower layouts.

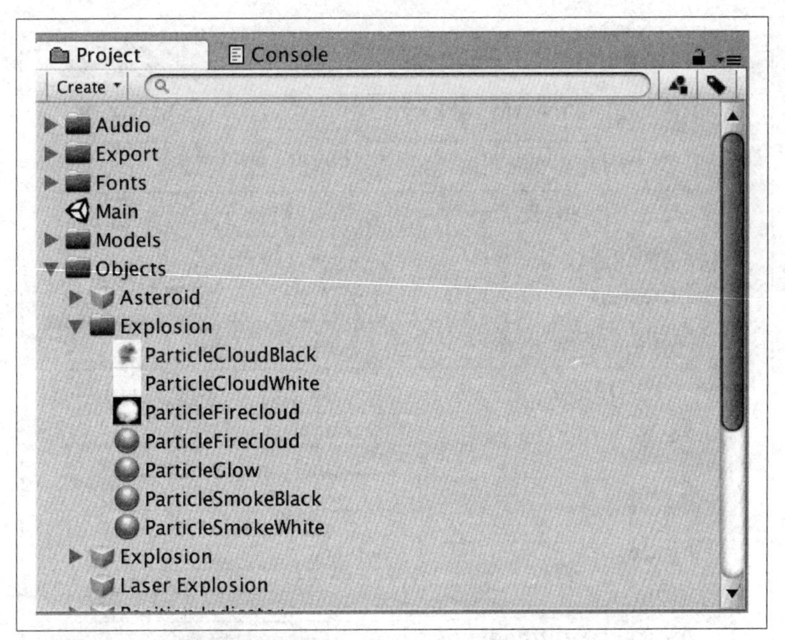

Figure 2-9. The Project view, in single-column mode

The Inspector

The Inspector (Figure 2-10) is one of the most important views in the entire editor, second only to the Scene view. The Inspector displays information about the currently selected objects, and it's where you'll go to configure your game objects. The Inspector appears to the righthand side of the window; by default, it's in the same tab group as the Services tab.

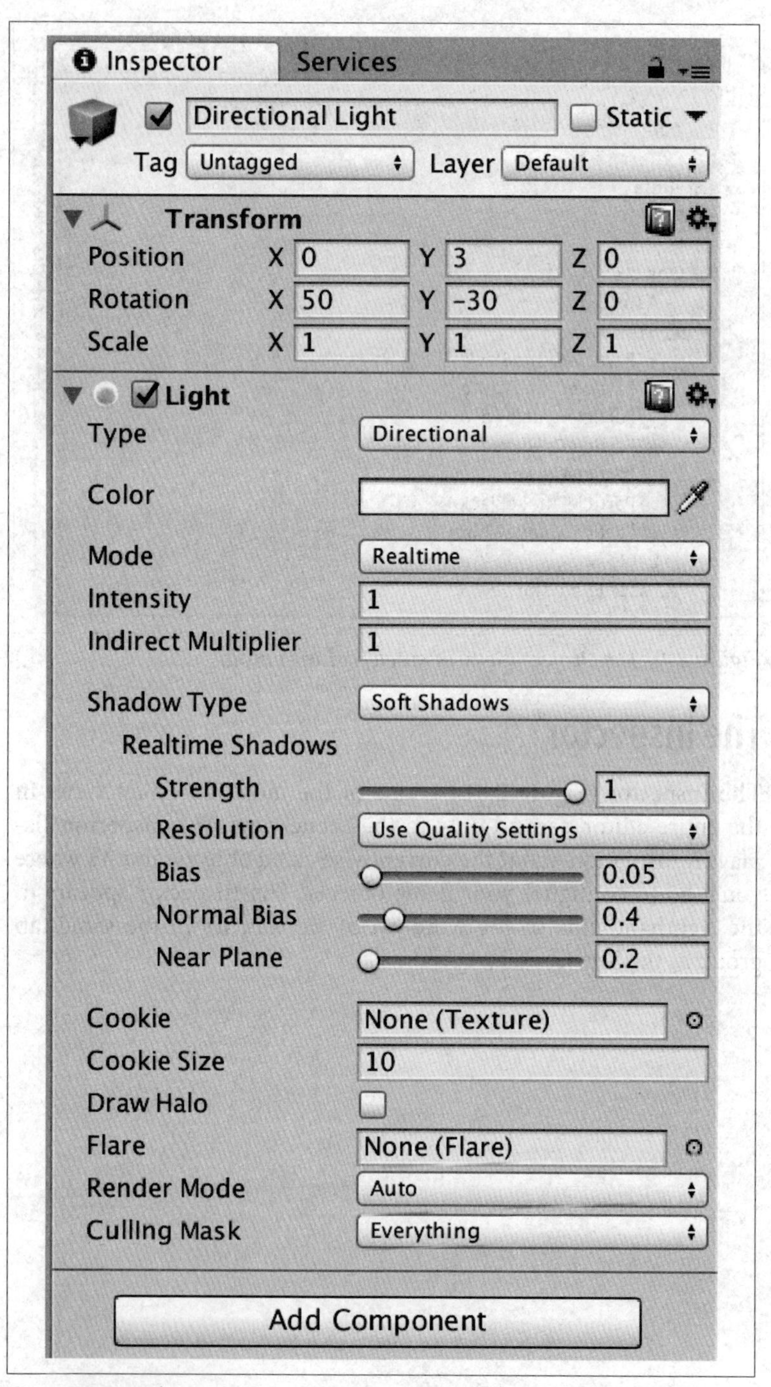

Figure 2-10. The Inspector, showing information about an object containing a Light component

The Inspector shows the list of all components attached to the selected object or asset. Each component shows different information; as we build the projects in Parts II and III, we'll be looking at a wide variety of them. This means that we'll get a lot more familiar with the Inspector and its contents as time goes on.

In addition to showing information about the current selection, the Inspector also shows your project's settings. which you can access via the Edit → Project Settings menu.

The Game View

The Game view, which is in the same tab group as the Scene view, displays the view from the game's currently active camera. When you enter Play Mode (see "Play Mode and Edit Mode" on page 10), the Game view automatically activates, letting you play your game.

The Game view isn't interactive on its own—all it does is show what the camera is rendering. This means that, when the editor is in Edit Mode, attempting to interact with the Game view won't do anything.

Wrapping Up

Now that you know how to get around Unity, you're ready to start making it do what you want. There's always more to explore in such a complex piece of software; take the time to poke around.

In the next chapter, we'll talk about how to work with game objects and scripts. From there, you'll be ready to start making your games.

CHAPTER 3

Scripting in Unity

In order for your game to work, you need to define what actually *happens* in your game. Unity provides you with the foundations of what you need, such as rendering graphics, getting input from the player, and playing audio; it's up to you to add the features that are unique to your game.

To make this happen, you write *scripts* that get added to your game's objects. In this chapter, we'll introduce you to Unity's scripting system, which uses the C# programming language.

Languages in Unity

You have a choice of languages when programming in Unity. Unity officially supports two different languages: C# and "JavaScript."

We put JavaScript in quotes because it's not actually the JavaScript language that you might be familiar with from the wider world. Instead, it's a language that *looks* like JavaScript, but has multiple differences from its namesake. It's different enough that it's often called "UnityScript," by both users of Unity and sometimes the Unity team themselves.

We don't use Unity's JavaScript in this book for a couple of reasons. The first is that Unity's reference material tends to show C# examples more than JavaScript, and we get the feeling that the use of C# is preferred by Unity's developers.

Secondly, when you use C# in Unity, it's the same language you'll find anywhere else, whereas Unity's version of JavaScript is very specific to Unity. This means that it's easier to find help about the language.

A Crash Course in C#

When writing scripts for Unity games, you write in a language called C#. We're not going to explain the fundamentals of programming in this book (we don't have the space!), but we'll highlight some main points to keep in mind.

A great general reference on the C# language is *C# in a Nutshell*, by Joseph and Ben Albahari (O'Reilly, 2015).

To give you a quick introduction, we'll take a chunk of C# code, and highlight some important elements:

```
using UnityEngine; ❶

namespace MyGame { ❷

    [RequireComponent(typeof(SpriteRenderer))] ❸
```

```
    class Alien : MonoBehaviour { ❹
        public bool appearsPeaceful; ❺

        private int cowsAbducted;

        public void GreetHumans() {
            Debug.Log("Hello, humans!");

            if (appearsPeaceful == false) {
                cowsAbducted += 1;
            }
        }
    }
}
```

❶ The `using` keyword indicates to the user which packages you'd like to use. The `UnityEngine` package contains the core Unity types.

❷ C# lets you put your types in *namespaces*, which means that you can avoid naming collisions.

❸ *Attributes* are placed between square brackets, and let you add additional information about a type or method.

❹ *Classes* are defined using the `class` keyword, and you specify the superclass after a colon. When you make a class a subclass of `MonoBehaviour`, it can be used as a script component.

❺ Variables attached to classes are called *fields*.

Mono and Unity

Unity's scripting system is powered by the Mono framework. Mono is an open source implementation of Microsoft's .NET Framework, which means that in addition to the libraries that come with Unity, you also have the complete set of libraries that come with .NET.

A common misconception is that Unity is built on top of Mono. Unity is not built on Mono; it merely uses Mono as its scripting engine. Unity supports scripting, through Mono, using both the C# language and the UnityScript language (what Unity calls "JavaScript;" see Languages in Unity).

The versions of C# and the .NET Framework available in Unity are older than the most current versions. At the time of writing in early 2017, the version of the C# language available is 4, while the version of the .NET Framework available is 3.5. The reason for this is that Unity uses its own fork of the Mono project, which diverged from the mainline branch several years ago. This has meant that Unity can add features that are specific to their uses, which are primarily mobile-oriented compiler features.

Unity is in the middle of updating its compiler tools to make the latest versions of the C# language and the .NET Framework available to users. Until that happens, your code will be a few versions behind.

For this reason, if you're looking for C# code or advice around the web, you should search for Unity-specific code most of the time. Similarly, when you're coding C# for Unity, you're going to be using a combination of Mono's API (for generic things that most platforms provide) and Unity's API (for game engine-specific things).

MonoDevelop

MonoDevelop is the development environment that's included with Unity. MonoDevelop's main role is to be the text editor that you write your scripts with; however, it contains some useful features that can make your life easier when programming.

When you double-click on any script file in your project, Unity will open the editor that's currently configured. By default, this will be MonoDevelop, though you can configure it to be any other text editor you like.

Unity will automatically update the project in MonoDevelop with the scripts in your project, and will compile your code when you return to Unity. This means that all you need to do to edit your scripts is to save your changes, and return to the editor.

There are several features in MonoDevelop that can save you a lot of time.

Code completion

In MonoDevelop, press Ctrl-Space (on both PC and Mac). Mono Develop will display a pop-up window that offers a list of suggestions for what to type next; for example, if you're halfway through typing a class name, MonoDevelop will offer to complete it. Press the up and down arrows to select from the list, and press Enter to accept the suggestion.

Refactoring

When you press Alt-Enter (Option-Enter on a Mac), MonoDevelop will offer to perform certain tasks that edit your source code. These tasks include things like adding or removing braces around `if` statements, automatically filling in the case labels for `switch` statements, or splitting a variable's declaration and assignment into two lines.

Building

Unity will automatically rebuild your code when you return to the editor. However, if you press Command-B (F7 on a PC), all of your code will be built in MonoDevelop. The files that result from this won't be used in this game, but doing this means that you're able to verify that there are no compilation errors in your code before you return to Unity.

Game Objects, Components, and Scripts

Unity scenes are composed of *game objects*. On their own, they're invisible objects, and have nothing but a name. Their behavior is defined by their *components*.

Components are the building blocks of your game, and anything you see in the Inspector is a component. Each component has a different responsibility; for example, Mesh Renderers display 3D meshes, while Audio Sources play sound to the user. Scripts that you write are components as well.

To create a script:

1. *Create the script asset*. Open the Assets menu, and choose Create → Script → C# Script.
2. *Name the script asset*. A new script file will appear in the folder you had selected in the Project panel, ready for you to name.
3. *Double-click the script asset*. The script will open in the script editor, which defaults to MonoDevelop. Most of your scripts will start off looking like this:

```
using UnityEngine;
using System.Collections;
using System.Collections.Generic;

public class AlienSpaceship : MonoBehaviour { ❶

    // Use this for initialization
    void Start () { ❷

    }

    // Update is called once per frame
    void Update () { ❸

    }
}
```

❶ The name of the class, in this case `AlienSpaceship`, must be the same as the asset filename.

❷ The `Start` function is called before the `Update` function is called for the first time, and is where you might put code that initializes variables, loads stored preferences, or sets up other scripts and GameObjects.

❸ The `Update` function is called every frame, and is an opportunity to include code that responds to input, triggers another script, or moves things around—anything that needs to happen.

You may be familiar with *constructors* from other programming environments. In Unity, you don't construct your `MonoBehaviour` subclasses yourself, because the construction of objects is performed by Unity itself, and does not necessarily take place when you think it might.

Script assets in Unity do not actually do anything—none of their code is executed—until they are attached to a GameObject (seen in Figure 3-1). There are two main ways to attach a script to a GameObject:

1. *Dragging the script asset onto the GameObject.* This can be done with either the Inspector or the Hierarchy panel.
2. *Using the Component menu.* You will find all the scripts that are in the project under Component → Scripts.

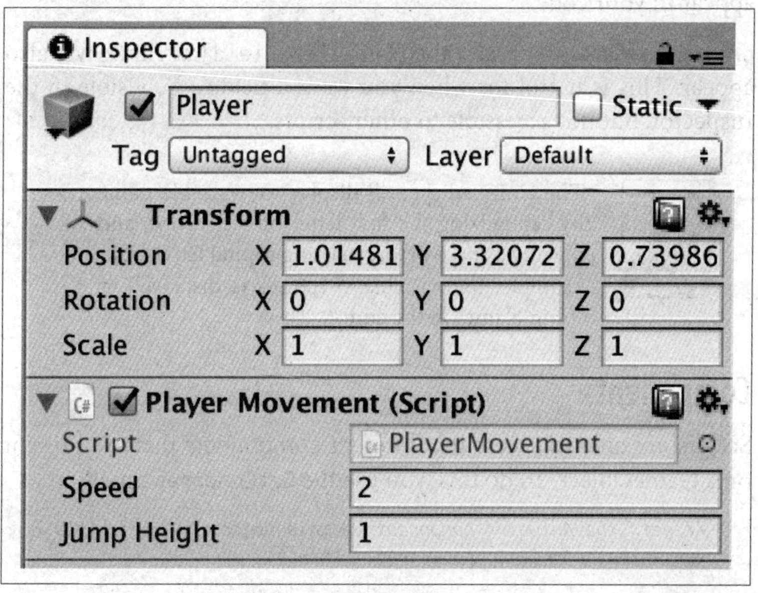

Figure 3-1. The Inspector for a GameObject, showing a script called "PlayerMovement" added as a component

Since scripts are primarily exposed in the Unity editor through being attached, as components, to GameObjects, Unity allows you to expose properties in your script as editable values in the Inspector. To do this, you create a public variable in your script. Everything specified as public will be visible in the editor; you can also set variables to private, though.

```
public class AlienSpaceship : MonoBehaviour {
    public string shipName;

    // "Ship Name" will appear in the Inspector
    // as an editable field
}
```

The Inspector

When your script is added as a component to a game object, it appears in the Inspector when that object is selected. Unity will automatically display all variables that are `public`, in the order they appear in your code.

`private` variables that have the `[SerializeField]` attribute will also appear. This is useful for when you want a field to be visible in the Inspector, but not accessible to other scripts.

The Unity editor will display the variable name by capitalizing the first letter of each word, and placing a space before existing capital letters. For example, the variable `shipName` is displayed as "Ship Name" in the editor.

Components

Scripts are able to access the different components that are present on a GameObject. To do this, you use the `GetComponent` method.

```
// gets the Animator component on this object, if it exists
var animator = GetComponent<Animator>();
```

You can also call `GetComponent` on other objects, to get components attached to them.

You can also get the components that are attached to parent or child objects, using the `GetComponentInChildren` or `GetComponentInParent` methods.

Important Methods

Your `MonoBehaviours` have several methods that are particularly important to Unity. These methods are called at different times during the component's life cycle, and are opportunities to run the right behavior at the right moment. This section lists the methods in the order that they're run.

Awake and OnEnable

`Awake` is run immediately after an object is instantiated in the scene, and is the first opportunity you have to run code in your script. `Awake` is called once in the object's lifetime.

By contrast, `OnEnable` is called each time an object becomes enabled.

Start

The `Start` method is called immediately before the first call to an object's `Update` method.

Start Versus Awake

You might wonder why there are two opportunities for setting up an object: `Awake` and `Start`. After all, doesn't that just mean that you'll pick one of them at random?

There's actually a very good reason for it. When you start a scene, all objects in it run their `Awake` and `Start` methods. Critically, however, Unity makes sure that all objects have finished running their `Awake` methods *before* any `Start` methods are run.

This means that any work that's done in an object's `Awake` method is guaranteed to have been done by the time another object runs its `Start` method. This can be useful, such as in this example, where object A uses a field set up by object B:

```
// In a file called ObjectA.cs
class ObjectA : MonoBehaviour {

    // A variable for other scripts to access
    public Animator animator;

    void Awake() {
        animator = GetComponent<Animator>();
    }
}

// In a file called ObjectB.cs
class ObjectB : MonoBehaviour {

    // Connected to the ObjectA script
    public ObjectA someObject;

    void Awake() {
```

```
        // Check to see if someObject has set its 'animator'
        // variable
        bool hasAnimator = someObject.animator == null;

        // May print 'true' OR 'false', depending on which
        // one happens to run first
        Debug.Log("Awake: " + hasAnimator.ToString());
    }

    void Start() {
        // Check to see if someObject has set its 'animator'
        // variable
        bool hasAnimator = someObject.animator == null;

        // Will *always* print 'true'
        Debug.Log("Start: " + hasAnimator.ToString());
    }
}
```

In this example, the `ObjectA` script is on an object that also has an Animator component attached. (The Animator itself does nothing in this example, and could just as easily be any other kind of component.) The `ObjectB` script has been set up so that its `someObject` variable is connected to the object containing the `ObjectA` script.

When the scene begins, the `ObjectB` script will log twice—once in its `Awake` method, and once in its `Start` method. In both cases, it will try to figure out if its `someObject` variable's `animator` field is not null, and print either "true" or "false."

If you were to run this example, the first log message, which runs in `ObjectB`'s `Awake` method, would be either "true" or "false," depending on which script's `Awake` method ran first. (Without manually setting up an execution order in Unity, it's impossible to know which runs first.)

However, the second log message, which runs in `ObjectB`'s `Start` method, is *guaranteed* to return "true." This is because, when a scene starts up, all existing objects will run their `Awake` methods before any `Start` methods are run.

Update and LateUpdate

The `Update` method is run every single frame, as long as the component is enabled, and the object that the script is attached to is active.

`Update` methods should do as little work as possible, because they're run every single frame. If you do some long-running work in an `Update` method, you'll slow down the rest of the game. If you need to do something that will take some time, you should use a coroutine (described in the following section).

Unity will call the `Update` method on all scripts that have one. Once that's done, it will call `LateUpdate` method on all scripts that have one. `Update` and `LateUpdate` have a similar relationship to that of `Awake` and `Start`: no `LateUpdate` methods will be called until all `Update` methods have been run.

This is useful for when you want to do work that relies on some other object to have done work in `Update`. You can't control which objects run their `Update` method first; however, when you write code that runs in `LateUpdate`, you're guaranteed that any work in any object's `Update` method will have completed.

In addition to `Update`, the `FixedUpdate` method can be used. While `Update` is called once per frame, `FixedUpdate` is called a fixed number of times each second. This can be useful when working with physics, where you need to apply forces at regular intervals.

Coroutines

Most functions do their work and return immediately. However, sometimes you need something to take place over time. For example, if you want an object to slide from one point to another, you need that movement to happen over multiple frames.

A *coroutine* is a function that runs over multiple frames. In order to create a coroutine, first create a method that has a return type of `IEnumerator`:

```
IEnumerator MoveObject() {

}
```

Next, use the `yield return` statement to make the coroutine temporarily stop, allowing the rest of the game to carry on. For example, to make an object move forward by a certain amount every frame,[1] you'd do this:

```
IEnumerator MoveObject() {
    // Loop forever
    while (true) {

        transform.Translate(0,1,0); // move 1 unit on the Y
                                    // axis every frame

        yield return null; // wait until the next frame

    }
}
```

If you include an infinite loop (such as the `while (true)` in the previous example), then you *must* yield during it. If you don't, it will loop forever without giving the rest of your code a chance to do any other work. Because your game's code runs inside Unity, you run the risk of causing Unity to freeze up if you enter an infinite loop. If that happens, you'll need to force Unity to quit, and may lose unsaved work.

When you `yield return` from a coroutine, you temporarily pause the execution of the function. Unity will resume execution later; the specifics of *when* it resumes depends on what value you `yield return` with.

For example:

`yield return null`
waits until the next frame

`yield return new WaitForSeconds(3)`
waits three seconds

`yield return new WaitUntil(() => this.someVariable == true)`
waits until `someVariable` equals `true`; you can also use any expression that evaluates to a `true` or `false` variable

1 This is actually not a great idea, for reasons that are explained in "Time in Scripts" on page 39, but it's the simplest example.

To stop a coroutine, you use the `yield break` statement:

```
// stop this coroutine immediately
yield break;
```

Coroutines will also automatically stop when execution reaches the end of the method.

Once you have a coroutine function, you can start it. To start a coroutine, you don't call it on its own; instead, you use it in conjunction with the `StartCoroutine` function:

```
StartCoroutine(MoveObject());
```

When you do this, the coroutine will start executing until it either reaches a `yield break` statement, or it reaches the end.

In addition to the `yield return` examples we just looked at, you can also `yield return` on another coroutine. This means that the coroutine you're `yielding` from will wait until the other coroutine ends.

It's also possible to stop a coroutine from outside of it. To do this, keep a reference to the return value of the `StartCoroutine` method, and pass it to the `StopCoroutine` method:

```
Coroutine myCoroutine = StartCoroutine(MyCoroutine());

// ... later ...

StopCoroutine(myCoroutine);
```

Creating and Destroying Objects

There are two ways to create an object during gameplay. The first is by creating an empty GameObject, and attaching components to it by using code; the second involves duplicating another object (called *instantiation*). The second method is more popular because you can do everything in a single line of code, so we'll discuss it first.

When you create new objects in Play Mode, those objects will disappear when you stop the game. If you want them to stick around, follow these steps:

1. Select the objects you want to save.
2. Copy them, either by pressing Command-C (Ctrl-C on a PC), or opening the Edit menu and choosing Copy.
3. Leave Play Mode. The objects will disappear from the scene.
4. Paste, either by pressing Command-V (Ctrl-V on a PC), or opening the Edit menu and choosing Paste. The objects will reappear; you can now work with them in Edit Mode.

Instantiation

In Unity, instantiating an object means that it, along with all of its components, child objects, and *their* components, are copied. This is particularly powerful when the object you're instantiating is a prefab. Prefabs are prebuilt objects that you save as assets. This means that you can create a single template of an object, and instantiate many copies of it across many different scenes.

To instantiate an object, you use the `Instantiate` method:

```
public GameObject myPrefab;

void Start() {
    // Create a new copy of myPrefab,
    // and position it at the same point as this object
    var newObject = (GameObject)Instantiate(myPrefab);

    newObject.transform.position = this.transform.position;
}
```

The `Instantiate` method's return type is `Object`, not `GameObject`. You'll need to do a cast in order to treat it as a `GameObject`.

Creating an Object from Scratch

The other way you can create objects is by building them up yourself through code. To do this, you use the `new` keyword to construct a new `GameObject`, and then call `AddComponent` on it to add new components.

```
// Create a new game object; it will appear as
// "My New GameObject" in the hierarchy
var newObject = new GameObject("My New GameObject");

// Add a new SpriteRenderer to it
var renderer = newObject.AddComponent<SpriteRenderer>();

// Tell the new SpriteRenderer to display a sprite
renderer.sprite = myAwesomeSprite;
```

The `AddComponent` method takes as a generic parameter the type of component you want to add. You can specify any class here that's a subclass of `Component`, and it will be added.

Destroying Objects

The `Destroy` method removes an object from the scene. Notice that we didn't say *game object*, but *object*! `Destroy` is used for removing both game objects and components.

To remove a game object from the scene, call `Destroy` on it:

```
// Destroy the game object that this script is attached to
Destroy(this.gameObject);
```

`Destroy` works on both components and game objects.

If you call `Destroy` and pass in `this`, which means *the current script component*, you won't remove the game object, but instead the script will end up removing itself from the game object it's attached to. The game object will stick around, but will no longer have your script attached.

Attributes

An *attribute* is a piece of information that you can attach to a class, variable, or method. Unity defines several useful attributes that you can use, which change the behavior of the class or how it's presented in the Editor.

RequireComponent

The `RequireComponent` attribute, when attached to a class, allows you to specify to Unity that the script requires that another type of component be present. This is useful when your script only makes sense when that kind of component is attached. For example, if your script only does one thing, such as changing the settings of an Animator, it makes sense that that class should require an Animator to be present.

To specify the type of component that your component requires, you provide the type of component as a parameter, like so:

```
[RequireComponent(typeof(Animator))]
class ClassThatRequiresAnAnimator : MonoBehaviour {
    // this class requires that an Animator also
    // be attached to the GameObject
}
```

If you add a script that requires a certain component to a GameObject, and that GameObject doesn't already have that component, Unity will automatically add one for you.

Header and Space

The `Header` attribute, when added to a field, causes Unity to draw a label above the field in the Inspector. `Space` works similarly, but adds empty space. Both are useful for visually organizing the contents of the Inspector.

For example, Figure 3-2 shows the Inspector's rendering of the following code:

```
public class Spaceship : MonoBehaviour {

    [Header("Spaceship Info")]

    public string name;
```

```
        public Color color;

        [Space]

        public int missileCount;
    }
```

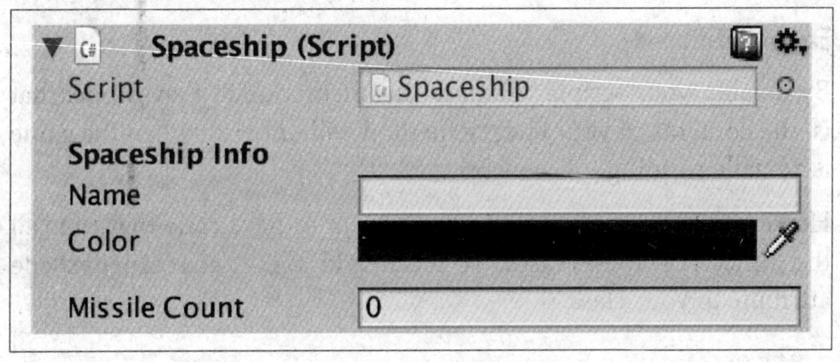

Figure 3-2. The Inspector, showing header labels and spaces

SerializeField and HideInInspector

Normally, only `public` fields are displayed in the Inspector. However, making variables `public` means that other objects can directly access them, which means that it can be difficult for an object to have full control over its own data. However, if you make the variable `private`, Unity won't display it in the Inspector.

To get around this, add the `SerializeField` attribute to `private` variables you want to appear in the Inspector.

If you want the *opposite* behavior (that is, the variable is `public`, but doesn't appear in the Inspector), then you can use the `HideInInspector` attribute:

```
class Monster : MonoBehaviour {

    // Appears in Inspector, because it's public
    // Accessible from other scripts
    public int hitPoints;

    // Doesn't appear in Inspector, because it's private
    // Not accessible from other scripts
    private bool isAlive;

    // Appears in Inspector, because of SerializeField
    // Not accessible from other scripts
    [SerializeField]
```

```
    private int magicPoints;

    // Doesn't appear in Inspector, because of HideInInspector
    // Accessible from other scripts
    [HideInInspector]
    public bool isHostileToPlayer;
}
```

ExecuteInEditMode

By default, your scripts will only run their code in Play Mode; that is, the contents of your `Update` method will only run when the game is actually running.

However, it can sometimes be convenient to have code that runs all the time. For these cases, you can add the `ExecuteInEditMode` attribute to your class.

The component life cycle performs differently in Edit Mode as compared to Play Mode. When in Edit Mode, Unity will only redraw itself when it has to, which generally means in response to user input events like mouse clicks. This means that the `Update` method will run only sporadically instead of continuously. Additionally, coroutines won't behave the way you expect.

Moreover, you can't call `Destroy` in Edit Mode, because Unity defers the actual removal of an object until the next frame. In Edit Mode, you should call `DestroyImmediate` instead, which removes an object right away.

For example, here's a script that makes an object always face its target, even when not in Play Mode:

```
[ExecuteInEditMode]
class LookAtTarget : MonoBehaviour {

    public Transform target;

    void Update() {
        // Don't continue if we don't have a target
        if (target != null) {
            return;
        }

        // Rotate to look at target
```

```
            transform.LookAt(target);
        }
    }
```

If you were to attach this script to an object, and set its `target` variable to another object, the first object would rotate to face its target in both Play Mode and Edit Mode.

Time in Scripts

The `Time` class is used to get information about the current time in your game. There are several variables available in the `Time` class (and we strongly recommend you look at the documentation (*https://docs.unity3d.com/Manual/TimeFrameManagement.html*) for it!), but the most important and commonly used variable is `deltaTime`.

`Time.deltaTime` measures the amount of time since the last frame was rendered. It's important to realize that this time can vary a lot. Doing this allows you to perform an action that's updated every frame, but needs to take a certain amount of time.

In "Coroutines" on page 31, the example we used was one that moves the object one unit every frame. This is a bad idea, because the number of frames in a second can vary quite a lot. For example, if the camera is looking at a very simple part of your scene, the frames per second could be very high, whereas looking at more visually complex scenes could result in very low framerates.

Because you can't be sure of the number of frames per second you're running at, the best thing to do is to take into account `Time.deltaTime`. The easiest way to explain this is with an example:

```
    IEnumerator MoveSmoothly() {
        while (true) {

            // move 1 unit per second
            var movement = 1.0f * Time.deltaTime;

            transform.Translate(0, movement, 0);

            yield return null;

        }
    }
```

Logging to the Console

As we saw in "Awake and OnEnable" on page 29, it's sometimes convenient to log some information to the Console, either for diagnostic purposes, or to warn you about some problem.

The `Debug.Log` function can be used to do this. There are three different levels of logging: info, warning, and error. There's no functional difference between the three types, except warnings and errors are each more visible and prominent than the last.

In addition to `Debug.Log`, you can also use `Debug.LogFormat`, which allows you to embed values in the string that's sent to the Console:

```
Debug.Log("This is an info message!");
Debug.LogWarning("This is a warning message!");
Debug.LogError("This is a warning message!");

Debug.LogFormat("This is an info message! 1 + 1 = {0}", 1+1);
```

Wrapping Up

Scripting in Unity is a critical skill, and becoming comfortable with both the C# language and the tools used for writing it will make building your game easier and more fun to do.

PART II

Building a 2D Game: Gnome on a Rope

Now that we've explored Unity in the abstract, we're going to put these skills to work. In both this part and the next, we'll be building entire games from scratch.

In the next several chapters, we'll build a side-scrolling action game called *Gnome's Well That Ends Well*. This game relies on the 2D graphics and physics features of Unity to a large degree, in addition to fairly heavy use of the UI system. It's going to be fun.

CHAPTER 4

Getting Started Building the Game

Knowing how to navigate Unity's interface is one thing. Creating an entire game with it is another. In this section of the book, you'll take what you've learned in Part I, and use it to create a 2D game. By the end of this part, you'll have *Gnome's Well That Ends Well*, a side-scrolling action game (see Figure 4-1 for a sneak peek at what the game looks like).

Figure 4-1. The finished product

Game Design

The gameplay of *Gnome's Well* is straightforward. The player controls a garden gnome, who's being lowered by an attached rope into a well. At the bottom of the well there's some treasure. The catch is that the well is filled with traps that kill the gnome if he touches them.

To begin with, we created a very rough sketch showing how the game would look. We used OmniGraffle, an excellent diagramming application, but the choice of tool doesn't especially matter—pen and paper is just as easy, and can often be better. The goal is to get a very rough sense of how the game will be put together, as quickly as possible. You can see our *Gnome's Well* sketch in Figure 4-2.

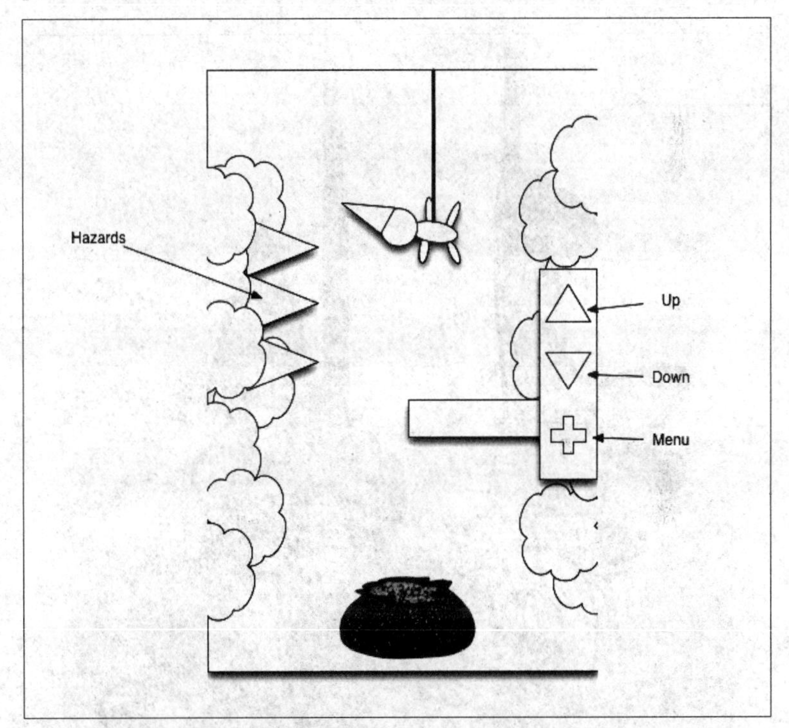

Figure 4-2. The rough concept sketch for the game

Once we decided on what the game was going to be, we started deciding the overall architecture. We started by working out what the visible objects were, and how they'd relate to one another. At the same time, we began thinking about how the "invisible" components would work—how input would be collected, and how the game's internal managers would communicate with each other.

Finally, we also considered the visuals of the game. We approached an artist friend of ours, and asked him to draw a picture of gnomes trying to get down a well and being menaced by traps. This gave us an idea of what the main character could look like, and set the overall tone of the game: a silly, cartoony, slightly violent game starring greedy gnomes. You can see this final sketch in Figure 4-3.

If you don't know any artists, draw it yourself! Even if you don't think that your skills are very good, *any* thoughts on what the game will look like are better than no thoughts at all.

Figure 4-3. The concept art for the gnome character

Once we'd done this preliminary design, it was possible to start working out the things that needed to be implemented: how the gnome would move around within the game, how the interface needed to be set up to make it work, and how the game objects would need to be linked together.

To get the gnome down the well, the player is given three buttons: one that increases the length of the rope, one that decreases it, and a third that displays the game's menu. By holding on the button that increases the rope length, the gnome is lowered down the well. To avoid the traps on the way down the well, the player tilts their device left and right. This moves the gnome left and right.

The gameplay is primarily the result of 2D physics simulation. The gnome is a "ragdoll"—a collection of pieces that are connected via joints, with each piece an independently simulated rigid body. This means that, when connected to the top of the well via the Rope object, the gnome will dangle correctly.

The rope is made in a similar way: it's a collection of rigid bodies, all connected to each other with joints. The first link in the chain is connected to the top of the well, and is connected to the second link via a rotating joint. This second joint is connected to the third, the third to the fourth, and so on, until the last link, which is connected to the gnome's ankle. To lengthen the rope, more links are added to the top of the rope, and to shorten it, links are removed.

The rest of the gameplay is handled through very straightforward collision detection:

- If any part of the gnome touches a trap object, the gnome is dead, and a new gnome is created. Additionally, a ghost sprite will be created that travels up the well.
- If the treasure is touched, the gnome's sprites are updated to show that he's holding the treasure.
- If the top of the well (an invisible object) is touched, and if the gnome is holding the treasure, the player wins the game.

In addition to the gnome, traps, and treasure, the game's camera has a script running that keeps its position linked to the vertical position of the gnome, but also keeps the camera from showing anything above the top of the well or below the bottom of the well.

The way that we'll build this game is as follows (don't worry—we'll walk you through it one step at a time):

1. First, we'll create the gnome, using temporary stick-figure images. We'll set up the ragdoll, and connect the sprites.
2. Next, we'll set up the rope. This will involve the first large amount of code, since the rope will be generated at runtime, and will need to support extending and contracting the rope.
3. Once the rope is set up, the input system will be created. This system will receive information about how the device is being tilted, and make it available to other parts of the game (in particular, the gnome.) At the same time, we'll also build the game's user interface, and create the buttons that extend and contract the rope.
4. With the rope, gnome, and input system in place, we can begin actually creating the game itself. We'll implement the traps and the treasure, and start playing the game proper.
5. From there, it's just a matter of polish: the gnome's sprites will be replaced with more complex ones, particle effects will be added, and audio will be added to the whole thing.

By the end of this chapter, the game will be functionally complete, but not all of the art will be in there. You'll be adding that later on, in Chapter 7. See Figure 4-4 for a look at where things will be at.

Over the course of this project, you'll end up adding lots of components to game objects, and tweaking the values of properties. There are a lot more components than the ones we're telling you to change, so feel free to play around with the settings for anything you wish to change; otherwise, you can leave everything with the default settings.

Let's get started!

Figure 4-4. The state the game will be in when we're done with the first, unpolished version

Creating the Project and Importing Assets

We'll start by creating the project in Unity, and do a little bit of setup. We'll also import some of the assets that will be needed in the early stages; as we progress, more will be imported.

1. *Create the project.* Choose File → New Project, and create a new project called GnomesWell. In the New Project dialog (see Figure 4-5), make sure you choose 2D, not 3D, and make sure that there are no asset packages marked for importing. You just want to make an empty project.

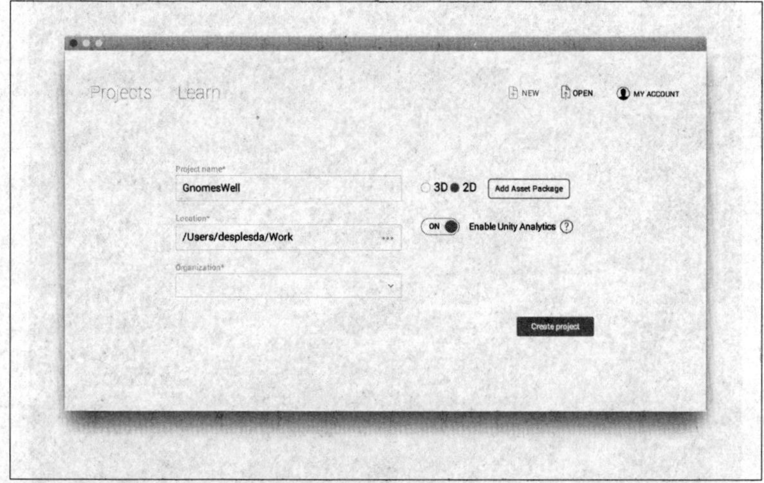

Figure 4-5. Creating the project

2. *Download the assets.* Download the art, sound, and other resources we've packaged for the project from *https://www.secretlab.com.au/books/unity*, and unzip them into an appropriate folder on your computer. You'll be importing these assets into your project as you continue.

3. *Save the scene as Main.scene.* You may as well save the scene now, so that hitting Command-S (or Ctrl-S on a PC) will instantly save your work. The first time you save, you'll be asked for the name of the scene and where to put it—put it in the *Assets* folder.

4. *Create the folders for your project.* To keep things organized, it's a good idea to create folders for different categories of assets. Unity is fine with you just keeping everything in one folder, but doing that can make finding assets more tedious than it needs to be. Create the following folders by right-clicking the *Assets* folder in the Project tab and choosing Create → Folder:

   Scripts
   : This folder will contain the C# code for the game. (By default, Unity likes to put any new code files in the root *Assets* directory; you'll need to move them into this *Scripts* folder yourself.)

   Sounds
   : This folder will contain both the music and the sound effects.

   Sprites
   : This folder will contain all of the sprite images. There are many of these, so they'll be stored in subfolders.

   Gnome
   : This folder will contain the prefabs needed for the gnome character, as well as additional, related objects like the rope, particle effects, and the ghost.

   Level
   : This folder will contain prefabs for the level itself, including the background, walls, decorative objects, and traps.

   App Resources
   : This folder will contain resources used by the app as a whole: its icon and its splash screen.

   When you're done, the *Assets* folder should look like Figure 4-6.

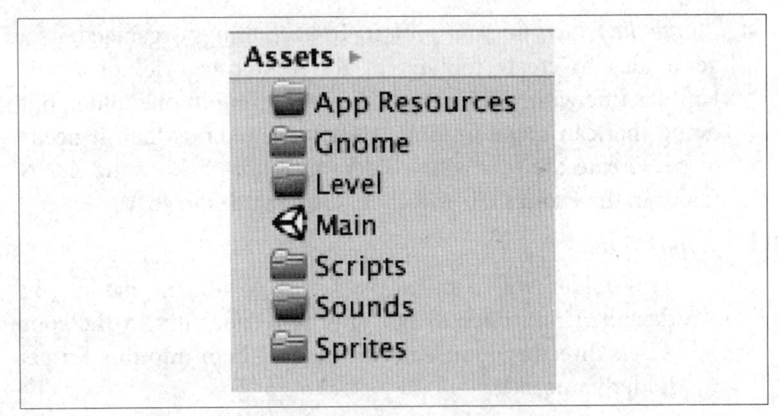

Figure 4-6. The Assets folder, with the folders created

5. *Import the prototype gnome assets.* The prototype gnome is the rough version of the gnome that we'll build first. We'll later replace it with more polished sprites.

   Locate the *Prototype Gnome* folder in the assets you downloaded, and drag it into the *Sprites* folder in Unity (see Figure 4-7).

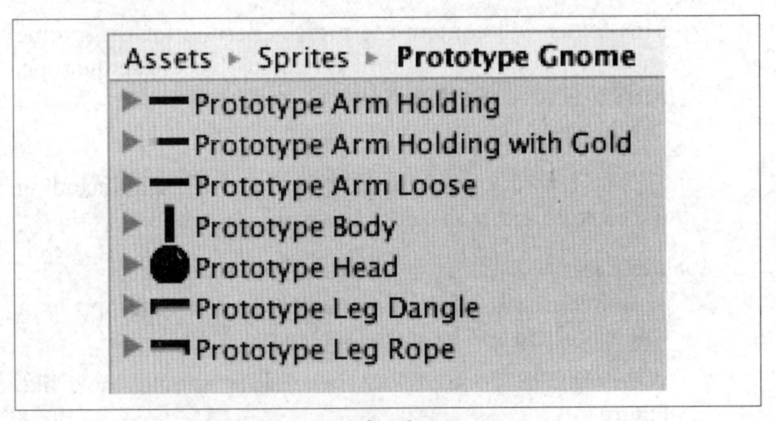

Figure 4-7. The prototype sprites for the Gnome

We're now ready to begin constructing the gnome.

Creating the Gnome

Because the gnome will be composed of multiple independently moving objects, we need to first create an object that will act as the container for each of the parts. This object will also need to be given the Player tag, because the collision detection system that's used to detect when the gnome touches a trap, treasure, or the level exit will need to know if that object was the special Player object. To build the gnome, follow these steps:

1. *Create the Prototype Gnome object.* Create a new empty game object by opening the GameObject menu and choosing Create Empty.

   Name this new object "Prototype Gnome", and then set the tag of the object to `Player`, by selecting "Player" from the Tag dropdown list at the top of the Inspector.

   The Prototype Gnome object's Position, which you can see in the Transform component near the top of the Inspector, should at zero on the X, Y and Z axes. If it's not, you can click on the gear menu at the top-right of the Transform and choose Reset Position.

2. *Add the sprites.* Locate the *Prototype Gnome* folder that you added earlier, and drag and drop each of the sprites into the scene, except for Prototype Arm Holding with Gold, which won't be used until later.

   You'll need to drag and drop each one individually—if you select all of the sprites and try to drag them all in at once, Unity will think that you're trying to drag a sequence of images in, and will create an animation.

   When you're done, you should have six new sprites in the scene: Prototype Arm Holding, Prototype Arm Loose, Prototype Body, Prototype Head, Prototype Leg Dangle, and Prototype Leg Rope.

3. *Set the sprites as children of the Prototype Gnome object.* In the Hierarchy, select all of the sprites that you just added, and drag and drop them onto the empty Prototype Gnome object. When

you're done, the hierarchy should look like that shown in Figure 4-8.

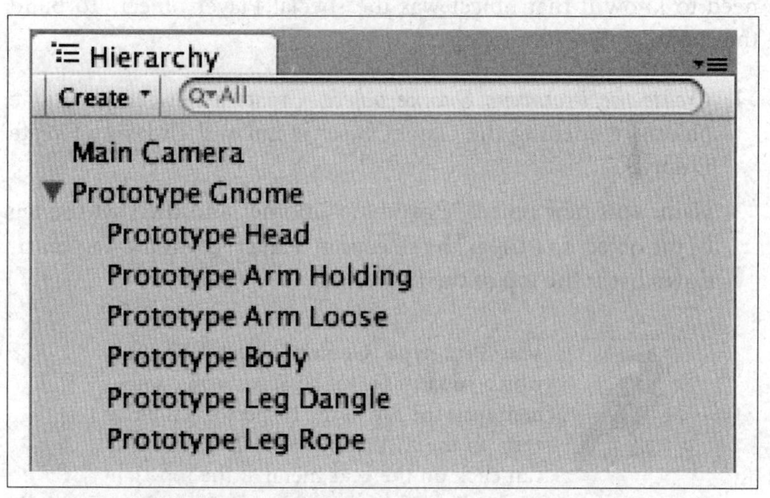

Figure 4-8. The hierarchy, with the gnome sprites attached as child objects of the Prototype Gnome object

4. *Position the sprites.* Once the sprites are added, they need to be positioned correctly—the arms, legs, and head need to be attached to the body. Within the Scene view, select the Move tool by either clicking on it in the toolbar or pressing T.

   Use the Move tool to rearrange the sprites so that they look like Figure 4-9.

   Additionally, make all of these sprites use the Player tag, just like the parent object. Finally, make sure that the Z position of each of the objects is zero. You can see the Z position in the Position field of each object's Transform component, at the top of the Inspector.

Figure 4-9. The prototype gnome's sprites

5. *Add Rigidbody 2D components to the body parts.* Select all of the body part sprites, and click Add Component in the Inspector. Type **`Rigidbody`** in the search field, and add a Rigidbody 2D (see Figure 4-10).

Make sure you add "Rigidbody 2D" components, and not the regular "Rigidbody." Regular rigidbody components do their simulation in 3D space, which isn't what you want for this game.

Additionally, make sure you only add the Rigidbody 2D on the sprites. Don't add a rigidbody to the parent Prototype Gnome object.

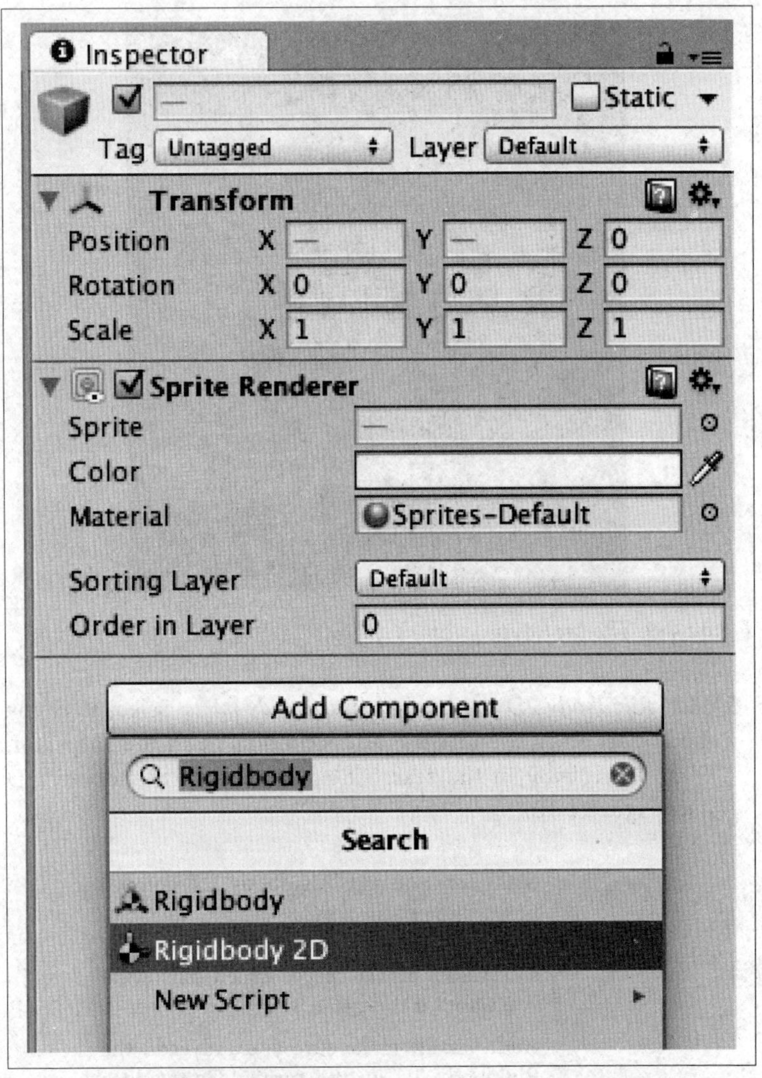

Figure 4-10. Adding Rigidbody 2D components to the sprites

6. *Add colliders to the body parts.* Colliders define the physical shape of an object. Because they're visually different shapes, different body parts will need different shaped colliders:

   a. Select the arm and leg sprites, and add `BoxCollider2D` components.

b. Select the head sprite, and add a `CircleCollider2D` component. Leave its radius as is.

c. Select the Body sprite, and add a `CircleCollider2D`. Once you've added it, go to the Inspector for the collider, and reduce the collider's `radius` by about half to fit the Body sprite.

The gnome and its body parts are now all ready to be linked together. The linkages between the body parts will all be done using the `HingeJoint2D` joint, which allows objects to rotate around a point, relative to one another. The legs, arms, and head will all be linked to the body. To configure the joints, follow these steps:

1. *Select all of the sprites,* except *for the body.* The body won't need any joint of its own—the other bodies will be connecting to it via *their* joints.
2. *Add a `HingeJoint2D` component to all of the selected sprites.* This is done by clicking the Add Component button at the bottom of the Inspector, and choosing Physics 2D → Hinge Joint 2D.
3. *Configure the joints.* While you still have the sprites selected, we'll set up a property that will be the same for all of the body parts: they'll all be connected to the Body sprite.

   Drag the Prototype Body from the Hierarchy pane into the "Connected Rigid Body" slot. This will make the objects linked to the body. When you're done, the settings for the hinge joints should look like Figure 4-11.

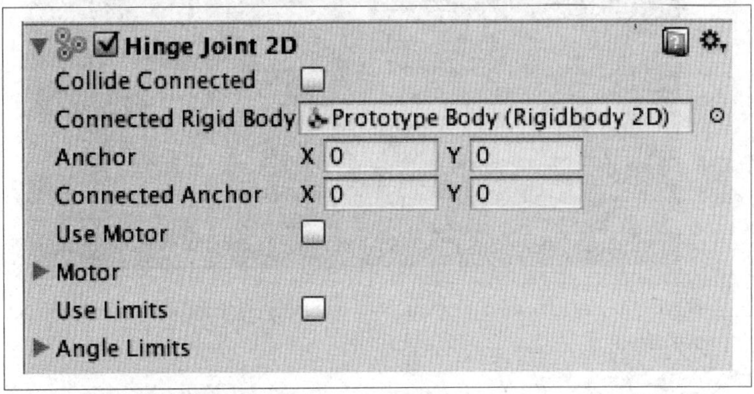

Hinge Joint 2D
Collide Connected
Connected Rigid Body Prototype Body (Rigidbody 2D)
Anchor X 0 Y 0
Connected Anchor X 0 Y 0
Use Motor
Motor
Use Limits
Angle Limits

Figure 4-11. The initial hinge joint settings

4. *Add limits to the joints.* We don't want the objects to rotate full circles, but instead want to add limits on how far they can rotate. This will prevent odd-looking behavior, like the leg appearing to move through the body.

   Select the arms and the head, and turn on Use Limits. Set the Lower Angle to -15, and the Upper Angle to 15.

   Next, select the legs, and also turn on Use Limits. Set the Lower Angle to -45, and the Upper Angle to 0.

5. *Update the pivot points for the joints.* We want the arms to rotate at the shoulder, and the legs to rotate at the hips. By default, the joints will rotate around the center of the object (see Figure 4-12), which will look odd.

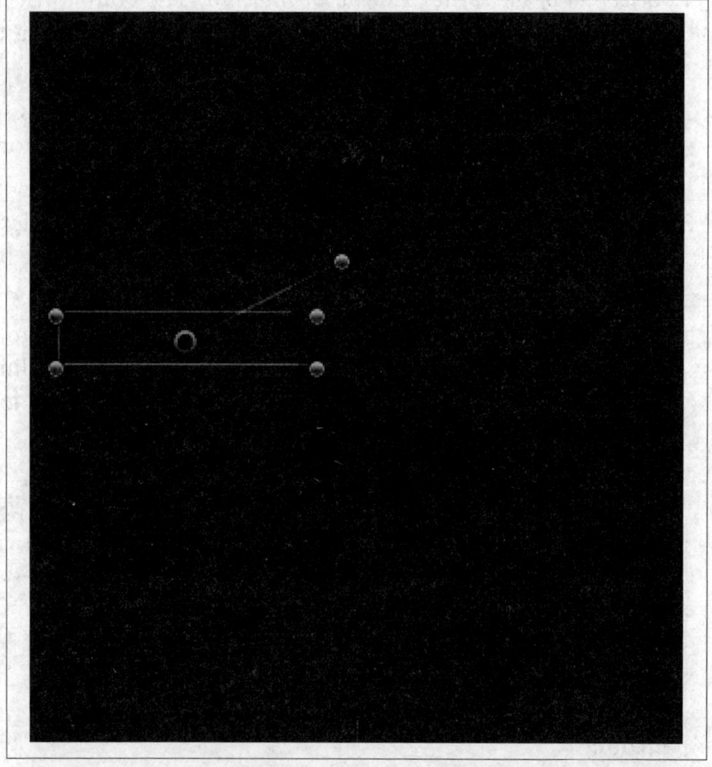

Figure 4-12. The anchor points of the hinge joints start at incorrect positions

To fix this, you need to update both the position of the joint's Anchor, as well as its Connected Anchor. The Anchor is the point at which the body that has the joint will rotate, and the Connected Anchor is the point at which the body that the joint is *connected to* will rotate. In the case of the gnome's joints, we want the Connected Anchor and the Anchor to both be at the same position.

When an object that has a hinge joint is selected, both the Anchor and Connected Anchor appear in the scene view: the Connected Anchor is shown as a blue dot, and the Anchor is shown as a blue circle.

Select each of the body parts that have hinge joints, and move both the Anchor and the Connected Anchor to the correct pivot point. For example, select the right arm, and drag the blue dot to the shoulder location to move the Connected Anchor.

Moving the Anchor is slightly trickier, since by default it's in the center, and dragging the center of the object makes Unity move the entire object. To move the Anchor, you first need to manually adjust the location of the Anchor by modifying the numbers in the Inspector—this will change the Anchor's location in the scene view. Once it's out of the center, you can drag it to the correct location, just like you do with the Connected Anchor (see Figure 4-13.)

Repeat this process for both arms (connecting at the shoulder), both legs (connecting at the hip), and the head (connecting at the base of the neck).

Now we'll add the joint that will connect to the Rope object. This will be a `SpringJoint2D` attached to the gnome's right leg, which will allow for free rotation around the joint's anchor point, and will limit the distance that the body will be allowed to be from the end of the rope. (We'll create the rope in the next section.) Spring joints work just like springs in the real world: they're bouncy, and can be stretched a little.

In Unity, they're controlled by two main properties: distance and frequency. The distance refers to the "preferred" length of the spring: the distance that the spring "wants" to return to after being squashed or stretched. The frequency refers to the amount of "stiffness" the string has. Lower values mean looser springs.

Figure 4-13. The anchor points of the left arm, in the correct position; notice how the dot has a ring surrounding it, indicating that both the Connected Anchor and the Anchor are in the same place

To set up springs for use in the Rope, follow these steps:

1. *Add the rope joint.* Select the "Prototype Leg Rope." This should be the top-right leg sprite.
2. *Add the spring joint to it.* Add a `SpringJoint2D` to it. Move its Anchor (the blue circle) so that it's near the end of the leg. Don't move the Connected Anchor (that is, move the blue circle, not the blue dot). The anchor positions on the Gnome can be seen in Figure 4-14.
3. *Configure the joint.* Turn off Auto Configure Distance, and, set the joint's Distance to 0.01, and the Frequency to 5.

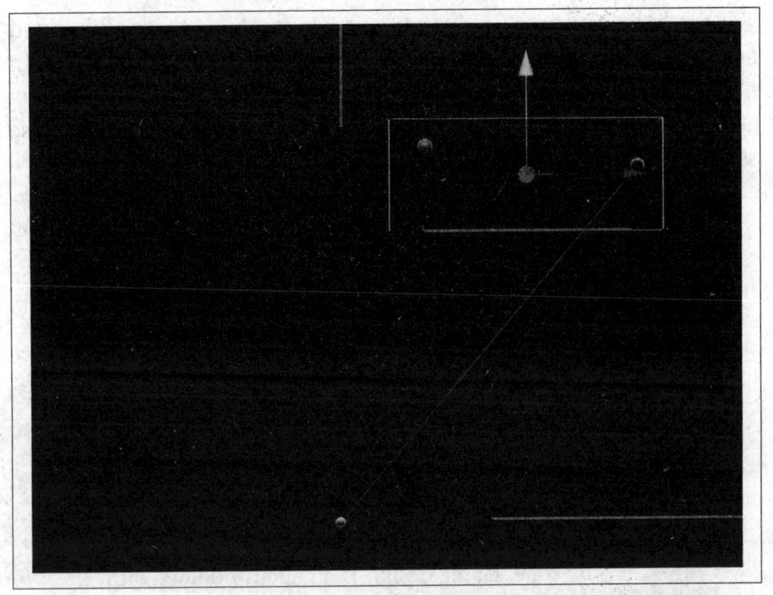

Figure 4-14. Adding the spring joint that will connect the leg to the rope—the joint's Anchor is near the toe

4. *Run the game.* When you do, the gnome will dangle from the middle of the screen.

The last step is to scale the gnome down, so that it will be shown at the right size alongside the other level objects.

5. *Scale the gnome.* Select the parent Gnome object, and change its X and Y scale values to 0.5. This will shrink the gnome by half.

The gnome's now ready to go. It's time to add the rope!

Rope

The rope is the first piece of the game that actually requires code. It works like this: the rope is a collection of game objects that each have rigid bodies and spring joints. Each spring joint is linked to the next Rope object, and that object is linked to the next, all the way up to the top of the rope. The top Rope object is linked to a rigidbody that's fixed in place, so that it stays put. The end of the rope will be attached to one of the gnome's components: the Rope Leg object.

To create the rope, we first need to create an object that will be used as the template for each of the rope segments. We'll then create an object that uses this segment object along with some code to generate the entire rope. To prepare the Rope Segments, follow these steps:

1. *Create the Rope Segment object.* Create a new empty game object, and name it `Rope Segment`.
2. *Add a body to the object.* Add a `Rigidbody2D` component. Set its Mass to 0.5, so that the rope has a bit of heft to it.
3. *Add the joint.* Add a `SpringJoint2D` component. Set its Damping Ratio to 1, and its Frequency to 30.

Feel free to play with other values as well. We found that these values lead to a decently realistic rope. Game design is all about fiddling with numbers.

4. *Create a prefab using the object.* Open the *Gnome* folder in the Assets pane, and drag the Rope Segment object from the Hierarchy Pane into the Assets pane. This will create a new prefab file in that folder.
5. *Delete the original Rope Prefab object.* It won't be needed anymore—you're about to write code that creates multiple instances of the Rope Segment, and connects them up into a rope.

We'll now create the Rope object itself:

1. *Create a new empty game object, and name it "Rope".*
2. *Change Rope's icon.* Because the rope won't have any visible presence in the scene view when the game's not running, you'll want to set an icon for it. Select the newly created Rope object, and click on the cube icon at the top-left of the Inspector (see Figure 4-15).

   Choose the red rounded-rectangle shape, and the Rope object will appear in the scene as a red pill-shaped object (see Figure 4-16).

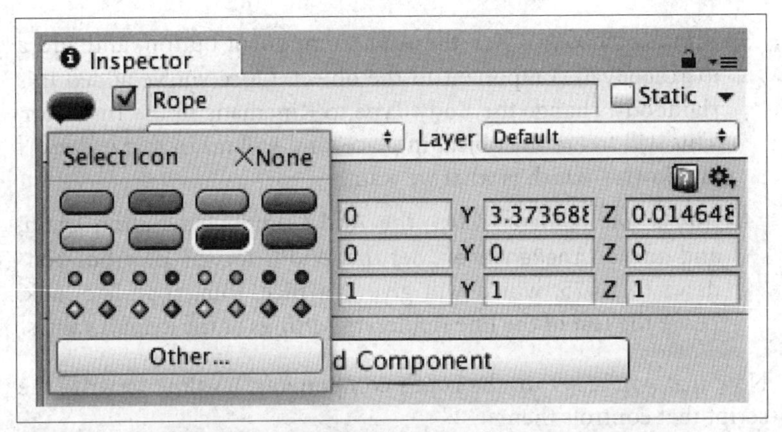

Figure 4-15. Selecting an icon for the Rope object

Figure 4-16. With an icon selected, the Rope object appears in the scene

3. *Add a rigidbody.* Click the Add Component button, and add a `Rigidbody2D` component to the object. Once you've added this rigidbody, change the Body Type to Kinematic in the Inspector. This will freeze the object in place, and will mean that it doesn't fall down—which is what we want.
4. *Add a line renderer.* Click the Add Component button again, and add a `LineRenderer`. Set the Width for the new line renderer to 0.075, which will give it a nice, thin, rope-like look. Leave the rest of the line renderer's settings as the default values.

Now that you've set up the rope's components, it's time to write the script that controls them.

Coding the Rope

Before we can write the code itself, we need to add a script component. To do so, follow these steps:

1. *Add a Rope script to it.* This script doesn't exist yet, but Unity will create the file for it. Select the Rope object, and click the Add Component button.

   Type **Rope**; you won't see any components appear, because Unity doesn't have any components named Rope. What you *will* see is a New Script option (see Figure 4-17). Select it.

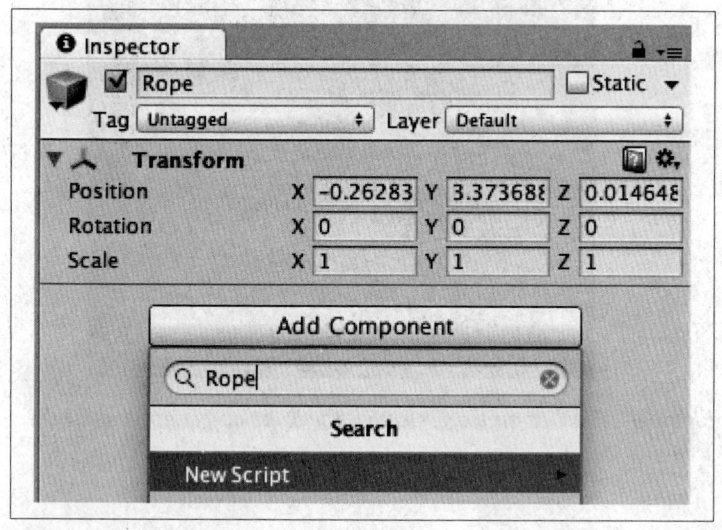

Figure 4-17. Creating the Rope.cs file

Unity will offer to create a new script file. Ensure that the language is set to C Sharp, and that Rope is spelled with a capital R. Click Create and Add. Unity will create the *Rope.cs* file, and will also attach a Rope script component to the Rope object.

2. *Move Rope.cs to the Scripts folder.* By default, Unity puts new scripts in the *Assets* folder; to keep things tidy, move it into *Scripts*.

3. *Add the code to the Rope.cs file.* Open *Rope.cs* by double-clicking on it, or by opening the file in the text editor of your choice.

   Add the following code to it (we'll explain what it does in a moment):

```
using UnityEngine;
using System.Collections;
using System.Collections.Generic;

// The connected rope.
public class Rope : MonoBehaviour {

    // The Rope Segment prefab to use.
    public GameObject ropeSegmentPrefab;

    // Contains a list of Rope Segment objects.
    List<GameObject> ropeSegments = new List<GameObject>();

    // Are we currently extending or retracting the rope?
    public bool isIncreasing { get; set; }
    public bool isDecreasing { get; set; }

    // The rigidbody object that the end of the rope
    // should be attached to.
    public Rigidbody2D connectedObject;

    // The maximum length a rope segment should be (if we
    // need to extend by more than this, create a new rope
    // segment).
    public float maxRopeSegmentLength = 1.0f;

    // How quickly we should pay out new rope.
    public float ropeSpeed = 4.0f;

    // The LineRenderer that renders the actual rope.
    LineRenderer lineRenderer;

    void Start() {

        // Cache the line renderer, so we don't have to look
```

```
        // it up every frame.
        lineRenderer = GetComponent<LineRenderer>();

        // Reset the rope, so that we're ready to go.
        ResetLength();

    }

    // Remove all rope segments, and create a new one.
    public void ResetLength() {

        foreach (GameObject segment in ropeSegments) {
            Destroy (segment);

        }

        ropeSegments = new List<GameObject>();

        isDecreasing = false;
        isIncreasing = false;

        CreateRopeSegment();

    }

    // Attaches a new rope segment at the top of the rope.
    void CreateRopeSegment() {

        // Create the new rope segment.
        GameObject segment = (GameObject)Instantiate(
        ropeSegmentPrefab,
        this.transform.position,
        Quaternion.identity);

        // Make the rope segment be a child of this object,
        // and make it keep its world position
        segment.transform.SetParent(this.transform, true);

        // Get the rigidbody from the segment
        Rigidbody2D segmentBody = segment
          .GetComponent<Rigidbody2D>();

        // Get the distance joint from the segment
        SpringJoint2D segmentJoint =
          segment.GetComponent<SpringJoint2D>();

        // Error if the segment prefab doesn't have a
        // rigidbody or spring joint - we need both
        if (segmentBody == null || segmentJoint == null) {
        Debug.LogError("Rope segment body prefab has no " +
          "Rigidbody2D and/or SpringJoint2D!");
```

```
            return;
        }

        // Now that it's checked, add it to the start of the
        // list of rope segments
        ropeSegments.Insert(0, segment);

        // If this is the *first* segment, it needs to be
        // connected to the gnome

        if (ropeSegments.Count == 1) {
            // Connect the joint on the connected object to
            // the segment
            SpringJoint2D connectedObjectJoint =
              connectedObject.GetComponent<SpringJoint2D>();

            connectedObjectJoint.connectedBody
              = segmentBody;

            connectedObjectJoint.distance = 0.1f;

            // Set this joint to already be at the max
            // length
            segmentJoint.distance = maxRopeSegmentLength;
        } else {
            // This is an additional rope segment. We now
            // need to connect the previous top segment to
            // this one

            // Get the second segment
            GameObject nextSegment = ropeSegments[1];

            // Get the joint that we need to attach to
            SpringJoint2D nextSegmentJoint =
              nextSegment.GetComponent<SpringJoint2D>();

            // Make this joint connect to us
            nextSegmentJoint.connectedBody = segmentBody;

            // Make this segment start at a distance of 0
            // units away from the previous one - it will
            // be extended.
            segmentJoint.distance = 0.0f;
        }

        // Connect the new segment to the
        // rope anchor (i.e., this object)
        segmentJoint.connectedBody =
          this.GetComponent<Rigidbody2D>();
    }
```

```
// Called when we've shrunk the rope, and
// we need to remove a segment.
void RemoveRopeSegment() {

    // If we don't have two or more segments, stop.
    if (ropeSegments.Count < 2) {
        return;
    }

    // Get the top segment, and the segment under it.
    GameObject topSegment = ropeSegments[0];
    GameObject nextSegment = ropeSegments[1];

    // Connect the second segment to the rope's anchor.
    SpringJoint2D nextSegmentJoint =
      nextSegment.GetComponent<SpringJoint2D>();

    nextSegmentJoint.connectedBody =
      this.GetComponent<Rigidbody2D>();

    // Remove the top segment and destroy it.
    ropeSegments.RemoveAt(0);
    Destroy (topSegment);

}

// Every frame, increase or decrease
// the rope's length if necessary
void Update() {

    // Get the top segment and its joint.
    GameObject topSegment = ropeSegments[0];
    SpringJoint2D topSegmentJoint =
        topSegment.GetComponent<SpringJoint2D>();

    if (isIncreasing) {

        // We're increasing the rope. If it's at max
        // length, add a new segment; otherwise,
        // increase the top rope segment's length.

        if (topSegmentJoint.distance >=
          maxRopeSegmentLength) {
              CreateRopeSegment();
        } else {
            topSegmentJoint.distance += ropeSpeed *
            Time.deltaTime;
        }

    }
```

```
if (isDecreasing) {

    // We're decreasing the rope. If it's near zero
    // length, remove the segment; otherwise,
    // decrease the top segment's length.

    if (topSegmentJoint.distance <= 0.005f) {
        RemoveRopeSegment();
    } else {
        topSegmentJoint.distance -= ropeSpeed *
            Time.deltaTime;
    }

}

if (lineRenderer != null) {
    // The line renderer draws lines from a
    // collection of points. These points need to
    // be kept in sync with the positions of the
    // rope segments.

    // The number of line renderer vertices =
    // number of rope segments, plus a point at the
    // top for the rope anchor, plus a point at the
    // bottom for the gnome.
    lineRenderer.positionCount
      = ropeSegments.Count + 2;

    // Top vertex is always at the rope's location.
    lineRenderer.SetPosition(0,
      this.transform.position);

    // For every rope segment we have, make the
    // corresponding line renderer vertex be at its
    // position.
    for (int i = 0; i < ropeSegments.Count; i++) {
        lineRenderer.SetPosition(i+1,
              ropeSegments[i].transform.position);
    }

    // Last point is at the connected
    // object's anchor.
    SpringJoint2D connectedObjectJoint =
      connectedObject.GetComponent<SpringJoint2D>();
    lineRenderer.SetPosition(
      ropeSegments.Count + 1,
      connectedObject.transform.
          TransformPoint(connectedObjectJoint.anchor)
    );
}
```

```
        }
    }
```

This is a large piece of code, so let's step through each part of it:

```
void Start() {

    // Cache the line renderer, so we don't
    // have to look it up every frame.
    lineRenderer = GetComponent<LineRenderer>();

    // Reset the rope, so that we're ready to go.
    ResetLength();

}
```

When the Rope object first appears, its `Start` method is called. This method calls `ResetLength`, which will also be called when the gnome dies. Additionally, the `lineRenderer` variable is set up to point toward the line renderer component attached to the object:

```
// Remove all rope segments, and create a new one.
public void ResetLength() {

    foreach (GameObject segment in ropeSegments) {
        Destroy (segment);
    }

    ropeSegments = new List<GameObject>();

    isDecreasing = false;
    isIncreasing = false;

    CreateRopeSegment();
}
```

The `ResetLength` method deletes all rope segments, resets its internal state by clearing the `ropeSegements` list and the `isDecreasing`/`isIncreasing` properties, and finally calls `CreateRopeSegment` to create a fresh new rope:

```
// Attaches a new rope segment at the top of the rope.
void CreateRopeSegment() {

    // Create the new rope segment.
    GameObject segment = (GameObject)Instantiate(
        ropeSegmentPrefab,
        this.transform.position,
        Quaternion.identity);

    // Make the rope segment be a child of this object,
```

```
// and make it keep its world position.
segment.transform.SetParent(this.transform, true);

// Get the rigidbody from the segment
Rigidbody2D segmentBody
  = segment.GetComponent<Rigidbody2D>();

// Get the distance joint from the segment
SpringJoint2D segmentJoint =
    segment.GetComponent<SpringJoint2D>();

// Error if the segment prefab doesn't have a
// rigidbody or spring joint - we need both
if (segmentBody == null || segmentJoint == null) {
    Debug.LogError(
      "Rope segment body prefab has no " +
      "Rigidbody2D and/or SpringJoint2D!"
      );

    return;
}

// Now that it's checked, add it to the start of
// the list of rope segments
ropeSegments.Insert(0, segment);

// If this is the *first* segment, it needs to be
// connected to the gnome

if (ropeSegments.Count == 1) {
    // Connect the joint on the connected object to
    // the segment
    SpringJoint2D connectedObjectJoint =
        connectedObject.GetComponent<SpringJoint2D>();

    connectedObjectJoint.connectedBody =
      segmentBody;
    connectedObjectJoint.distance = 0.1f;

    // Set this joint to already be at the max
    // length
    segmentJoint.distance = maxRopeSegmentLength;
} else {
    // This is an additional rope segment. We now
    // need to connect the previous top segment
    // to this one

    // Get the second segment
    GameObject nextSegment = ropeSegments[1];

    // Get the joint that we need to attach to
```

```
            SpringJoint2D nextSegmentJoint =
                nextSegment.GetComponent<SpringJoint2D>();

            // Make this joint connect to us
            nextSegmentJoint.connectedBody = segmentBody;

            // Make this segment start at a distance of
            // 0 units away from the previous one - it
            // will be extended.
            segmentJoint.distance = 0.0f;
        }

        // Connect the new segment to the rope
        // anchor (i.e., this object)
        segmentJoint.connectedBody =
            this.GetComponent<Rigidbody2D>();
    }
```

`CreateRopeSegment` creates a new copy of the Rope Segment object, and adds it to the top of the rope chain. As part of doing this, it disconnects the current top of the rope (if one exists), and reconnects it to the newly created segment. It then connects the new segment to the `Rigidbody2D` attached to the Rope object itself.

If this new segment is the only rope segment created so far, it attaches itself to the `connectedObject` rigidbody. This variable will be set up to be the gnome's leg:

```
    // Called when we've shrunk the rope, and
    // we need to remove a segment.
    void RemoveRopeSegment() {

        // If we don't have two or more segments, stop.
        if (ropeSegments.Count < 2) {
            return;
        }

        // Get the top segment, and the segment under it.
        GameObject topSegment = ropeSegments[0];
        GameObject nextSegment = ropeSegments[1];

        // Connect the second segment to the rope's anchor.
        SpringJoint2D nextSegmentJoint =
            nextSegment.GetComponent<SpringJoint2D>();

        nextSegmentJoint.connectedBody =
            this.GetComponent<Rigidbody2D>();

        // Remove the top segment and destroy it.
        ropeSegments.RemoveAt(0);
```

```
        Destroy (topSegment);

    }
```

RemoveRopeSegment works in the opposite way. The top segment is deleted, and the segment underneath it is connected to the Rope rigidbody. Note that RemoveRopeSegment doesn't do anything if there's only a single rope segment, which means that the rope will not vanish entirely if retracted all the way:

```
    // Every frame, increase or decrease
    // the rope's length if necessary
    void Update() {

        // Get the top segment and its joint.
        GameObject topSegment = ropeSegments[0];
        SpringJoint2D topSegmentJoint =
            topSegment.GetComponent<SpringJoint2D>();

        if (isIncreasing) {

            // We're increasing the rope. If it's at max
            // length, add a new segment; otherwise,
            // increase the top rope segment's length.

            if (topSegmentJoint.distance >=
                maxRopeSegmentLength) {

                CreateRopeSegment();

            } else {

                topSegmentJoint.distance += ropeSpeed *
                    Time.deltaTime;

            }

        }

        if (isDecreasing) {

            // We're decreasing the rope. If it's near zero
            // length, remove the segment; otherwise,
            // decrease the top segment's length.

            if (topSegmentJoint.distance <= 0.005f) {
                RemoveRopeSegment();
            } else {
                topSegmentJoint.distance -= ropeSpeed *
                    Time.deltaTime;
            }
```

```
        }

        if (lineRenderer != null) {
            // The line renderer draws lines from a
            // collection of points. These points need to
            // be kept in sync with the positions of the
            // rope segments.

            // The number of line renderer vertices =
            // number of rope segments, plus a point at the
            // top for the rope anchor, plus a point at the
            // bottom for the gnome.
            lineRenderer.positionCount =
              ropeSegments.Count + 2;

            // Top vertex is always at the rope's location.
            lineRenderer.SetPosition(0,
              this.transform.position);

            // For every rope segment we have, make the
            // corresponding line renderer vertex be at its
            // position.
            for (int i = 0; i < ropeSegments.Count; i++) {
                lineRenderer.SetPosition(
                    i+1,
                    ropeSegments[i].transform.position
                );
            }

            // Last point is at the connected
            // object's anchor.
            SpringJoint2D connectedObjectJoint =
              connectedObject.GetComponent<SpringJoint2D>();

            var lastPosition = connectedObject
                .transform
                .TransformPoint(
                    connectedObjectJoint.anchor
                );

            lineRenderer.SetPosition(
              ropeSegments.Count + 1,
              position
            );
        }
    }
```

Every time the Update method is called (that is, every time the game redraws the screen), the rope checks to see if isIncreasing or isDecreasing is true.

If the check reveals that `isIncreasing` is true, then the rope gradually increases the `distance` property of the top rope segment's spring joint. If this property is greater than or equal to the `maxRopeSegment` variable, then a new rope segment is created.

Conversely, if `isDecreasing` is true, the `distance` property is decreased. If this value is near zero, then the top rope segment is removed.

Finally, the `LineRenderer` is updated so that the vertices that define the visual position of the line match the location of the rope segment objects.

Configuring the Rope

Now that the Rope's code has been set up, we can now make the objects in the scene use it. To do so, follow these steps:

1. *Configure the Rope object.* Select the Rope game object. Drag the Rope Segment prefab into the rope's Rope Segment Prefab slot, and drag the gnome's Rope Leg object into the rope's Connected Object slot. Leave everything else as the default values, which were defined in the *Rope.cs* file. When you're done, the Rope's inspector should look like Figure 4-18.
2. *Run the game.* The gnome will now be dangling from the Rope object, and you'll see the line connecting the gnome to a point slightly above it.

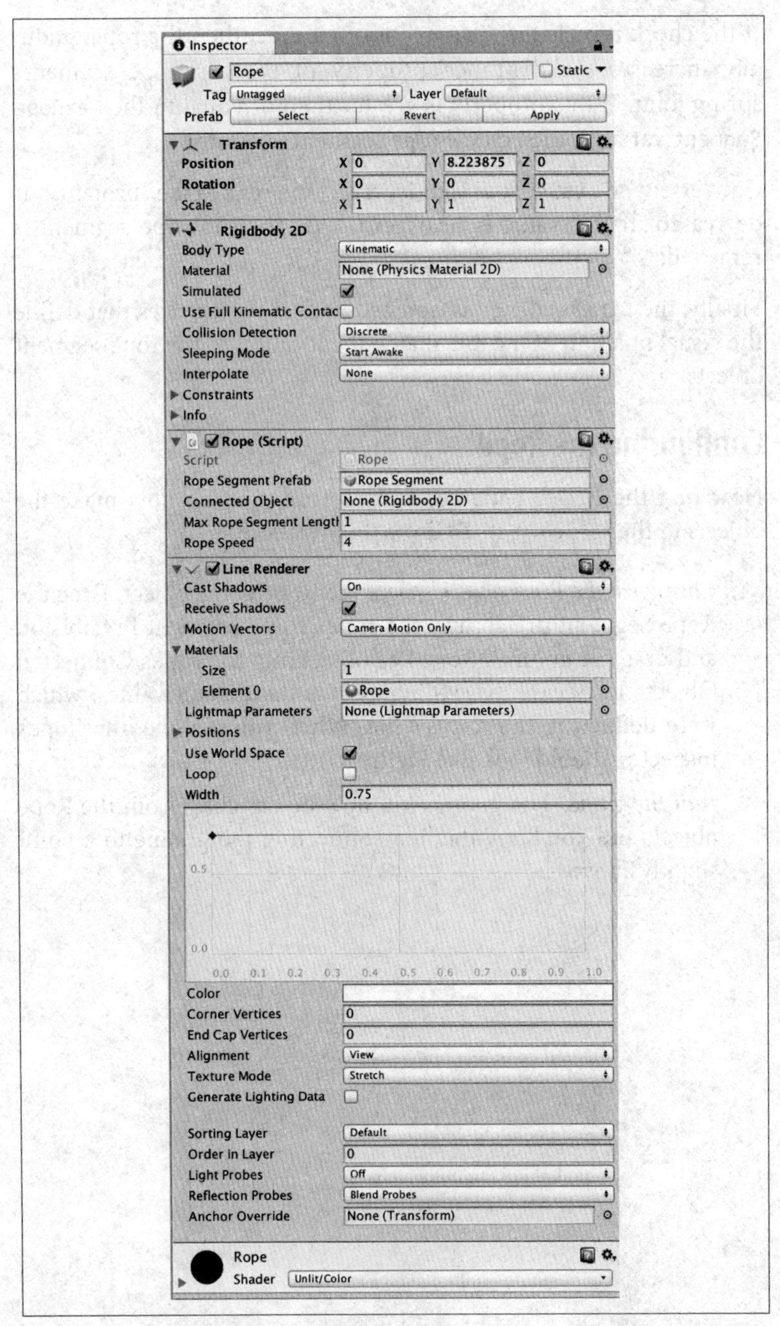

Figure 4-18. The configured Rope object

There's one step left for the Rope—we need to set up a material for the Line Renderer to use:

1. *Create the material.* Open the Assets menu, and choose Create → Material. Name the new material Rope.
2. *Set up the Rope material.* Select the new Rope material, and open the Shader menu in the Inspector. Choose Unlit → Color.

   The inspector will change to show the parameters for the new shader, which will be a single color slot. Change this color to be a dark brown by clicking on the color and picking a new one from the pop-up window.
3. *Make the Rope use the new material.* Select the Rope object, and open the Materials property. Drag and drop the Rope material you just created into the Element 0 slot.
4. *Run the game again.* The rope will now be brown.

Wrapping Up

At this point, the bare-bones structure of the game is starting to take shape. We've got the two most important parts of the game functioning: a ragdoll gnome, and the rope from which it is suspended.

In the next chapter, we'll start creating the systems that implement gameplay using these objects. It's going to be great.

CHAPTER 5

Preparing for Gameplay

Now that the gnome and the rope have both been created, it's time to set up the system that lets the user provide input to the game.

We'll do this in two parts: first, we'll add the script that makes the gnome swing from side to side when the phone is tilted. After that, we'll add the buttons that lengthen and retract the rope.

Once that's done, we'll begin implementing the code that drives the game itself: first, we'll do a bunch of setup work that the gnome will end up using, and then we'll implement a manager object that keeps track of some important game states.

Input

Because we're now at the point where we need to get input from the device, it's time to make sure that the Unity Editor can receive input. Without this, the only way to test the game is to build the game and install it on a device, which can take a while. Unity's all about being able to rapidly test your changes, and waiting for a build to finish will slow you down a lot.

Unity Remote

To allow quickly providing input to the Unity Editor, Unity has an app on the App Store called the Unity Remote. Unity Remote connects to the Unity Editor through your phone's cable; when the game is playing in the Editor, the phone displays a copy of what's being shown in the Game window, and sends back all touch and sensor information to your script. This allows you to test the game without having to do a build—all you need to do is launch the app on your phone, and play the game as if it were already installed.

There are a couple of downsides to the Unity Remote:

- In order to display the game on your phone, Unity compresses the image down quite a bit. In addition to reducing the visual quality of the picture, transferring the image to the phone adds some latency and reduces the framerate.
- Because the game is running on your computer, the framerate won't be the same as if it were running on the phone. If you've got a very graphics-intensive scene, or if your scripts take a long time to run every frame, then you won't get the same performance as if it were running on the phone.
- Finally, of course, it will only work when the phone is connected to your computer.

To get the Unity Remote working, download it from your device's app store, launch it, and connect your phone to your computer using your USB cable. Then click the Play button. The game will appear on your device.

If you don't see anything on the device, open the Edit menu, and choose Project Settings → Editor. The Editor settings will open in the Inspector. Change the Device setting to your phone.

For up-to-the-minute instructions on installing Unity Remote for your device, check Unity's documentation (*http://bit.ly/unity-remote-5*).

Adding Tilt Control

This will be driven by two scripts: `InputManager` (which reads information from the accelerometer) and `Swinging` (which gets the input from the `InputManager` and applies a sideways force to a rigidbody —the rigidbody in question will be the gnome's body).

Creating a Singleton class

InputManager will be a singleton object. This means that there will be precisely one InputManager in the scene, and all other objects will access it. There will be other types of singletons that we'll eventually add to the code, so it makes sense to create a class that multiple parts of our code can reuse. To prepare the Singleton class that the InputManager uses, follow these steps:

1. *Create the Singleton script.* Create a new C# script asset in the *Scripts* folder by opening the Assets menu, and choosing Create → C# Script. Name the script "Singleton".
2. *Add the Singleton code.* Open *Singleton.cs*, and replace its contents with the following code:

```
using UnityEngine;
using System.Collections;

// This class allows other objects to refer to a single
// shared object. The GameManager and InputManager classes
// use this.

// To use this, subclass like so:
// public class MyManager : Singleton<MyManager>  { }

// You can then access the single shared instance of the
// class like so:
// MyManager.instance.DoSomething();

public class Singleton<T> : MonoBehaviour
  where T : MonoBehaviour {

  // The single instance of this class.
  private static T _instance;

  // The accessor. The first time this is called, _instance
  // will be set up. If an appropriate object can't be found,
  // an error will be logged.
  public static T instance {
    get {
      // If we haven't already set up _instance...
      if (_instance == null)
      {
        // Try to find the object.
        _instance = FindObjectOfType<T>();

        // Log if we can't find it.
        if (_instance == null) {
```

```
                Debug.LogError("Can't find " +
                  typeof(T) + "!");
            }
        }

        // Return the instance so that it can be used!
        return _instance;
      }
    }
  }
```

The Singleton class works like this: other classes will subclass this template class, and will gain a static property called instance. This property will always point to the shared instance of this class. This means that when other scripts ask for InputManager.instance, they'll always get the single InputManager.

The advantage of doing it like this is that scripts that need the InputManager won't need to have variables that connect to it.

Implementing an InputManager Singleton

Now that you've created the Singleton class, it's time to create the InputManager.

1. *Create the InputManager game object.* Make a new game object, and name it InputManager.
2. *Create and add the InputManager script.* Select the InputManager object, and click Add Component. Type **InputManager**, and choose to create a new script. Make sure that the name of the script is "InputManager", and that the language is C Sharp.
3. *Add the code to InputManager.cs.* Open the *InputManager.cs* file that was just created, and add the following code to it:

```
using UnityEngine;
using System.Collections;

// Translates the accelerometer data into sideways motion
// info.
public class InputManager : Singleton<InputManager> {

  // How much we're moving. -1.0 = full left, +1.0 = full
  // right
  private float _sidewaysMotion = 0.0f;

  // This property is declared as read-only, so that other
  // classes can't change it
```

```
public float sidewaysMotion {
  get {
    return _sidewaysMotion;
  }
}

// Every frame, store the tilt
void Update () {
  Vector3 accel = Input.acceleration;

  _sidewaysMotion = accel.x;
}
}
```

Every frame, the `InputManager` class samples data from the accelerometer via the built-in `Input` class, and stores the X component (which measures the amount of force being applied to the left and right sides of the device) in a variable. This variable is exposed using the public read-only property `sidewaysMotion`.

A read-only property is used to prevent other classes from accidentally writing to this value.

In short, if any other class wants to find out how much the phone is tilting along the left-right axis, all it needs to do is simply ask for `InputManager.instance.sidewaysMotion`.

Now it's time to write the `Swinging` code:

1. *Select the gnome's Body object.*
2. *Create and add a new C# script called Swinging.cs.* Add the following code to it:

```
using UnityEngine;
using System.Collections;

// Uses the input manager to apply sideways forces to an
// object. Used to make the gnome swing side-to-side.
public class Swinging : MonoBehaviour {

  // How much should we swing by? Bigger numbers = more
  // swing
  public float swingSensitivity = 100.0f;
```

```
    // Use FixedUpdate instead of Update, in order to play
    // better with the physics engine
    void FixedUpdate() {

      // If we have no ridigbody (anymore), remove this
      // component
      if (GetComponent<Rigidbody2D>() == null) {
        Destroy (this);
        return;
      }

      // Get the tilt amount from the InputManager
      float swing = InputManager.instance.sidewaysMotion;

      // Calculate a force to apply
      Vector2 force =
        new Vector2(swing * swingSensitivity, 0);

      // Apply the force
      GetComponent<Rigidbody2D>().AddForce(force);
    }

  }
```

The Swinging class runs code every time the physics system updates. First, it checks to see if the object still has a `Rigidbody2D` component. If it doesn't, then it immediately returns. If it still does, then it takes the `swidewaysMotion` from the InputManager, uses it to create a `Vector2`, and applies that as a force to the object's rigidbody.

3. *Run the game.* Launch Unity Remote on your phone, and tilt the phone side to side; the gnome will move left and right.

If you rotate your phone too far, Unity Remote may rotate to landscape mode, stretching the picture. You may need to turn on your device's rotation lock feature to prevent this..

Controlling the Rope

We'll now add buttons that make the rope lengthen and shorten. These will be implemented using Unity GUI buttons; when the user starts holding the Down button down, it will signal the Rope to begin extending, and when the user stops holding it down, the Rope will stop extending. The Up button works in a similar way, and will make the Rope start and stop contracting.

1. *Add the button.* Open the GameObject menu, and choose UI → Button. This will add the button, as well as a Canvas for it to appear in, and an EventSystem that will handle its input. (You don't need to worry about these additional objects.) Name the button's game object "Down".
2. *Position the button at the bottom-right.* Select the Down button, and click on the Anchor button, which appears at the top-left of the Inspector. Hold down the Shift and Alt keys (Option on a Mac), and click the "'bottom-right" option (see Figure 5-1). Doing this means that we're setting the anchor and position of the button to the bottom-right; as a result, when you do this, the button will move to the bottom-right of the screen.

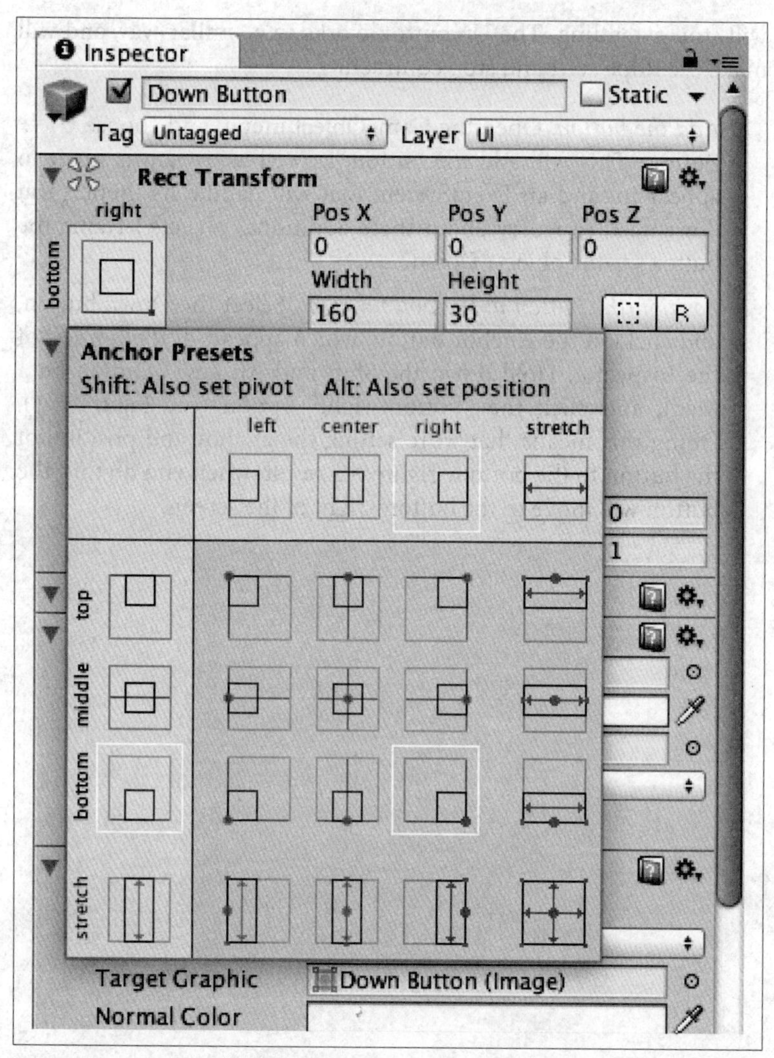

Figure 5-1. Setting the Down button's anchor to bottom-right; in this screenshot, the Shift and Alt keys are being held down, which means clicking the bottom-right anchor point will also set the pivot point and its position

3. *Make the button's text "Down".* The Button has a single child object, named Text. This object is the label that's included inside the button. Select it, and locate the `Text` component that's attached to it in the Inspector. Change the Text property to `Down`. The button will change to read "Down".

4. *Remove the Button component from the Button object.* Click the gear icon at the top-right of the component, and choose "Remove Component."

> This might seem a little unexpected, but we don't actually want this UI element to behave like a "regular" button.
>
> Regular buttons send an event when they're "clicked"—that is, when the user puts a finger on the button, and then lifts that finger. The event is only sent when the finger is lifted, which won't suit our needs—what we want is for an event to be sent when the finger lands on the button, and a second event when the finger is lifted from the button.
>
> So, what we'll do instead is manually add components that will send messages to the rope.

5. *Add an Event Trigger component to the Button object.* This component watches for interactions, and sends messages when those interactions happen.
6. *Add a Pointer Down event.* Click the Add New Event Type button, and choose Pointer Down from the list that appears.
7. *Connect the Rope's `isIncreasing` property to the event.* Click the + button in the Pointer Down list, and a new entry will appear (see Figure 5-2).

   Drag the Rope object from the Hierarchy pane into the object slot that appears.

   Change the Function from "No Function" to Rope → isIncreasing. (When you select this, the drop-down menu will display `Rope.isIncreasing`.) This will make the button modify the `isIncreasing` property on the rope when the finger lands on the button.

   Change the checkbox that appears from unchecked to checked. This will make the `isIncreasing` property change to `true`.

   When you're done, the new item in the Pointer Down event should look like Figure 5-3.

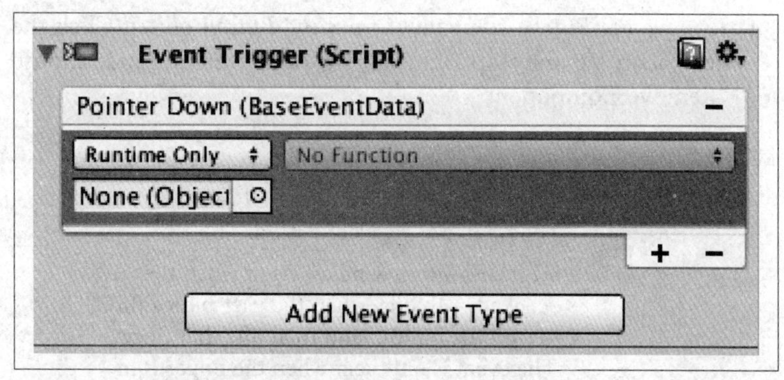

Figure 5-2. A new event in the list

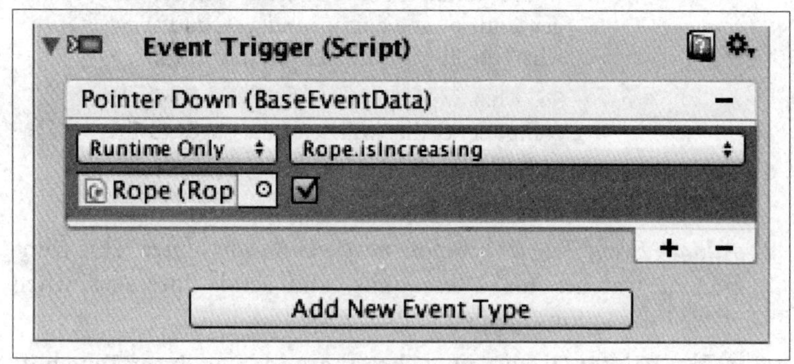

Figure 5-3. The configured Pointer Down event

8. *Add a Pointer Up event, and make it set the Rope's `isIncreasing` property to `false`.* When the finger lifts up off the button, we want the rope to stop increasing.

   Add a new event, Pointer Up, to the Event Trigger by clicking Add New Event Type, and uncheck the checkbox for the Rope's `isIncreasing` property. This will make the `isIncreasing` property change to `false` when the finger lifts.

   When you're done, the Inspector for the Event Trigger should look like Figure 5-4.

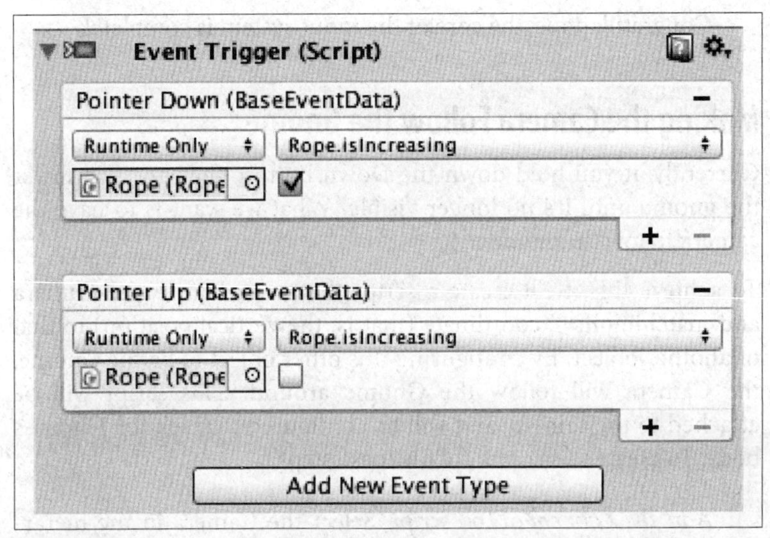

Figure 5-4. The event trigger for the Down button, fully configured

9. *Test out the Down button.* Play the game, and click and hold on the Down button. The rope should start lengthening, and should stop when you release the mouse button. If it doesn't, double-check the events you've configured on the Down button; Pointer Down should set `isIncreasing` to `true`, and Pointer Up should set `isIncreasing` to `false`.

10. *Add the Up button.* You now need to repeat the same process, but for the button that retracts the rope. Add a new button, position it at the bottom-right just like you did for the Down button, and then move it up a little.

    Make its label say "Up," remove the Button component, and add an Event Trigger (with two event types; Pointer Down and Pointer Up). Make the two Event Triggers affect the Rope's `isDecreasing` property.

    The only difference between the two buttons is the text of the label, and the property that they affect. Otherwise, they're identical.

11. *Test out the Up button.* Play the game again. You should now be able to extend and retract the rope.

    You can also use Unity Remote, running on your phone, to swing the gnome from side to side at the same time as changing the rope.

Congratulations: the core of the input system is complete!

Making the Camera Follow the Gnome

Currently, if you hold down the Down button, the rope will lower the gnome until it's no longer visible. What we want is to have the camera follow the gnome.

To achieve this, we'll create a script that's attached to the Camera and matching its Y coordinate (that is, the vertical position) to that of another object. By configuring this other object to be the Gnome, the Camera will follow the Gnome around. This script will be attached to the camera, and will be configured to track the Gnome's body. To create this script, follow these steps:

1. *Add the `CameraFollow` script.* Select the Camera in the hierarchy, and add a new C# component called `CameraFollow`.
2. Add the following code to *CameraFollow.cs*:

```
// Adjusts the camera to always match the Y-position of a
// target object, within certain limits.
public class CameraFollow : MonoBehaviour {

  // The object we want to match the Y position of.
  public Transform target;

  // The highest point the camera can go.
  public float topLimit = 10.0f;

  // The lowest point the camera can go.
  public float bottomLimit = -10.0f;

  // How quickly we should move toward the target.
  public float followSpeed = 0.5f;

  // After all objects have updated position, work out where
  // this camera should be
  void LateUpdate () {

    // If we have a target...
    if (target != null) {

      // Get its position
      Vector3 newPosition = this.transform.position;

      // Work out where this camera should be
      newPosition.y = Mathf.Lerp (newPosition.y,
```

```
              target.position.y, followSpeed);

            // Clamp this new location to within our
            // limits
            newPosition.y =
              Mathf.Min(newPosition.y, topLimit);
            newPosition.y =
              Mathf.Max(newPosition.y, bottomLimit);

            // Update our location
            transform.position = newPosition;
        }

    }

    // When selected in the editor, draw a line from the top
    // limit to the bottom.
    void OnDrawGizmosSelected() {
        Gizmos.color = Color.yellow;

        Vector3 topPoint =
          new Vector3(this.transform.position.x,
            topLimit, this.transform.position.z);
        Vector3 bottomPoint =
          new Vector3(this.transform.position.x,
            bottomLimit, this.transform.position.z);

        Gizmos.DrawLine(topPoint, bottomPoint);
    }
}
```

The `CameraFollow` code uses the `LateUpdate` method, which runs after all other objects have run their `Update` method. `Update` is often used to update the position of objects, which means that using `LateUpdate` means that your code will run *after* these position updates are done.

`CameraFollow` matches the Y-position of the transform of the object that it's attached to, but also ensures that that position doesn't go above or below certain thresholds. This means that when the rope is fully retracted, the camera won't show the empty space above the top of the well. In addition, the code uses the `Mathf.Lerp` function to calculate a position that's close to the target's position. This makes it "loosely" follow the object—the closer the `followSpeed` parameter is to 1, the faster the camera will move.

To visualize these thresholds, the `OnDrawGizmosSelected` method is implemented. This method, which is used by the Unity Editor itself, draws a line from the top threshold to the bottom whenever the camera is selected. If you use the Inspector to change the `topLimit` and `bottomLimit` properties, you'll see the line change length.

2. *Configure the CameraFollow component.* Drag the gnome's Body object into the Target slot (see Figure 5-5), and leave the other properties as they are.

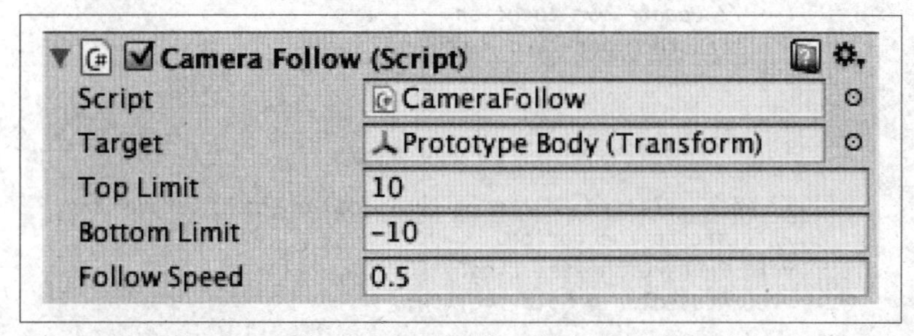

Figure 5-5. Setting up the CameraFollow script

3. *Test the camera.* Run the game, and lower the gnome down using the Down button. The camera will follow the gnome.

Scripts and Debugging

This is a good point to discuss how to find and fix problems in your scripts, since the code will only get more complex from here on out.

Sometimes scripts don't behave the way you want them to, either due to typos or otherwise because of logic errors. To track down and solve these problems in your scripts, you can use the debugging features available in MonoDevelop. You can set breakpoints in your code, inspect the state of a program, and precisely control the execution of your scripts.

While you can use any text editor you want to edit your scripts, you need to use a dedicated development environment app to do your development work. This means using MonoDevelop or Visual Studio. In this book, we'll use MonoDevelop; if you want to use Visual Studio, Microsoft has some excellent documentation (*http://bit.ly/ms-debugger-basics*).

Setting breakpoints

To explore this feature, we'll set a breakpoint in the Rope script that we just wrote, and use it to get a closer look at the behavior of the script. To do so, follow these steps:

1. *Open Rope.cs in MonoDevelop.*
2. *Locate the `Update` method.* Specifically, find the following line:

    ```
    if (topSegmentJoint.distance >= maxRopeSegmentLength) {
    ```

3. *Click in the thin gray line at the left of this line.* A breakpoint will be added (Figure 5-6).

```
// Every frame, increase or decrease the rope's length if neccessary
void Update() {

    // Get the top segment and its joint.
    GameObject topSegment = ropeSegments[0];
    SpringJoint2D topSegmentJoint =
        topSegment.GetComponent<SpringJoint2D>();

    if (isIncreasing) {

        // We're increasing the rope. If it's at max length,
        // add  a new segment; otherwise, increase the top
        // rope segment's length.

        if (topSegmentJoint.distance >= maxRopeSegmentLength) {
            CreateRopeSegment();
        } else {
            topSegmentJoint.distance += ropeSpeed *
                Time.deltaTime;
        }

    }
```

Figure 5-6. Adding a breakpoint

Next, we'll connect MonoDevelop to Unity. This means that when the breakpoint is hit, MonoDevelop will jump in and pause Unity.

4. *Click the Play button at the top left of the window in MonoDevelop (Figure 5-7).*

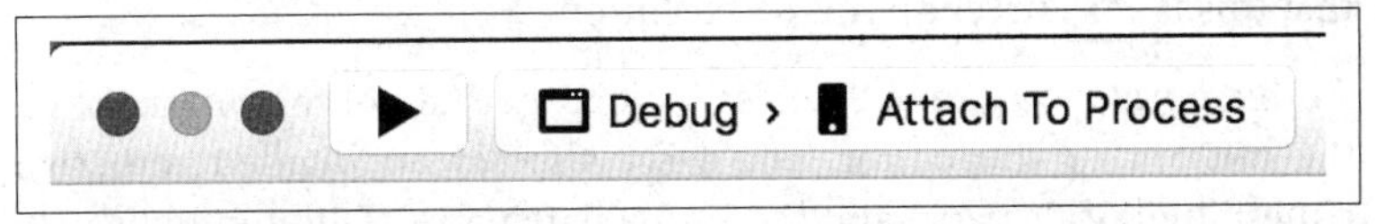

Figure 5-7. The Play button at the top left of the MonoDevelop window

5. *In the window that appears, click Attach (Figure 5-8).*

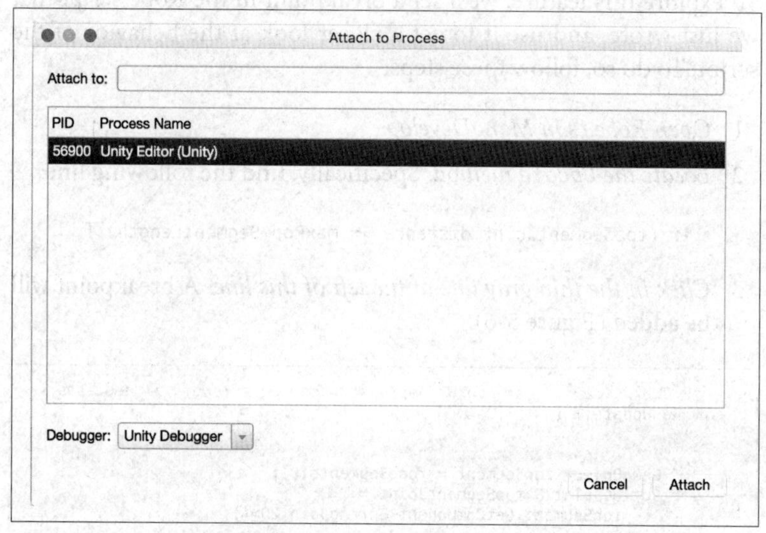

Figure 5-8. The Attach to Process window

MonoDevelop is now attached to Unity. When the breakpoint is hit, Unity will pause, allowing you to debug the code.

When we say Unity will pause, we don't mean that the *game* inside Unity will pause, such as what happens when you click the Pause button. Rather, *the entire Unity application* will freeze, and will not continue executing until you tell MonoDevelop to continue running. If it looks like Unity is hanging, don't panic.

6. *Run the game, and click the Down button.*

The moment you do so, Unity will freeze, and MonoDevelop will appear. The line with the breakpoint will be highlighted, indicating that this is the current point of execution.

At this point, you can get a close look at the state of the program. At the bottom of the editor, you'll see the screen divided into two panes: the Locals pane, and the Immediate pane. (Different tabs may be open, depending on your circumstances; if they are, just click on them to open them.)

The Locals pane lets you see the list of the variables that are currently in scope.

7. *Open the `topSegmentJoint` variable in the Locals pane.* A list of fields inside this variable will appear, allowing you to inspect them (Figure 5-9).

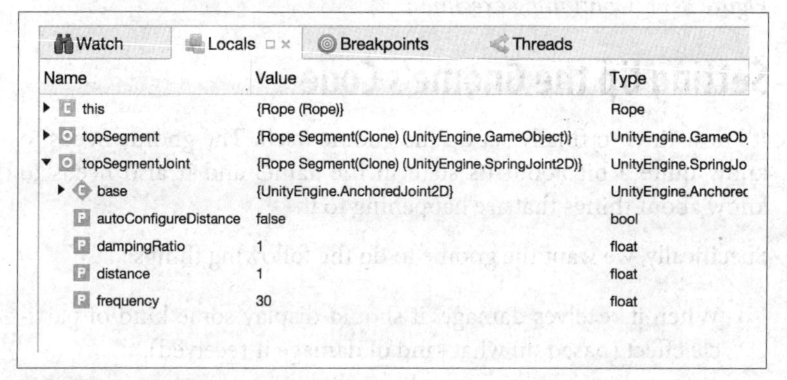

Figure 5-9. The Locals pane, showing the data inside topSegmentJoint

The Immediate pane lets you type C# code that you want to see the result of. For example, you can access the same information about the `distance` property on `topSegmentJoint` seen in Figure 5-9 by typing **`topSegmentJoint.distance`**.

When you're done debugging your code, you'll need to indicate to the debugger that Unity should continue operating. There are two ways you can do this: you can detach the debugger, or you can keep the debugger attached and signal that execution should continue.

If you detach the debugger, breakpoints will stop being hit, and you'll need to reattach the debugger. If you keep the debugger attached, the next breakpoint that's hit will pause the game again.

- *To detach the debugger, click the Stop button (Figure 5-10).*

Figure 5-10. Stopping the debugger

- *To keep the debugger attached and continue executing, click the Continue button (Figure 5-11).*

Figure 5-11. Continuing execution

Setting Up the Gnome's Code

It's now time to finally set up the gnome itself. The gnome needs to know quite a bit about its state in the game, and it also needs to know about things that are happening to it.

Specifically, we want the gnome to do the following things:

- When it receives damage, it should display some kind of particle effect (based on what kind of damage it received).
- When it dies, it should do several things:
  - It should update the sprites for the different body parts (again based on damage), and detach some of them.
  - It should create a Ghost object a short time after dying, which will travel upward.
  - It should spawn blood fountains from the body when a limb detaches; we'll need to know where to spawn these fountains for each limb.
  - When a detached limb stops moving, it should lose all physics so that it doesn't interfere with the player (we don't want dead gnomes piling up at the bottom that stop you from reaching the treasure).
- It should keep track of whether it's holding the treasure; when that changes, it should swap out the sprite of the *holding* arm to one that shows it's holding treasure.
- It should store some important information, like which object the camera should follow and which rigidbody the rope should attach to.
- It should track whether it's dead or not.

Keep in mind this is separate from the overall game state—the gnome doesn't track if you've won the game or not, it just manages the state of the gnome itself. We'll also (eventually) create an object that manages the overall game state, and causes the gnome to die.

To implement this system, we need a script to manage the gnome as a whole. Additionally, we need to add a script to each body part (to manage their sprite and to stop physics after they move.)

We also need to add some additional info to track where the blood fountains should emit from. These positions will be represented by game objects (since they can be positioned in the scene); each body part will have a reference to its corresponding "blood fountain" position.

We'll start with the body part script, and then move on to the gnome script. The reason we're doing it in this order is because the main gnome script will need to know about the `BodyPart` script, while the `BodyPart` script doesn't need to know about the gnome.

1. *Create the BodyPart.cs file.* Create a new C# script, called *BodyPart.cs*. Add the following code to it:

```
[RequireComponent (typeof(SpriteRenderer))]
public class BodyPart : MonoBehaviour {

  // The sprite to use when ApplyDamageSprite is called with
  // damage type 'slicing'
  public Sprite detachedSprite;

  // The sprite to use when ApplyDamageSprite is called with
  // damage type 'burning'
  public Sprite burnedSprite;

  // Represents the position and rotation that a blood
  // fountain will appear at on the main body
  public Transform bloodFountainOrigin;

  // If true, this object will remove its collision, joints,
  // and rigidbody when it comes to rest
  bool detached = false;

  // Decouple this object from the parent, and flag it as
  // needing physics removal
  public void Detach() {
    detached = true;

    this.tag = "Untagged";

    transform.SetParent(null, true);
```

```
}

// Every frame, if we're detached, remove physics if the
// rigidbody is sleeping. This means this detached body
// part will never get in the way of the gnome.
public void Update() {

  // If we're not detached, do nothing
  if (detached == false) {
    return;
  }

  // Is our rigidbody sleeping?
  var rigidbody = GetComponent<Rigidbody2D>();

  if (rigidbody.IsSleeping()) {

    // If so, destroy all joints..
    foreach (Joint2D joint in
      GetComponentsInChildren<Joint2D>()) {
      Destroy (joint);
    }

    // ...rigidbodies...
    foreach (Rigidbody2D body in
      GetComponentsInChildren<Rigidbody2D>()) {
      Destroy (body);
    }

    // ...and the collider.
    foreach (Collider2D collider in
      GetComponentsInChildren<Collider2D>()) {
      Destroy (collider);
    }

    // Finally, remove this script.
    Destroy (this);
  }
}

// Swaps out the sprite for this part based on what kind
// of damage was received
public void ApplyDamageSprite(
  Gnome.DamageType damageType) {

  Sprite spriteToUse = null;

  switch (damageType) {

  case Gnome.DamageType.Burning:
    spriteToUse = burnedSprite;
```

```
            break;

        case Gnome.DamageType.Slicing:
            spriteToUse = detachedSprite;

            break;
        }

        if (spriteToUse != null) {
            GetComponent<SpriteRenderer>().sprite =
                spriteToUse;
        }

    }

}
```

This code won't compile yet, because it makes use of the `Gnome.DamageType` type that hasn't been written yet. We'll add it when we write the Gnome class shortly.

The `BodyPart` script works with two different types of damage: burning and slicing. These are represented by the `Gnome.DamageType` enumeration, which we'll write shortly and which will be used by the damage-related methods in several different classes. `Burning` damage, which will be applied by some kinds of traps, will cause a burning visual effect, while `Slicing` damage will be applied by other traps and will cause a (fairly bloody) slicing effect that involves a stream of red blood particles emitting from the gnome's body.

The `BodyPart` class itself is marked as requiring a `SpriteRenderer` to be attached to the game object for it to work. Because different types of damage will result in changing the sprite for the body part, it's reasonable to require that any object that has the `BodyPart` script also has a `SpriteRenderer`.

The class stores a few different properties: the `detachedSprite` is the sprite that should be used when the gnome receives `Slicing` damage, while the `burnedSprite` is the sprite that should be used when the gnome receives `Burning` damage. Additionally, the `bloodFountainOrigin` is a `Transform` that the main Gnome component will use to add a blood fountain object; it isn't used by this class, but the information is stored in it.

Additionally, the BodyPart script detects if the RigidBody2D component has fallen asleep (that is, it has stopped moving for a few moments, and no new forces are acting on it). When this happens, the BodyPart script removes everything but the sprite renderer from it, effectively turning it into decoration. This is necessary to keep the level from getting filled up with Gnome limbs, which might block player movement.

The blood fountain feature is something that we'll be returning to in "Particle Effects" on page 182; what we're doing here is a bit of initial setup that will make it a lot quicker to add later.

Nextr, it's time to add the Gnome script itself. This script is mostly in preparation for later, when we'll make the gnome actually die, but it's good to have it set up ahead of time.

2. *Create the Gnome script.* Create a new C# script called *Gnome.cs*.
3. *Add the code for the Gnome component.* Add the following code to *Gnome.cs*:

```
public class Gnome : MonoBehaviour {

  // The object that the camera should follow.
  public Transform cameraFollowTarget;

  public Rigidbody2D ropeBody;

  public Sprite armHoldingEmpty;
  public Sprite armHoldingTreasure;

  public SpriteRenderer holdingArm;

  public GameObject deathPrefab;
  public GameObject flameDeathPrefab;
  public GameObject ghostPrefab;

  public float delayBeforeRemoving = 3.0f;
  public float delayBeforeReleasingGhost = 0.25f;

  public GameObject bloodFountainPrefab;
```

```
bool dead = false;

bool _holdingTreasure = false;

public bool holdingTreasure {
  get {
    return _holdingTreasure;
  }
  set {
    if (dead == true) {
      return;
    }

    _holdingTreasure = value;

    if (holdingArm != null) {
      if (_holdingTreasure) {
        holdingArm.sprite =
          armHoldingTreasure;
      } else {
        holdingArm.sprite =
          armHoldingEmpty;
      }
    }

  }
}

public enum DamageType {
  Slicing,
  Burning
}

public void ShowDamageEffect(DamageType type) {
  switch (type) {

  case DamageType.Burning:
    if (flameDeathPrefab != null) {
      Instantiate(
          flameDeathPrefab,cameraFollowTarget.position,
          cameraFollowTarget.rotation
      );
    }
    break;

  case DamageType.Slicing:
    if (deathPrefab != null) {
      Instantiate(
          deathPrefab,
          cameraFollowTarget.position,
          cameraFollowTarget.rotation
```

```
      );
    }
    break;
  }
}

public void DestroyGnome(DamageType type) {

  holdingTreasure = false;

  dead = true;

  // find all child objects, and randomly disconnect
  // their joints
  foreach (BodyPart part in
    GetComponentsInChildren<BodyPart>()) {

    switch (type) {

    case DamageType.Burning:
      // 1 in 3 chance of burning
      bool shouldBurn = Random.Range (0, 2) == 0;
      if (shouldBurn) {
        part.ApplyDamageSprite(type);
      }

      break;

    case DamageType.Slicing:
      // Slice damage always applies a damage sprite
      part.ApplyDamageSprite (type);

      break;
    }

    // 1 in 3 chance of separating from body
    bool shouldDetach = Random.Range (0, 2) == 0;

    if (shouldDetach) {

      // Make this object remove its rigidbody and
      // collider after it comes to rest
      part.Detach ();

      // If we're separating, and the damage type was
      // Slicing, add a blood fountain

      if (type == DamageType.Slicing) {

        if (part.bloodFountainOrigin != null &&
          bloodFountainPrefab != null) {
```

```
                // Attach a blood fountain for
                // this detached part
                GameObject fountain = Instantiate(
                    bloodFountainPrefab,
                    part.bloodFountainOrigin.position,
                    part.bloodFountainOrigin.rotation
                ) as GameObject;

                fountain.transform.SetParent(
                    this.cameraFollowTarget,
                    false
                );
            }
        }

        // Disconnect this object
        var allJoints = part.GetComponentsInChildren<Joint2D>();
        foreach (Joint2D joint in allJoints) {
            Destroy (joint);
        }
      }
    }

    // Add a RemoveAfterDelay component to this object
    var remove = gameObject.AddComponent<RemoveAfterDelay>();
    remove.delay = delayBeforeRemoving;

    StartCoroutine(ReleaseGhost());
  }

  IEnumerator ReleaseGhost() {

    // No ghost prefab? Bail out.
    if (ghostPrefab == null) {
      yield break;
    }

    // Wait for delayBeforeReleasingGhost seconds
    yield return new WaitForSeconds(delayBeforeReleasingGhost);

    // Add the ghost
    Instantiate(
        ghostPrefab,
        transform.position,
        Quaternion.identity
    );
  }

}
```

When you add this code, you'll see a couple of compiler errors, including one or more lines of "The type or namespace name `RemoveAfterDelay` could not be found." This is expected, and we'll address it in a moment by adding the `RemoveAfterDelay` class!

The `Gnome` script is primarily in charge of holding important data to do with the gnome, and for handling what happens when the gnome receives damage. Many of these properties aren't used by the gnome itself, but are used by the Game Manager (which we'll be writing shortly) to set up the game when a new gnome needs to be created.

Some of the highlights of the `Gnome` script:

- The `holdingTreasure` property is set up with an overridden setter. When the `holdingTreasure` property changes, the gnome needs to visually change: if the gnome is now holding treasure (that is, the `holdingTreasure` property is set to `true`), then the "Arm Holding" sprite renderer needs to change to use a sprite that contains the treasure. Conversely, if the property changes to `false`, then the sprite renderer needs to use a sprite that *doesn't*.
- When the gnome receives damage, a "damage effect" object will be created. The specific object will depend on the specific type of damage—if it's `Burning`, then we want a puff of smoke to appear, and if it's `Slicing`, we want a burst of blood. We use the `ShowDamageEffect` to depict this.

In this book, we'll implement the blood effect. The burning effect is left as a challenge for you!

- The `DestroyGnome` method is in charge of telling all connected `BodyPart` components that the gnome has received damage, and that they should detach. Additionally, if the damage type was `Slicing`, then blood fountains should be created.

The method also creates a `RemoveAfterDelay` component, which is something we'll add momentarily. This will remove the entire gnome from the game.

Finally, the method kicks off the `ReleaseGhost` coroutine, which waits for a certain amount of time and then creates a Ghost object. (We're leaving the creation of the Ghost prefab itself as a challenge for you.)

4. *Add the `BodyPart` script component to all of the gnome's body parts.* Do this by selecting all of the body parts (the head, legs, arms, and body), and adding a BodyPart component to it.
5. *Add the container for the blood fountains.* Create an empty game object, and name it "Blood Fountains". Make it a child of the main Gnome object (that is, not any of the body parts, but rather the parent object).
6. *Add the markers for the blood fountains.* Create five new empty game objects, and make them be children of the Blood Fountains object.

   Name them the same as the attached parts: Head, Leg Rope, Leg Dangle, Arm Holding, Arm Loose.

   Move them so that each of the objects is positioned where you want the blood fountain to appear for each limb (e.g., move the Head object to the gnome's neck); then, rotate the object so that its Z axis (the blue arrow) is aiming in the direction you want the blood fountain to shoot from. See Figure 5-12 for an example—the Head object is selected, and the blue arrow is pointing down. This will make the blood fountain shoot upward, out of the gnome's neck.

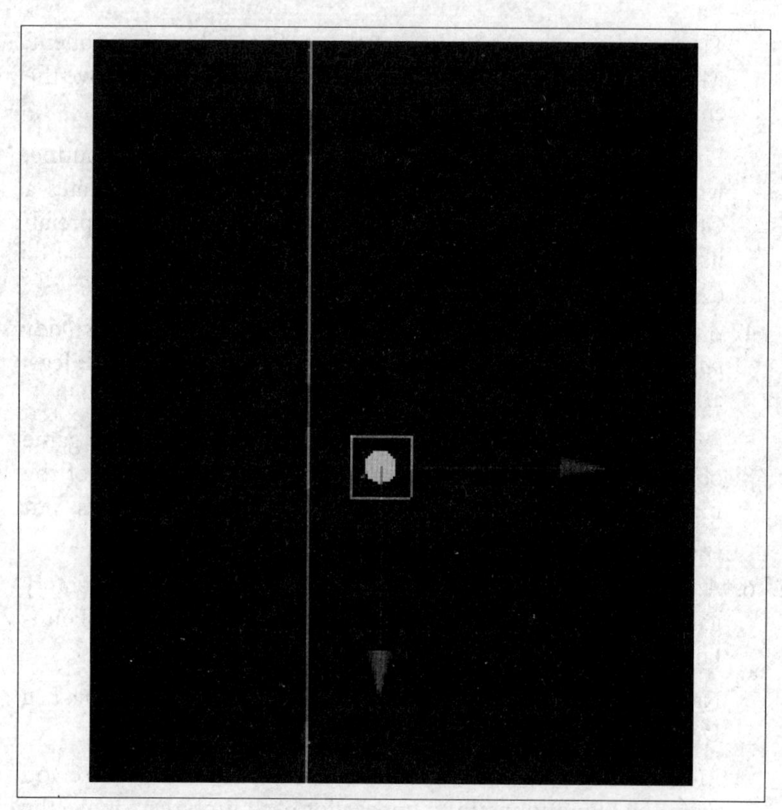

Figure 5-12. The position and rotation of the Head blood fountain

7. *Connect the blood fountain markers to each body part.* For each body part, drag the blood fountain for each part to the Blood Fountain Origin slot. For example, drag the game object for the Head's blood fountain origin to the Head body part (see Figure 5-13). Note that the Body doesn't have one—it's not a part that you detach. Don't drag the body part itself into the slot! Drag the new game object that you just created.

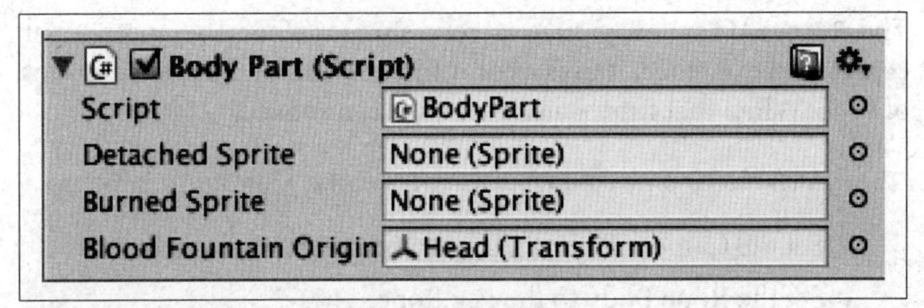

Figure 5-13. Connecting the Head blood fountain object

The gnome body needs to disappear after a certain delay. So, we'll create a script that removes an object after a while. This will be useful for the main game, too—fireballs need to vanish after a while, as well as the ghost.

1. *Create the `RemoveAfterDelay` script.* Create a new C# script called *RemoveAfterDelay.cs*. Add the following code to it:

```
// Removes an object after a certain delay.
public class RemoveAfterDelay : MonoBehaviour {

  // How many seconds to wait before removing.
  public float delay = 1.0f;

  void Start () {
    // Kick off the 'Remove' coroutine.
    StartCoroutine("Remove");
  }

  IEnumerator Remove() {
    // Wait 'delay' seconds, and then destroy the
    // gameObject attached to this object.
    yield return new WaitForSeconds(delay);
    Destroy (gameObject);

    // Don't say Destroy(this) - that just destroys this
    // RemoveAfterDelay script.
  }
}
```

Once you add this code, the compiler error mentioned earlier will go away—the `Gnome` class needs the `RemoveAfterDelay` class to exist in order to compile correctly.

The RemoveAfterDelay class is very simple: when the component comes into existence, it uses a coroutine to wait for a certain amount of time. When that time is up, the object is removed.

2. *Attach the Gnome component to the gnome.* Configure it like so:

   - Set the Camera Follow Target to be the gnome's body.
   - Set the Rope Body to the Leg Rope.
   - Set the Arm Holding Empty sprite to the Prototype Arm Holding sprite.
   - Set the Arm Holding Treasure sprite to the Prototype Arm Holding with Gold sprite.
   - Set the Holding Arm object to the gnome's Arm Holding body part.

   When you're done, the scripts settings should look like Figure 5-14.

   These properties are used by the Game Manager, which we'll add in a moment, so that the Camera Follow is aiming at the correct object, and the Rope is connected to the correct body.

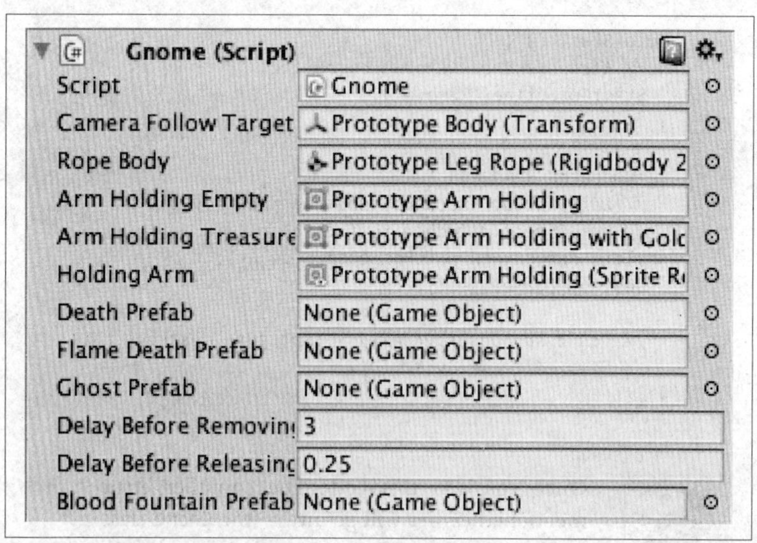

Figure 5-14. The configured Gnome component

Setting Up the Game Manager

The Game Manager is the object that's responsible for managing the entire game. It's in charge of creating gnomes when the game starts, dealing with the gnome touching important objects like traps, treasure, or the level exit, and generally dealing with everything that lasts longer than an individual gnome will.

Specifically, the Game Manager needs to do the following things:

1. When the game starts or restarts:
    a. Create an instance of the gnome.
    b. Remove the old gnome, if necessary.
    c. Position it at the level start.
    d. Attach the Rope to it.
    e. Make the Camera start following it.
    f. Reset all objects that need resetting, like the treasure.
2. When the gnome touches treasure:
    a. Tell the gnome that it has claimed the treasure by changing its `holdingTreasure` property.
3. When the gnome touches a trap:
    a. Tell it to show damage by calling `ShowDamageEffect`.
    b. Kill it by calling `DestroyGnome`.
    c. Reset the game.
4. When the gnome touches the exit:
    a. If it's holding treasure, show a Game Over view.

Before we add the code for the Game Manager, we need to add a class that the Game Manager will depend upon: the `Resettable` class.

We want a general way to run code when the game resets. One way to do this is with Unity Events—we'll make a script called *Resettable.cs* that has a Unity Event, which can be attached to any object that needs resetting. When the game resets, the Game Manager will find all objects that have a `Resettable` component, and invoke the Unity Event.

Doing it this way means that individual objects can be configured to reset themselves, without having to write code for each of them. For example, the `Treasure` object, which we'll add later, will need to change its sprite to indicate that there's no longer any treasure; we'll be adding a `Resettable` object to it that will change the sprite back to its original, treasure-present sprite.

Create the `Resettable` script. Add a new C# script called *Resettable.cs*, and add the following code to it:

```
using UnityEngine.Events;

// Contains a UnityEvent that can be used to reset the state
// of this object.
public class Resettable : MonoBehaviour {

  // In the editor, connect this event to methods that should
  // run when the game resets.
  public UnityEvent onReset;

  // Called by the GameManager when the game resets.
  public void Reset() {
    // Kicks off the event, which calls all of the
    // connected methods.
    onReset.Invoke();
  }
}
```

The `Resettable` code is incredibly simple. All it contains is a `UnityEvent` property, which allows you to add method calls and property changes in the inspector. When the `Reset` method is called, the event is invoked, and all of these methods and property changes are performed.

We can now create the Game Manager:

1. *Create the Game Manager object.* Create a new empty game object, and name it "Game Manager".
2. *Create and add the `GameManager` code to it.* Add a new C# script to it called *GameManager.cs*, and add the following code to it:

```
// Manages the game state.
public class GameManager : Singleton<GameManager> {

  // The location where the gnome should appear.
  public GameObject startingPoint;

  // The rope object, which lowers and raises the gnome.
```

```
public Rope rope;

// The follow script, which will follow the gnome
public CameraFollow cameraFollow;

// The 'current' gnome (as opposed to all those dead ones)
Gnome currentGnome;

// The prefab to instantiate when we need a new gnome
public GameObject gnomePrefab;

// The UI component that contains the 'restart' and 'resume'
// buttons
public RectTransform mainMenu;

// The UI component that contains the 'up', 'down' and
// 'menu' buttons
public RectTransform gameplayMenu;

// The UI component that contains the 'you win!' screen
public RectTransform gameOverMenu;

// If true, ignore all damage (but still show damage
// effects) The 'get; set;' make this a property, to make
// it show up in the list of methods in the Inspector for
// Unity Events
public bool gnomeInvincible { get; set; }

// How long to wait after dying before creating a new gnome
public float delayAfterDeath = 1.0f;

// The sound to play when the gnome dies
public AudioClip gnomeDiedSound;

// The sound to play when the game is won
public AudioClip gameOverSound;

void Start() {
  // When the game starts, call Reset to set up the
  // gnome.
  Reset ();
}

// Reset the entire game.
public void Reset() {

  // Turn off the menus, turn on the gameplay UI
  if (gameOverMenu)
      gameOverMenu.gameObject.SetActive(false);

  if (mainMenu)
```

```
        mainMenu.gameObject.SetActive(false);

    if (gameplayMenu)
        gameplayMenu.gameObject.SetActive(true);

    // Find all Resettable components and tell them to
    // reset
    var resetObjects = FindObjectsOfType<Resettable>();

    foreach (Resettable r in resetObjects) {
        r.Reset();
    }

    // Make a new gnome
    CreateNewGnome();

    // Un-pause the game
    Time.timeScale = 1.0f;
}

void CreateNewGnome() {

    // Remove the current gnome, if there is one
    RemoveGnome();

    // Create a new Gnome object, and make it be our
    // currentGnome
    GameObject newGnome =
      (GameObject)Instantiate(gnomePrefab,
        startingPoint.transform.position,
        Quaternion.identity);

    currentGnome = newGnome.GetComponent<Gnome>();

    // Make the rope visible
    rope.gameObject.SetActive(true);

    // Connect the rope's trailing end to whichever
    // rigidbody the Gnome object wants (e.g., his foot)
    rope.connectedObject = currentGnome.ropeBody;

    // Reset the rope's length to the default
    rope.ResetLength();

    // Tell the cameraFollow to start tracking the new
    // Gnome object
    cameraFollow.target = currentGnome.cameraFollowTarget;

}

void RemoveGnome() {
```

```
    // Don't actually do anything if the gnome is invincible
    if (gnomeInvincible)
      return;

    // Hide the rope
    rope.gameObject.SetActive(false);

    // Stop tracking the gnome
    cameraFollow.target = null;

    // If we have a current gnome, make that no longer be
    // the player
    if (currentGnome != null) {

      // This gnome is no longer holding the treasure
      currentGnome.holdingTreasure = false;

      // Mark this object as not the player (so that
      // colliders won't report when the object
      // hits them)
      currentGnome.gameObject.tag = "Untagged";

      // Find everything that's currently tagged
      // "Player" and remove that tag
      foreach (Transform child in
        currentGnome.transform) {
          child.gameObject.tag = "Untagged";
      }

      // Mark ourselves as not currently having a
      // gnome
      currentGnome = null;
    }
  }

  // Kills the gnome.
  void KillGnome(Gnome.DamageType damageType) {

      // If we have an audio source, play "gnome died"
      // sound
      var audio = GetComponent<AudioSource>();
      if (audio) {
          audio.PlayOneShot(this.gnomeDiedSound);
      }

      // Show the damage effect
      currentGnome.ShowDamageEffect(damageType);

      // If we're not invincible, reset the game and make
      // the gnome not be the current player.
```

```
    if (gnomeInvincible == false) {

        // Tell the gnome that it died
        currentGnome.DestroyGnome(damageType);

        // Remove the Gnome
        RemoveGnome();

        // Reset the game
        StartCoroutine(ResetAfterDelay());

    }
}

// Called when gnome dies.
IEnumerator ResetAfterDelay() {

    // Wait for delayAfterDeath seconds, then call Reset
    yield return new WaitForSeconds(delayAfterDeath);
    Reset();
}

// Called when the player touches a trap
public void TrapTouched() {
  KillGnome(Gnome.DamageType.Slicing);
}

// Called when the player touches a fire trap
public void FireTrapTouched() {
  KillGnome(Gnome.DamageType.Burning);
}

// Called when the gnome picks up the treasure.
public void TreasureCollected() {
    // Tell the currentGnome that it should have the
    // treasure.
    currentGnome.holdingTreasure = true;
}

// Called when the player touches the exit.
public void ExitReached() {
  // If we have a player, and that player is holding
  // treasure, game over!
  if (currentGnome != null &&
    currentGnome.holdingTreasure == true) {

    // If we have an audio source, play the "game
    // over" sound
    var audio = GetComponent<AudioSource>();
    if (audio) {
      audio.PlayOneShot(this.gameOverSound);
```

```
            }

            // Pause the game
            Time.timeScale = 0.0f;

            // Turn off the Game Over menu, and turn on the
            // "game over" screen!
            if (gameOverMenu) {
              gameOverMenu.gameObject.SetActive(true);
            }

            if (gameplayMenu) {
              gameplayMenu.gameObject.SetActive(false);
            }

        }
    }

    // Called when the Menu button is tapped, and when the
    // Resume Game button is tapped.
    public void SetPaused(bool paused) {

        // If we're paused, stop time and enable the menu (and
        // disable the game overlay)
        if (paused) {
            Time.timeScale = 0.0f;
            mainMenu.gameObject.SetActive(true);
            gameplayMenu.gameObject.SetActive(false);
        } else {
            // If we're not paused, resume time and disable
            // the menu (and enable the game overlay)
            Time.timeScale = 1.0f;
            mainMenu.gameObject.SetActive(false);
            gameplayMenu.gameObject.SetActive(true);
        }
    }

    // Called when the Restart button is tapped.
    public void RestartGame() {

        // Immediately remove the gnome (instead of killing it)
        Destroy(currentGnome.gameObject);
        currentGnome = null;

        // Now reset the game to create a new gnome.
        Reset();
    }

}
```

The Game Manager is primarily designed to deal with creating new gnomes, and connecting the other systems to the correct objects. When a new gnome needs to appear, the Rope needs to be connected to the gnome's leg, and the CameraFollow needs to be pointed toward the gnome's body. The Game Manager is also in charge of handling the display of the menus, and for responding to buttons from these menus. (We'll be implementing the menus later.)

Because this is such a large chunk of code, we'll step through what it does in some detail.

Setting Up and Resetting the Game

The `Start` method, which is called the first time the object appears, immediately calls the `Reset` method. `Reset`'s job is to reset the entire game to its initial state, so calling it from `Start` is a quick way to merge the "initial setup" and "reset the game" code into one place.

The `Reset` method itself ensures that the appropriate menu elements, which we'll be setting up later, are visible. All `Resettable` components in the scene are told to reset, and a new gnome is created by calling the `CreateNewGnome` method. Finally, the game is unpaused (just in case the game *was* paused).

```
void Start() {
  // When the game starts, call Reset to set up the gnome.
  Reset ();
}

// Reset the entire game.
public void Reset() {

  // Turn off the menus, turn on the gameplay UI
  if (gameOverMenu)
      gameOverMenu.gameObject.SetActive(false);

  if (mainMenu)
      mainMenu.gameObject.SetActive(false);

  if (gameplayMenu)
      gameplayMenu.gameObject.SetActive(true);

  // Find all Resettable components and tell them to
  // reset
  var resetObjects = FindObjectsOfType<Resettable>();

  foreach (Resettable r in resetObjects) {
```

```
        r.Reset();
    }

    // Make a new gnome
    CreateNewGnome();

    // Un-pause the game
    Time.timeScale = 1.0f;
}
```

Creating a New Gnome

The CreateNewGnome method replaces the gnome with a freshly constructed gnome. It does this by first removing the current gnome, if one exists, and creating a new one; it also enables the rope, and connects the gnome's ankle (its ropeBody) to the end of the rope. The rope is then told to reset its length to the initial value, and finally the camera is made to track the new gnome:

```
void CreateNewGnome() {

    // Remove the current gnome, if there is one
    RemoveGnome();

    // Create a new Gnome object, and make it be our
    // currentGnome
    GameObject newGnome =
      (GameObject)Instantiate(gnomePrefab,
        startingPoint.transform.position,
        Quaternion.identity);

    currentGnome = newGnome.GetComponent<Gnome>();

    // Make the rope visible
    rope.gameObject.SetActive(true);

    // Connect the rope's trailing end to whichever
    // rigidbody the Gnome object wants (e.g., his foot)
    rope.connectedObject = currentGnome.ropeBody;

    // Reset the rope's length to the default
    rope.ResetLength();

    // Tell the cameraFollow to start tracking the new
    // Gnome object
    cameraFollow.target = currentGnome.cameraFollowTarget;

}
```

Removing the Old Gnome

There are two cases where we need to cut the gnome from the rope: when the gnome dies, and when the player decides to start a new game. In both situations, the old gnome gets detached and is no longer treated as the player. It still remains in the level, but if it hits traps, the game won't interpret it as a signal to restart the level anymore.

To remove the active gnome, we disable the rope and stop tracking the current gnome with the camera. We then mark the gnome as not holding the treasure, which switches the sprite back to the regular version, and tags the object as "Untagged." This is done because the traps, which we'll be adding shortly, will be looking for an object tagged "Player"; if the old gnome was still tagged "Player," the trap would end up signaling the Game Manager to restart the level.

```
void RemoveGnome() {

  // Don't actually do anything if the gnome is
  // invincible
  if (gnomeInvincible)
    return;

  // Hide the rope
  rope.gameObject.SetActive(false);

  // Stop tracking the gnome
  cameraFollow.target = null;

  // If we have a current gnome, make that no longer be
  // the player
  if (currentGnome != null) {

    // This gnome is no longer holding the treasure
    currentGnome.holdingTreasure = false;

    // Mark this object as not the player (so that
    // colliders won't report when the object
    // hits them)
    currentGnome.gameObject.tag = "Untagged";

    // Find everything that's currently tagged
    // "Player", and remove that tag
    foreach (Transform child in
      currentGnome.transform) {
        child.gameObject.tag = "Untagged";
    }
```

```
            // Mark ourselves as not currently having a
            // gnome
            currentGnome = null;
        }
    }
```

Killing a gnome

When the gnome is killed, we need to show the appropriate in-game effects. These include sounds and special effects; in addition, if the gnome is not currently invincible, we should indicate to the gnome that it died, remove the gnome, and then reset the game after a certain delay. Here's the code we'll use to accomplish this:

```
void KillGnome(Gnome.DamageType damageType) {

    // If we have an audio source, play "gnome died" sound
    var audio = GetComponent<AudioSource>();

    if (audio) {
        audio.PlayOneShot(this.gnomeDiedSound);
    }

    // Show the damage effect
    currentGnome.ShowDamageEffect(damageType);

    // If we're not invincible, reset the game and make
    // the gnome not be the current player.
    if (gnomeInvincible == false) {

        // Tell the gnome that it died
        currentGnome.DestroyGnome(damageType);

        // Remove the gnome
        RemoveGnome();

        // Reset the game
        StartCoroutine(ResetAfterDelay());

    }
}
```

Resetting the Game

When the gnome dies, we want the camera to linger on the point where that death happened. This will allow the player to watch the gnome fall down for a bit, before returning to the top of the screen.

To do this, we use a coroutine to wait for a number of seconds (stored in `delayAfterDeath`), and then call `Reset` to reset the game state:

```
// Called when gnome dies.
IEnumerator ResetAfterDelay() {

    // Wait for delayAfterDeath seconds, then call Reset
    yield return new WaitForSeconds(delayAfterDeath);
    Reset();

}
```

Dealing with Touching

The next three methods all deal with reacting to the gnome touching certain objects. If the gnome touches a trap, we call `KillGnome` and indicate that slicing damage was done. If the gnome touches a fire trap, we indicate that burning damage was done. Finally, if the treasure was collected, we make the gnome start holding that treasure. Here's the code we'll use to accomplish this:

```
// Called when the player touches a trap
public void TrapTouched() {
  KillGnome(Gnome.DamageType.Slicing);
}

// Called when the player touches a fire trap
public void FireTrapTouched() {
  KillGnome(Gnome.DamageType.Burning);
}

// Called when the gnome picks up the treasure.
public void TreasureCollected() {
  // Tell the currentGnome that it should have the
  // treasure.
  currentGnome.holdingTreasure = true;
}
```

Reaching the Exit

When the gnome touches the exit at the top of the level, we need to check to see if the current gnome is holding treasure. If that's the case, the player wins! As a result, we play a "game over" sound (we'll be setting that up in "Audio" on page 196), pause the game by setting the time scale to zero, and show the Game Over screen (which will include a button that resets the game):

```
// Called when the player touches the exit.
public void ExitReached() {
  // If we have a player, and that player is holding
  // treasure, game over!
  if (currentGnome != null &&
    currentGnome.holdingTreasure == true) {

    // If we have an audio source, play the "game
    // over" sound
    var audio = GetComponent<AudioSource>();
    if (audio) {
      audio.PlayOneShot(this.gameOverSound);
    }

    // Pause the game
    Time.timeScale = 0.0f;

    // Turn off the Game Over menu, and turn on the
    // Game Over screen!
    if (gameOverMenu) {
      gameOverMenu.gameObject.SetActive(true);
    }

    if (gameplayMenu) {
      gameplayMenu.gameObject.SetActive(false);
    }
  }
}
```

Pausing and Unpausing

Pausing the game involves doing three things: first, time is stopped by setting the time scale to zero. Next, the main menu is made visible, and the gameplay UI is hidden. To unpause the game, we simply do the reverse—set time moving again, hide the main menu, and show the gameplay UI:

```
// Called when the Menu button is tapped, and when the
// Resume Game is tapped.
public void SetPaused(bool paused) {

  // If we're paused, stop time and enable the menu (and
  // disable the game overlay)
  if (paused) {
    Time.timeScale = 0.0f;
    mainMenu.gameObject.SetActive(true);
    gameplayMenu.gameObject.SetActive(false);
  } else {
    // If we're not paused, resume time and disable
    // the menu (and enable the game overlay)
```

```
        Time.timeScale = 1.0f;
        mainMenu.gameObject.SetActive(false);
        gameplayMenu.gameObject.SetActive(true);
    }
}
```

Handling the Reset Button

The `RestartGame` method will be called when the user clicks on certain buttons in the UI. This method immediately restarts the game:

```
// Called when the Restart button is tapped.
public void RestartGame() {

  // Immediately remove the gnome (instead of killing it)
  Destroy(currentGnome.gameObject);
  currentGnome = null;

  // Now reset the game to create a new gnome.
  Reset();
}
```

Preparing the Scene

Now that the code is written, we can set up the scene to use it:

1. *Create the start point.* This is an object that the Game Manager will use to position freshly created gnomes. Create a new game object, and call it "Start Point." Position it where you want the gnome to start (somewhere near the Rope will do; see Figure 5-15), and change its icon to be a yellow capsule (in the same way as you set up the icon of the Rope).

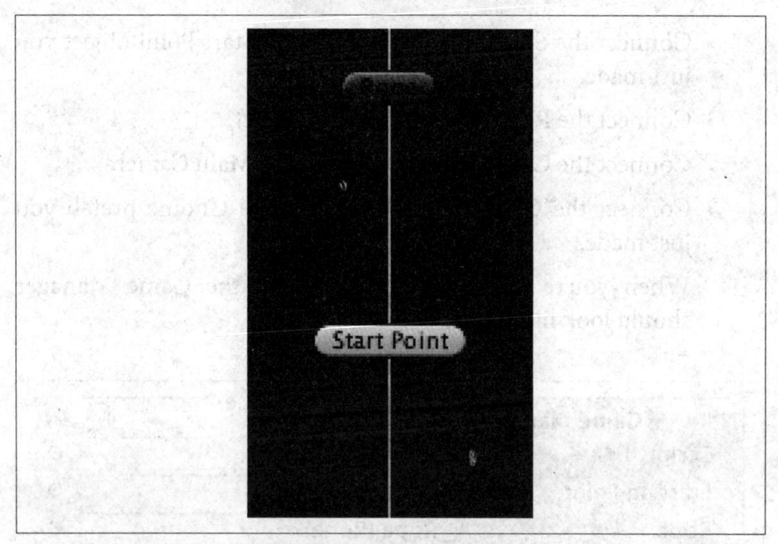

Figure 5-15. Positioning the start point

2. *Turn the gnome into a prefab.* Gnomes will now be created by the Game Manager, which means that the gnome that's currently in the scene needs to be removed. Before you remove it, you need to turn it into a prefab, so that the Game Manager can create instances of it at runtime.

   Drag the gnome into the *Gnome* folder in the Project pane. A new prefab object will be created (see Figure 5-16), which will be a complete copy of the origial Gnome object.

   Now that you've created a prefab, you don't need the object in the scene anymore. Delete the gnome from the scene.

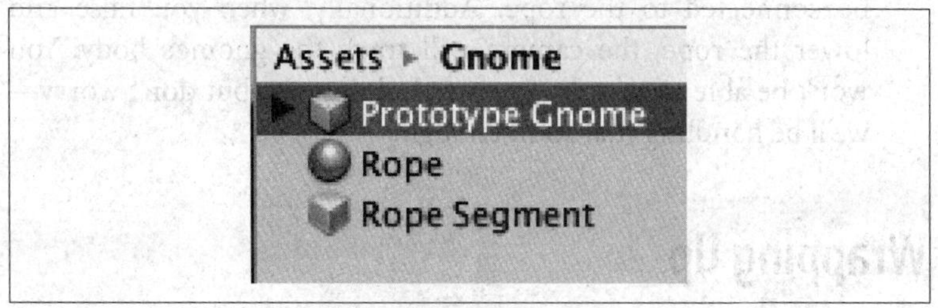

Figure 5-16. The gnome, as a prefab in the Gnome folder

3. *Configure the Game Manager.* There are several connections we need to set up for the Game Manager:

- Connect the Starting Point field to the Start Point object you just made.
- Connect the Rope field to the Rope object.
- Connect the Camera Follow field to the Main Camera.
- Connect the Gnome Prefab field to the Gnome prefab you just made.

  When you're done, the inspector for the Game Manager should look like Figure 5-17.

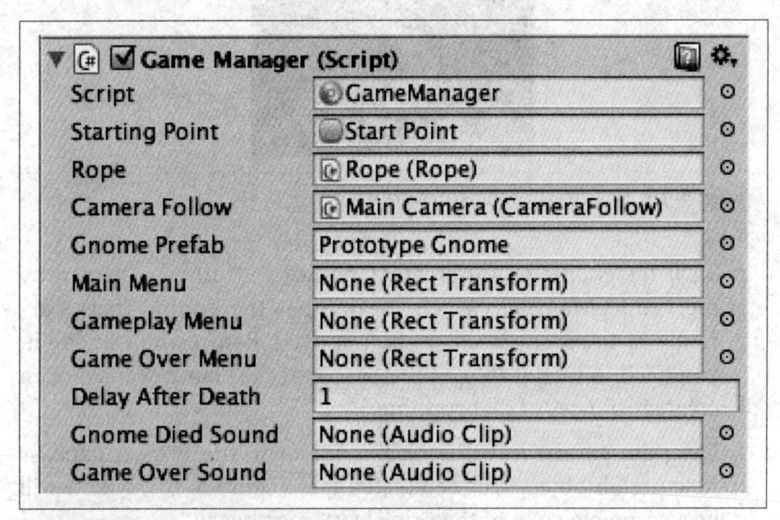

Figure 5-17. The Game Manager setup

4. *Test the game.* The gnome will appear at the start point, and will be connected to the rope. Additionally, when you raise and lower the rope, the camera will track the gnome's body. You won't be able to test the treasure holding yet, but don't worry—we'll be handling that soon enough!

Wrapping Up

Now that the Game Manager is in place, it's time to add actual gameplay. In Chapter 6, we'll start adding elements that the gnome will interact with: the treasure and the traps.

CHAPTER 6
Building Gameplay with Traps and Objectives

Now that the foundations of the gameplay are set up, we can start adding in game elements like traps and treasure. From that point, much of the rest of the game is simply level design.

Simple Traps

Most of this game is about when the player hits things—traps, the treasure, the exit, and so on. Because detecting when the player hits certain objects is so important, we'll create a generic script that triggers a Unity Event when any object that's tagged with "Player" collides with it. This event can then be set up in different ways for different objects: the traps can be configured to tell the Game Manager that the gnome has received damage, the treasure can be configured to tell the Game Manager that the gnome has collected treasure, and the exit can be configured to tell the Game Manager that the gnome has reached the exit.

Now, create a new C# script called *SignalOnTouch.cs*, and add the following code to it:

```
using UnityEngine.Events;

// Invokes a UnityEvent when the Player collides with this
// object.
[RequireComponent (typeof(Collider2D))]
public class SignalOnTouch : MonoBehaviour {
```

```
    // The UnityEvent to run when we collide.
    // Attach methods to run in the editor.
    public UnityEvent onTouch;

    // If true, attempt to play an AudioSource when we collide.
    public bool playAudioOnTouch = true;

    // When we enter a trigger area, call SendSignal.
    void OnTriggerEnter2D(Collider2D collider) {
      SendSignal (collider.gameObject);
    }

    // When we collide with this object, call SendSignal.
    void OnCollisionEnter2D(Collision2D collision) {
      SendSignal (collision.gameObject);
    }

    // Checks to see if this object was tagged as Player, and
    // invoke the UnityEvent if it was.
    void SendSignal(GameObject objectThatHit) {

      // Was this object tagged Player?
      if (objectThatHit.CompareTag("Player")) {

        // If we should play a sound, attempt to play it
        if (playAudioOnTouch) {
          var audio = GetComponent<AudioSource>();

          // If we have an audio component,
          // and this component's parents
          // are active, then play
          if (audio &&
            audio.gameObject.activeInHierarchy)
            audio.Play();
        }

        // Invoke the event
        onTouch.Invoke();
      }
    }

  }
```

The `SignalOnTouch` class's main code is handled in the `SendSignal` method, which is called by `OnCollisionEnter2D` and `OnTriggerEnter2D`. These latter two methods are called by Unity when an object touches a collider, or when an object enters a trigger. The `SendSignal` method checks the tag of the object that collided, and if it was "Player," it invokes the Unity Event.

Now that the `SignalOnTouch` class is ready, we can add the first trap:

1. *Import the level object sprites.* Import the contents of the *Sprites/Objects* folder into your project.
2. *Add the brown spikes.* Locate the SpikesBrown sprite, and drag it into the scene.
3. *Configure the spike object.* Add a `PolygonCollider2D` component to the spikes, as well as a `SignalOnTouch` component.

   Add a new function to the `SignalOnTouch`'s event. Drag the Game Manager into the object slot, and make the function be `GameManager.TrapTouched`. See Figure 6-1.

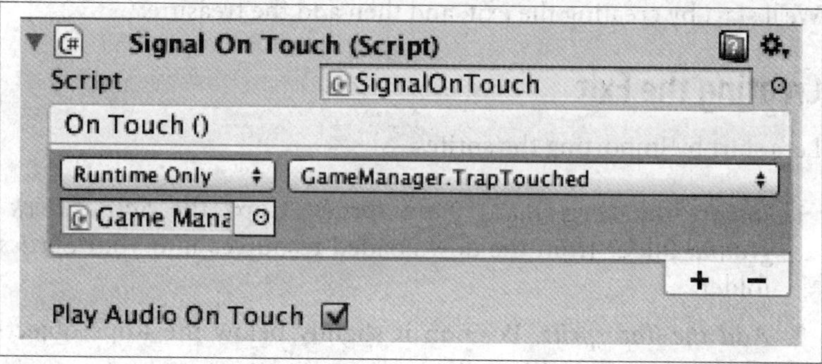

Figure 6-1. Setting up the spike

4. *Turn the spikes into a prefab.* Drag the SpikesBrown object from the Hierarchy into the *Level* folder. This will create a prefab, which means that you can make multiple copies of the objects.
5. *Test it out.* Run the game. Make the gnome hit the spikes. He'll fall off-camera, and respawn!

Treasure and Exit

Now that you've successfully added a way to kill the gnome, it's time to add a way to win the game. You'll do this by adding two new items: the treasure and the exit.

The treasure is a sprite at the bottom of the well that detects when the player has touched it, and signals the Game Manager. When that happens, the Game Manager will inform the gnome that it is carrying treasure, which will make the gnome's arm sprite change to look like it's carrying treasure.

The exit is another sprite, positioned at the top of the well. Like the treasure, it detects when the player has touched it, and notifies the Game Manager. If the gnome happens to be carrying treasure, the player has won the game.

The majority of the work for both of these objects is handled by the `SignalOnTouch` component—when the exit is reached, the Game Manager's `ExitReached` method needs to be called, and when the treasure is touched, the Game Manager's `TreasureCollected` method needs to be called.

We'll start by creating the exit, and then add the treasure.

Creating the Exit

Let's start by importing the sprites:

1. *Import the Level Background sprites.* Copy the *Sprites/Background* folder from the downloaded resources into your *Sprites* folder.
2. *Add the Top sprite.* Position it slightly below the Rope object. This sprite will be the Exit.
3. *Configure the sprite.* Add a `BoxCollider2D` component to the sprite, and set its Is Trigger property to on. Click the Edit Collider button, and resize the box so that it's short and wide (see Figure 6-2).

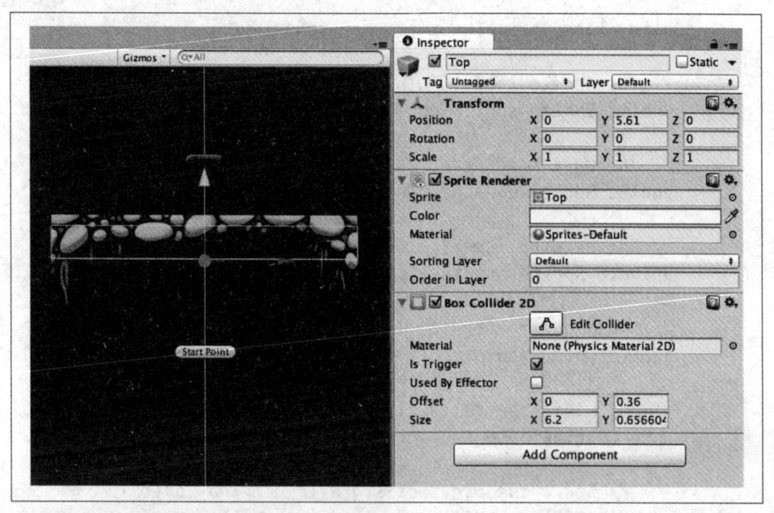

Figure 6-2. Setting up the exit's collider to be wide and short, and positioning it above the level

4. *Make the sprite signal the game controller when it's touched.* Add a `SignalOnTouch` component to the sprite. Add an entry in the component's event, and connect it to the Game Manager. Set the function to GameManager.ExitReached. This will make the Game Manager's `ExitReached` method run when the gnome touches it.

Next up, we need to add the treasure.

The way the treasure works is this: by default, the Treasure object shows a treasure sprite. When the player touches it, the Game Manager's `TreasureCollected` method is called, and the treasure's sprite will change to show that the treasure has been collected. If the gnome dies, the Treasure object will reset to show the sprite that contains the treasure.

Because swapping out one sprite for another is going to be a common thing in the rest of the game, especially when we get to the polishing stage, it makes sense to create a generic sprite-swapping class, and set up the treasure using it.

Create a new C# script called *SpriteSwapper.cs*. Add the following code to it: <<<

```
// Swaps out a sprite for another. For example, the treasure
// switches from 'treasure present' to 'treasure not present'.
```

```
public class SpriteSwapper : MonoBehaviour {

    // The sprite that should be displayed.
    public Sprite spriteToUse;

    // The sprite renderer that should use the new sprite.
    public SpriteRenderer spriteRenderer;

    // The original sprite. Used when ResetSprite is called.
    private Sprite originalSprite;

    // Swaps out the sprite.
    public void SwapSprite() {

      // If this sprite is different than the current sprite...
      if (spriteToUse != spriteRenderer.sprite) {

        // Store the previous store in originalSprite
        originalSprite = spriteRenderer.sprite;

        // Make the sprite renderer use the new sprite.
        spriteRenderer.sprite = spriteToUse;
      }
    }

    // Reverts back to the old sprite.
    public void ResetSprite() {

      // If we have a previous sprite...
      if (originalSprite != null) {
        // ...make the sprite renderer use it.
        spriteRenderer.sprite = originalSprite;
      }
    }
}
```

The `SpriteSwapper` class is designed to do two things: when the `SwapSprite` method is called, the `SpriteRenderer` attached to the game object is told to change its sprite. Additionally, the original sprite is stored in a variable. When the `ResetSprite` method is called, the sprite renderer is restored to its original sprite.

We can now create and set up the Treasure object:

1. *Add the treasure sprite.* Locate the TreasurePresent sprite, and add it to the scene. Put it somewhere near the bottom, but make sure that the gnome can still reach it.
2. *Add a collider to the treasure.* Select the treasure sprite, and add a Box Collider 2D. Make this collider be a trigger.

3. *Add and configure a sprite swapper.* Add a `SpriteSwapper` component. Drag the treasure sprite itself onto the Sprite Renderer field. Next, locate the TreasureAbsent sprite, and drag it onto the sprite swapper's Sprite To Use field.

4. *Add and configure a signal-on-touch component.* Add a `SignalOnTouch` component. Add two entries into the On Touch list:

   - First, connect the Game Manager object, and make the event's method be GameManager.TreasureCollected.
   - Next, connect the treasure sprite (that is, the object you're currently configuring), and make the method SpriteSwapper.SwapSprite.

5. *Add and configure a `Resettable` component.* Add a `Resettable` component to the object. Add a single entry to the On Touch method, make the method be SpriteSwapper.ResetSprite, and connect the Treasure object to it.

When you're done, the Treasure object's Inspector should look like Figure 6-3.

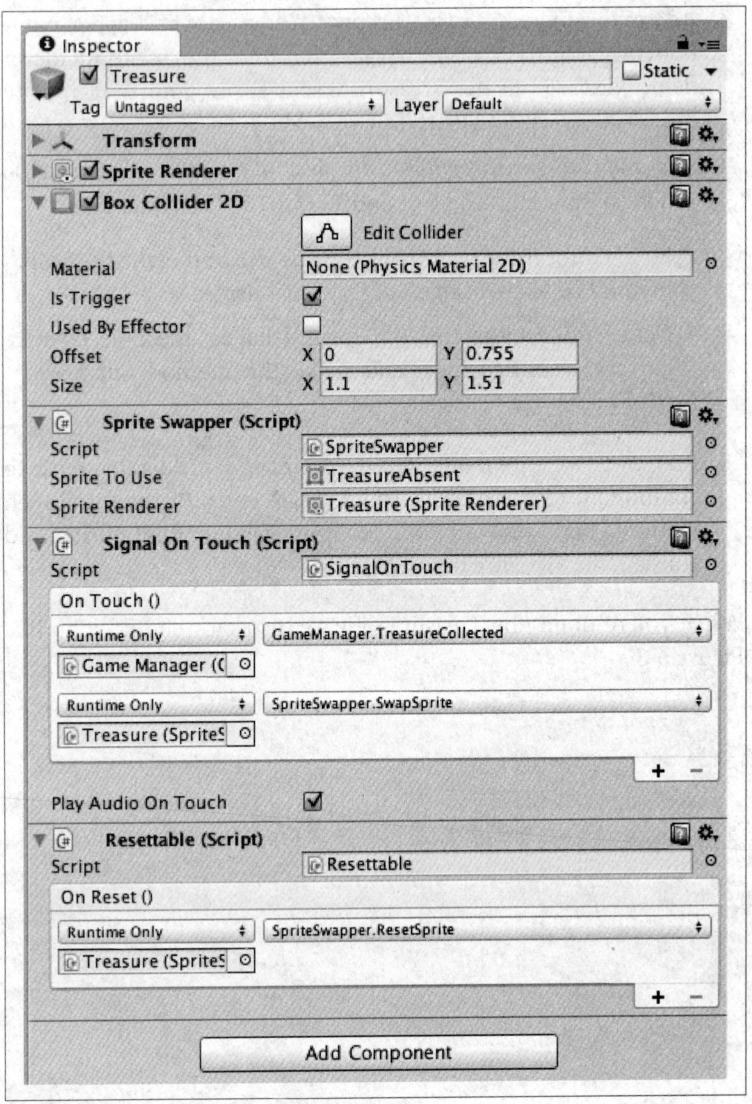

Figure 6-3. The configured Treasure object

6. *Test the game.* Run the game, and touch the treasure. When you touch it, the treasure will disappear; if you die, the treasure will reappear when the gnome respawns.

Adding a Background

Currently, the gnome is set against the default Unity background, which is kind of an ugly blue. We'll add a temporary background, which will be eventually replaced with a background sprite when we start to polish the art.

1. *Add the background quad.* Open the GameObject menu, and choose 3D Object → Quad. Name the new object "Background."
2. *Move the background further back.* To avoid a situation where the background quad is drawn in front of the game's sprites, we'll move it further back from the camera. Set the Z value of the background quad's position to 10.

Even though this is a 2D game, Unity is still a 3D engine. This means that we can take advantage of the fact that the concept of things being "behind" other objects still exists, such as what we're doing here.

3. *Position the background quad.* Switch to the Rect tool by pressing T, and then use the resizing handles to resize the background quad. Make the top edge of the background line up with the sprite at the top of the level, and the bottom edge line up with the treasure (see Figure 6-4).
4. *Test the game.* When you play the game, the level will have a gray background color behind it.

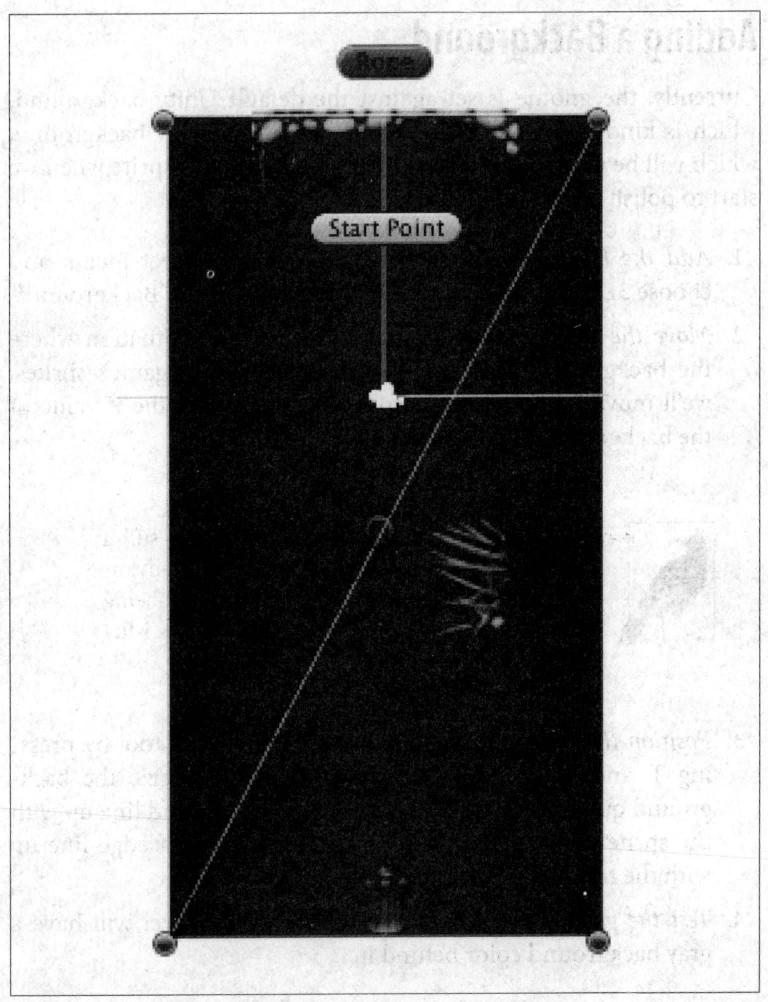

Figure 6-4. Sizing the background quad

Wrapping Up

At this point in the building of the game, the core gameplay functionality is present. Quite a lot of gameplay has been added:

- The gnome is physically simulated, and is attached to a physically simulated rope.
- The rope is controllable via on-screen buttons, which means that the gnome can be lowered and raised.

- The camera is set up to track the gnome, so that it stays in sight the entire time.
- The gnome responds to the phone tilting, and moves left to right.
- The gnome is able to be killed by colliding with traps, and is able to collect the treasure.

You can see a screenshot of the game in its current state in Figure 6-5.

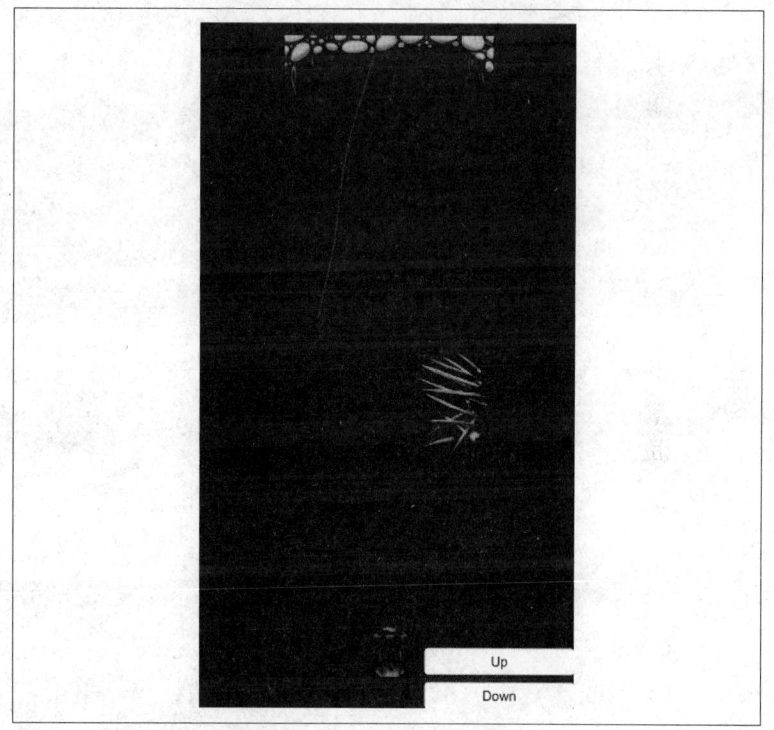

Figure 6-5. The game, at the end of this chapter

While functionally complete, it's not the best-looking game at this point. The gnome is still a stick figure, and the levels are quite barebones. In Chapter 7, we'll continue building the game and improving the visuals of every on-screen element.

CHAPTER 7

Polishing the Game

By the end of this chapter, you'll have applied a number of tweaks to *Gnome's Well That Ends Well*, and the final result will be something like Figure 7-1.

There are three main areas of polish that we'll add to the game:

Visual polish
: We'll be adding new sprites for the gnome, improving the look of the backgrounds, and adding particle effects that improve the look of the game.

Gameplay polish
: We'll be adding different kinds of traps, a title screen, and also a way to make the gnome invincible, which will help with gameplay testing.

Audio polish
: We'll also be adding sound effects to the game, which react to what the player's doing.

The resources used in this chapter can be found in the assets package available at *https://www.secretlab.com.au/books/unity*.

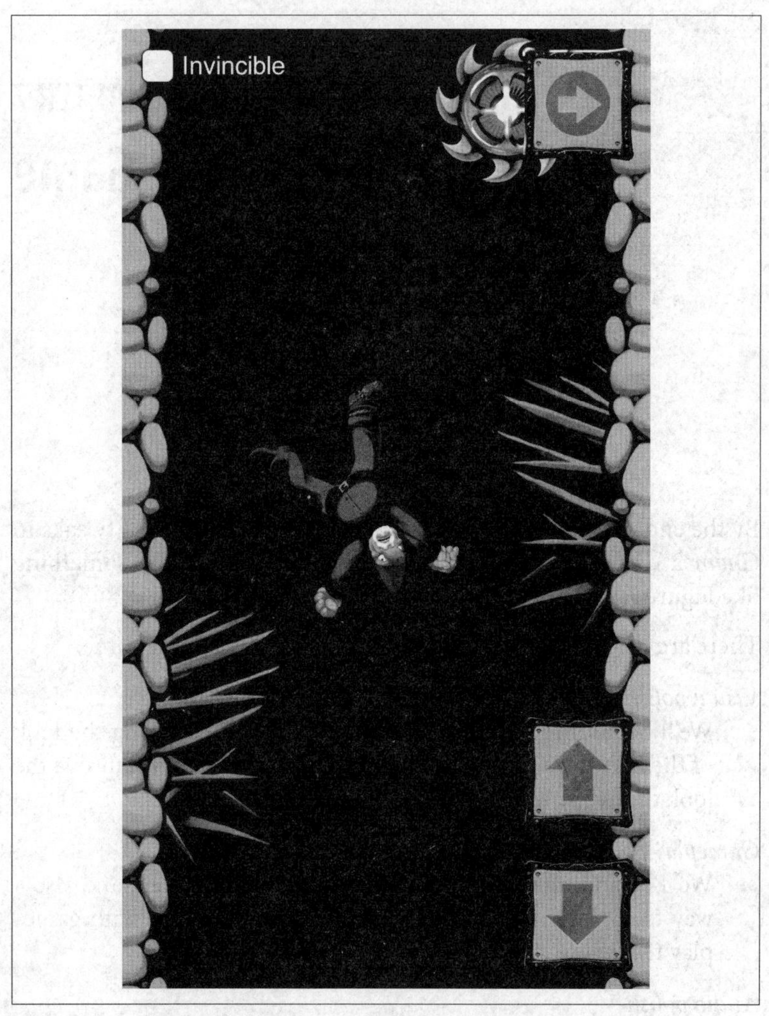

Figure 7-1. The final game

Updating the Gnome's Art

The first thing that we'll do to polish up the game is to change the gnome's sprites from their current stick-figure state, and swap them for a hand-painted replacement set of sprites.

To get started with this, copy the *GnomeParts* folder from the original resources into the *Sprites* folder. This folder contains two subfolders: *Alive* contains the new parts for the gnome, and *Dead* contains sprites that are used for when the gnome is dead (see

Figure 7-2). We'll start with the alive sprites, but we'll be using the dead sprites later.

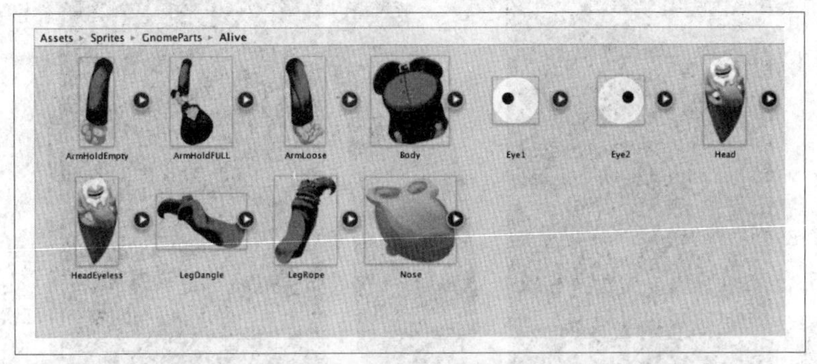

Figure 7-2. The Gnome's Alive sprites

There are more resources in the downloaded assets than the ones we'll be using, including a version of the head sprite with no eyes, designed to be used with the separated eye sprites. If you want to take the game even further than what we cover in this book, you might like to use these bonus assets!

The first step is to configure the sprites so that they're ready for use in the Gnome object. In particular, we need to ensure that they're imported as sprites, and that the pivot points for these sprites are in the right place. Here are the steps you'll need to follow:

1. *Convert the images into sprites, if they aren't already.* Select the sprites in the *Alive* folder, and ensure that the texture type is set to "Sprite (2D and UI)."
2. *Update the pivot points for the sprites.* For each sprite except the Body sprite, do the following:
   a. Select the sprite.
   b. Click the Sprite Editor button.
   c. Drag the pivot point icon (that is, the little blue circle) to the point around which the body part should rotate. For example, Figure 7-3 shows the position for the ArmHoldEmpty sprite.

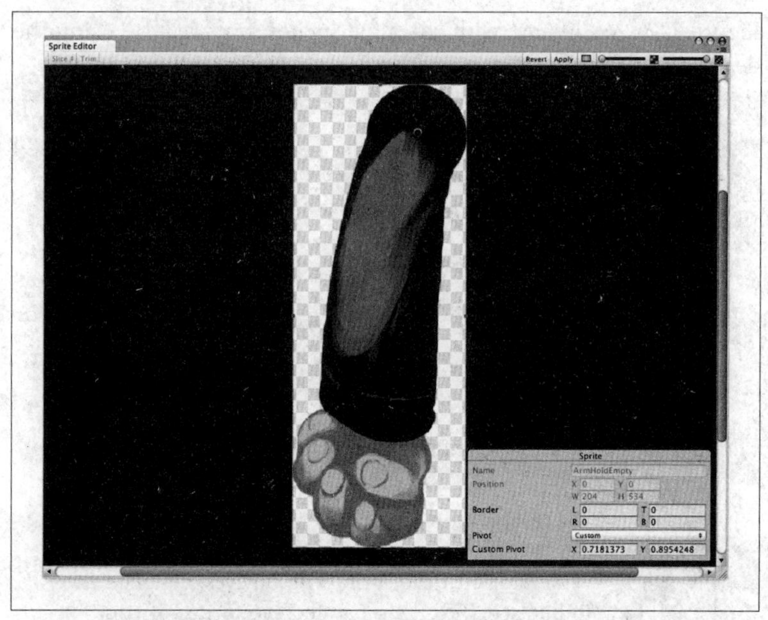

Figure 7-3. Setting the pivot point for the ArmHoldEmpty sprite; notice the position of the pivot point, at the top right

Once you've set up the sprites, it's time to add them to the gnome. To keep things tidy, and to allow you to keep the older version of the gnome around, we'll duplicate the gnome prefab, and make changes to the new version. We'll then tell the game to use this new and improved gnome.

Once we've done this, we'll add it to the scene, and start replacing the various components of the gnome's body with fresh art. Good plan? Great plan. Here's what to do:

1. *Duplicate the prototype gnome prefab.* Find the prototype gnome prefab, and duplicate it by pressing Ctrl-D (Command-D on a Mac). Name the new object "Gnome".
2. *Add the new gnome to the scene.* Drag the new gnome prefab into the Scene window to create a new instance of it.
3. *Replace the art.* Select all of the gnome's body parts and replace the sprite for it with the corresponding new sprite. For example, select the head and replace its sprite with the Head sprite in the *Alive* folder.

When you're done, the gnome should look something close to Figure 7-4. The body parts won't be in precisely the right position, but that's OK—we'll fix that in a moment.

Figure 7-4. The Gnome object, with updated sprites

Next, we need to adjust the position of the gnome's body parts. The new sprites are different shapes and sizes, and we need the various bits and pieces to line up correctly. Follow these steps to set it up:

1. *Reposition the head, arms, and legs.* Select the Head object, and adjust its position so that the neck lines up at the right point between the shoulders. Repeat this process with the arms (matching them with the shoulders) and the legs (matching them with the waist).

   Note that, for your convenience, the pivot points for the shoulders are the purple dots on the Body sprite.

Once you've repositioned them, it's important to ensure that the sprites are always in the right order—the legs should never be drawn on top of the body, the body should never be on top of the arms, and the head should be on top of everything.

2. *Adjust the sorting order for the body parts.* Select the head and both the arms, and change the Order in Layer property of the Sprite Renderer to 2.

   Next, select the body and change its order to 1.

When you're done, the Gnome should look like Figure 7-5.

Figure 7-5. The gnome, with properly positioned sprites

Updating the Physics

Now that the gnome's sprites have been updated, we need to update the physical components. There are two changes that need to be made: the colliders need to be updated so that they're the right shape, and the joints need to be adjusted so that the body parts will pivot at the right point.

We'll start by adjusting the colliders. Because the sprites aren't horizontal or vertical lines, we need to replace the simple box and circle colliders with *polygon colliders*.

There are two ways to create polygon colliders: you can either let Unity generate a shape for you or you can specify it yourself. We'll do it ourselves because it's more efficient (Unity tends to generate complex shapes, which isn't great for performance), and it also allows you much better control over the final result.

When you add a polygon collider to an object that also has a sprite renderer, Unity will use the sprite to build a polygon shape by drawing a line around all of the nontransparent parts of the image. If you want to define your own collision shape, the polygon collider component needs to be added to a game object that *doesn't* have a sprite renderer. The easiest way to do that is to create an empty child object, and add a polygon collider to that. To do so, follow these steps:

1. *Remove the existing colliders.* Select all legs and arms, and remove the Box Collider 2D. Next, select the head and remove the Circle Collider 2D.
2. Repeat these steps for each arm, each leg, and the head:
   a. *Add the child object for the collider.* Create a new empty game object called Collider. Make it a child of the body part, and ensure its position is 0,0,0.
   b. *Add the polygon collider.* Select this new Collider object, and add a Polygon Collider 2D component to it. A green collider shape will appear (see Figure 7-6); by default, Unity will create a pentagon shape for it, and you'll need to adjust it to make it fit the object.
   c. *Edit the shape of the polygon collider.* Click Edit Collider (see Figure 7-7) and you'll enter edit mode.

      While you're in edit mode, you can drag the individual points of the shape around. You can also click and drag on the lines connecting each point to create new points, and hold Ctrl (Command on a Mac) and click a point to remove it.

      Drag the points around so that they roughly match the shape of the body part (see Figure 7-8).

When you're done, click the Edit Collider button again.

The colliders that you just added should now look roughly like Figure 7-9.

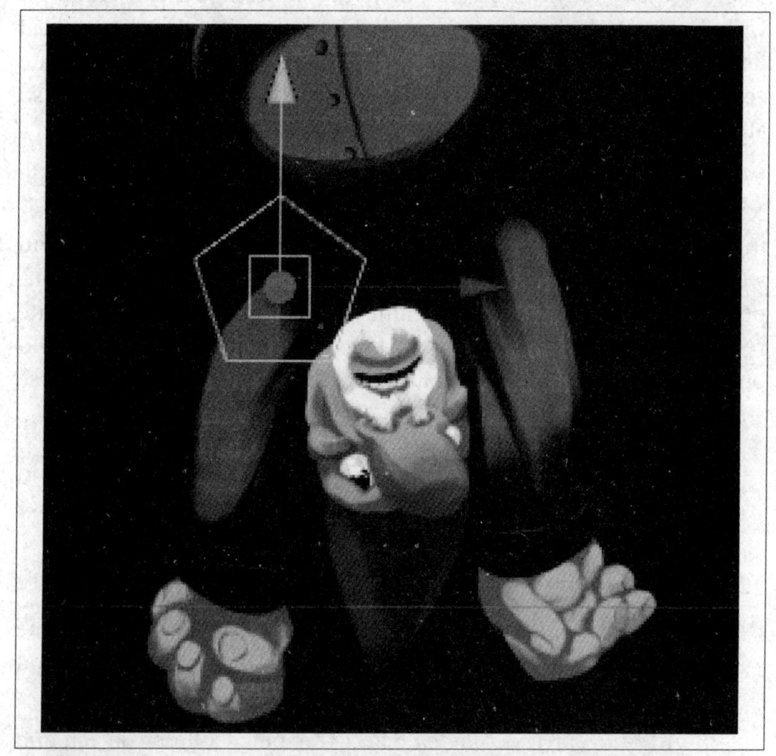

Figure 7-6. A newly added Polygon Collider 2D

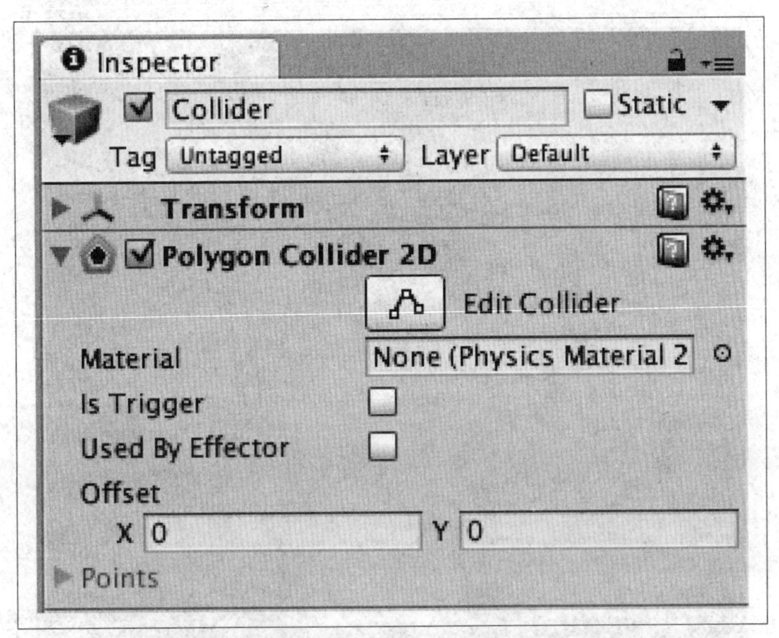

Figure 7-7. The Edit Collider button

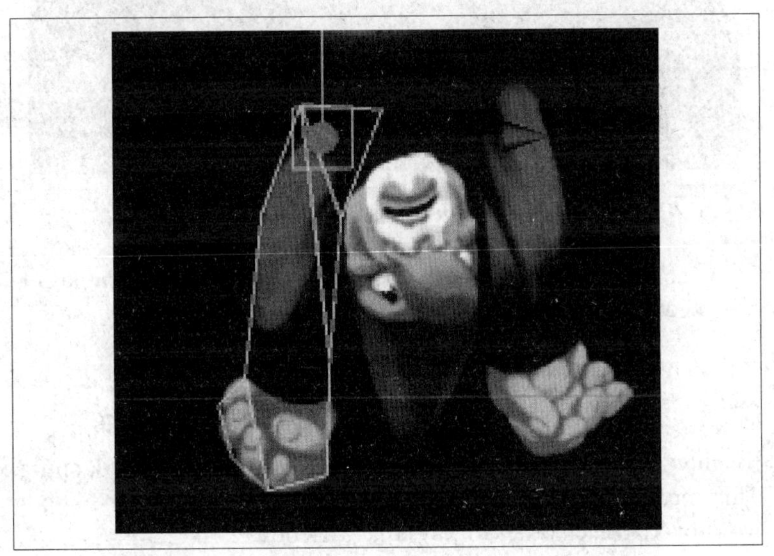

Figure 7-8. The updated polygon collider for the gnome's arm

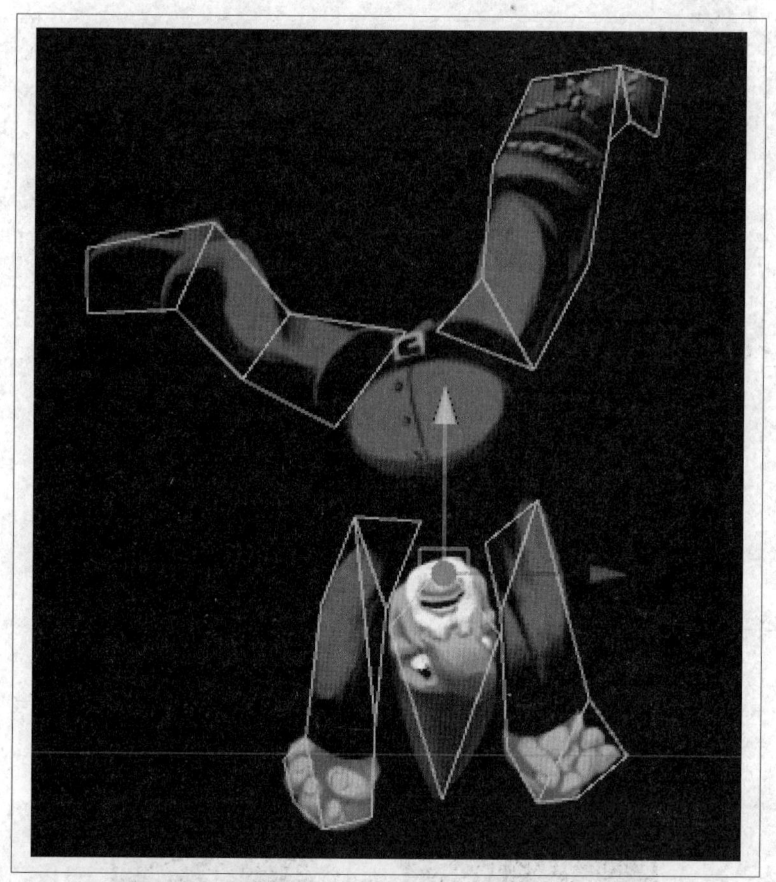

Figure 7-9. The colliders for the arms, legs, and head

There's one more change to the colliders to do: the body's circle collider needs to be enlarged slightly, to match the bulkier body.

3. *Increase the radius of the Body's circle collider to 1.2.*

The effect you're aiming to accomplish here is for the colliders to roughly match the shape of the sprites, but without overlapping. This means that, during gameplay, the parts of the gnome's body won't overlap each other in ways that look odd.

Now that the colliders are the right shape, it's time to update the joints. Recall that the head, arms, the legs all have a hinge joint attached to them, which connects them to the body. You'll need to make sure that the pivot point is correct to avoid weirdnesses like the gnome's arms appearing to rotate around the upper arm.

4. *Update the Connected Anchor and Anchor position of the gnome's joints.* For each body part except the body, drag the position of the Connected Anchor and the Anchor to the pivot point. The legs should pivot at the hips, the arms at the shoulders, and the head at the neck.

If you drag the anchor and connected anchor near the center point of the sprite, it will snap to that point.

Don't forget that Leg Rope has two joints on it: one that connects it to the body, and one that's used for the rope. Move the Anchor of this second joint to the ankle.

There are a couple of changes we need to make to the gnome's `Gnome` script. Remember from earlier that the gnome's arm sprite changes when he touches the treasure? Currently, that's still set to use the old prototype art, which doesn't fit the new art at all.

5. *Update the sprites used by the `Gnome` script.* Select the parent Gnome object.

   Drag the ArmHoldEmpty sprite into the Gnome's Arm Holding Empty slot, and drag the ArmHoldFull sprite into the Gnome's Arm Holding Full slot.

Now, when the gnome picks up the treasure, the arm's sprite will change to the correct image. Additionally, when the gnome drops the treasure (which happens when the gnome touches a trap and dies), the gnome's arm won't change into a stick-figure arm.

Lastly, we need to scale the gnome a bit, to make it fit in the world, and then save your changes to the prefab.

6. *Scale the Gnome.* Select the parent Gnome object, and change the X and Y scale from 0.5 to 0.3.
7. *Apply the changes to the prefab.* Select the parent Gnome object, and click Apply at the top of the Inspector.

8. *Remove the gnome from the world.* There's no need to keep it in the scene now that it's saved, so delete it.

Now that you've finished updating the gnome, it's time to update the Game Manager so that it uses this newly updated object.

9. *Make the Game Manager use the object.* Select the Game Manager, and drag the gnome prefab that you just updated into the Gnome Prefab slot.
10. *Test the game.* The updated gnome is now in the world! See Figure 7-10 for an example of how it should look.

Figure 7-10. The updated gnome, in game

Background

Currently, the background is a flat, gray quad, and doesn't look at all like the inside of a well. Let's change that!

To address this issue, we're going to add a more sophisticated set of objects that represent both the background and the side walls of the well. Before you continue, ensure that you've got the sprites from the *Background* folder added to your project.

Layers

Before adding the images, we need to first work out how they will be ordered in the scene. When you're making a 2D game, getting the right sprites to appear on top of other sprites is an important, and sometimes tricky, thing to maintain. Thankfully, Unity has a built-in solution for making it a little easier: *sorting layers*.

A sorting layer is a group of objects which are all drawn together. Sorting layers, as the name implies, are able to be sorted into the order that you want. This means that you can group certain objects into the "Background" layer, other objects into the "Foreground" layer, and so on. Additionally, each object can be ordered within its own layer, so that you can ensure that certain pieces of the background will always be drawn behind other pieces.

You always have at least one sorting layer, which is called "Default." All new objects go into this layer unless you change them.

We'll be adding multiple sorting layers to this project. Specifically, we'll add the following:

- The *Level Background* layer, which contains level background objects and will always be displayed behind everything else.
- The *Level Foreground* layer, which contains the foreground objects, like the walls.
- The *Level Objects* layer, which contains things like traps.

To create the layers, follow these steps:

1. *Open the Tags & Layers inspector.* Open the Edit menu, and select Project Settings → Tags & Layers.

2. *Add the Level Background sorting layer.* Open the Sorting Layers section, and add a new layer. Call it "Level Background".

   Drag this to the top of the list (above "Default"). This will make any object on this layer appear *behind* anything on the Default layer.

3. *Add the Level Foreground layer.* Repeat the process, and add a new layer called "Level Foreground". Place this *below* the Default layer. This makes objects on this layer appear *in front* of anything on the Default layer.

4. *Add the Level Objects layer.* Finally, repeat this process one more time, adding a new layer called "Level Objects". Place this below "Default" and above "Level Foreground". This is where the traps and treasure will go, which means that they need to be behind the foreground.

Creating the Backgrounds

Now that the layers are set up, it's time to start building the background itself. The background has three different themes—namely, brown, blue, and red—and each theme is composed of multiple sprites: a background, a side wall sprite, and a background version of the side wall sprite.

Because you'll want to lay out the content of the level to suit your own tastes, the best thing to do is to create prefabs for each of the three different themes. We'll start with the Brown background theme, build up the object, then save it as a prefab; next, you'll do the same thing with the Blue and Red themes.

Before we start any of *that*, however, we'll want to create an object to contain all of the level background objects, just to keep things tidy. Here are the steps you'll need to follow:

1. *Create the Level container object.* Create a new empty game object by opening the GameObject menu, and choosing "Create Empty." Name this new object "Level", and set its position to (0,0,1).

2. *Create the container for the Background Brown object.* Create another game object, and name this one "Background Brown". Make it a child of the Level object, and make sure that its posi-

tion is (0,0,0). This will make the object's position not offset from the position of the Level object.

3. *Add the main background sprite.* Drag the BrownBack sprite into the scene, and then make it a child of the Background Brown object.

   Select this new sprite, and change its Sorting Layer to "Level Background." Finally, set its X position 0 so that it's centered.

4. *Add the background side object.* Drag the BrownBackSide sprite into the scene, and make it a child of the Background Brown object.

   Set its Sorting Layer to Level Background, and set its Order In Layer to 1. This will make the object appear above the main background, while still being behind all objects in the other layers.

   Set its X position to -3, so that it's pushed to the left.

5. *Add the foreground side object.* Drag in the BrownSide sprite, and make it a child of the Background Brown sprite. Set the Sorting Layer to Level Foreground.

   Set its X position to -3.7, and set its Y position to the same as the BrownBackSide sprite. You want them to line up horizontally, with the foreground object a little further to the left.

Because the side objects are only half as high as the main background image, we'll now create a second row of these side objects.

To duplicate the side objects, select both the BrownBackSide and the BrownSide sprite, and duplicate them by pressing Ctrl-D (Command-D on a Mac).

Move these new side objects down so that the bottom edge of the upper row is at the same point as the top edge of the lower row. When you're done, the background should look like Figure 7-11.

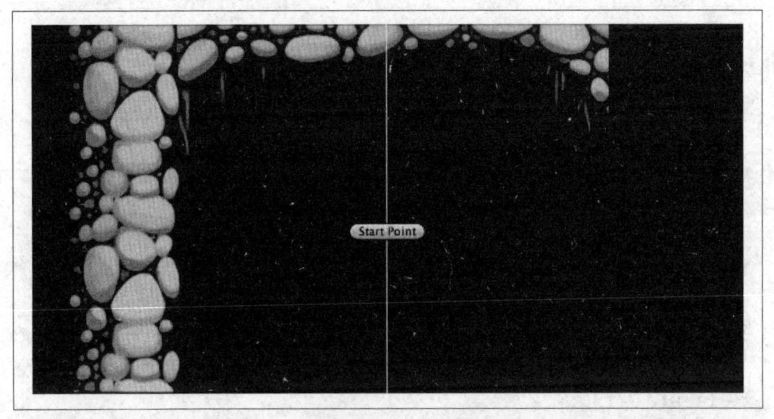

Figure 7-11. The partially updated background

We've now set up the side objects for the lefthand side, and it's time to set up the righthand side. To do this, we'll duplicate the existing sprites, and adjust them to suit the righthand side:

1. *Duplicate the side objects again.* Select all of the BrownSide and BrownBackSide objects, and press Ctrl-D (or Command-D on a Mac.)
2. *Ensure that the pivot mode is set to Center.* If the Pivot Mode button is currently set to *Pivot*, click it and it will change to Center.
3. *Rotate the objects.* Using the Rotate tool, rotate the righthand objects to 180 degrees. Hold down the Ctrl (Command on a Mac) to snap the rotation.

Don't use the Inspector to change their rotation value, since that will rotate them around their individual origins. What we want is for the objects to rotate around their common center.

4. *Flip the objects vertically.* Do this by changing their Y scale to -1. If you don't do this, the lighting will look incorrect when upside down.

When you're done, the transform inspector for these objects should look like Figure 7-12.

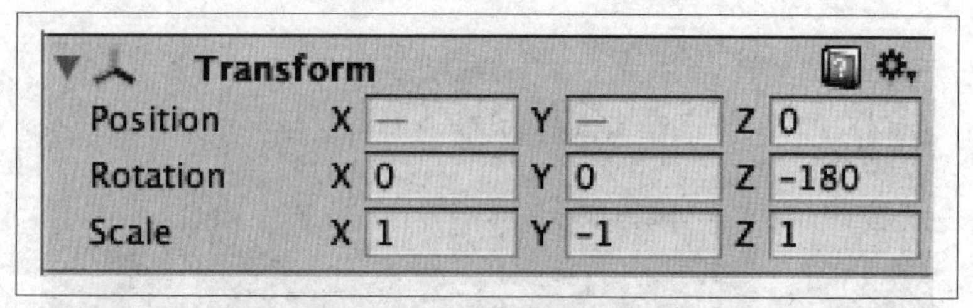

Figure 7-12. The transform of the righthand background elements

5. *Move the new objects to the righthand side of the level.* It's where they belong, after all. When you're done it should look like Figure 7-13.

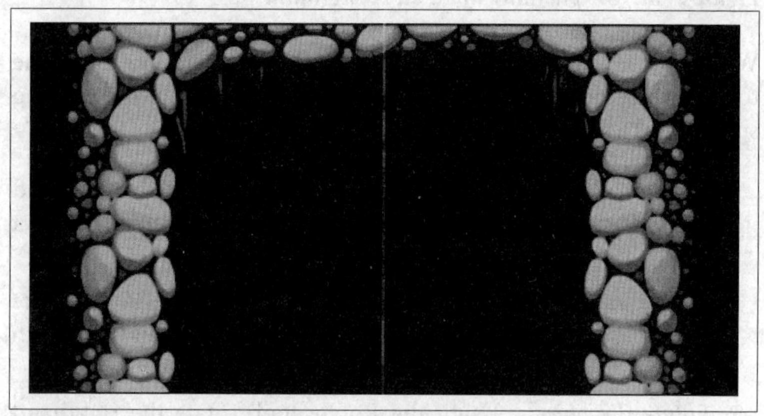

Figure 7-13. The updated background

Now that the Background Brown object has been set up, it's time to turn it into a prefab. To do so, follow these steps:

1. *Create a prefab from the Background Brown object.* Drag the Background Brown object into the Project tab, and a prefab will be created. Move this prefab into the *Level* folder.
2. *Duplicate the Background Brown object.* Select the Background Brown object, and press Ctrl-D (Command-D on a Mac) a few times. Move each of these new objects down, until you've made a decently long region of background.

Different Backgrounds

Now that the first background has been created, you can follow the exact same steps for the other two background themes:

1. *Create the Background Blue theme.* Make a new empty object called "Background Blue", and make it a child of the Level object.

   Follow the same steps you followed to create the Background Brown object, but this time, use the BlueBack, BlueBackSide, and BlueSide sprites.

   Don't forget to make a prefab out of the Background Blue object when you're done.

2. *Create the Background Red theme.* Again, follow the same steps, using the RedBack, RedBackSide, and RedSide sprites.

When you're all done, the level should look something like Figure 7-14.

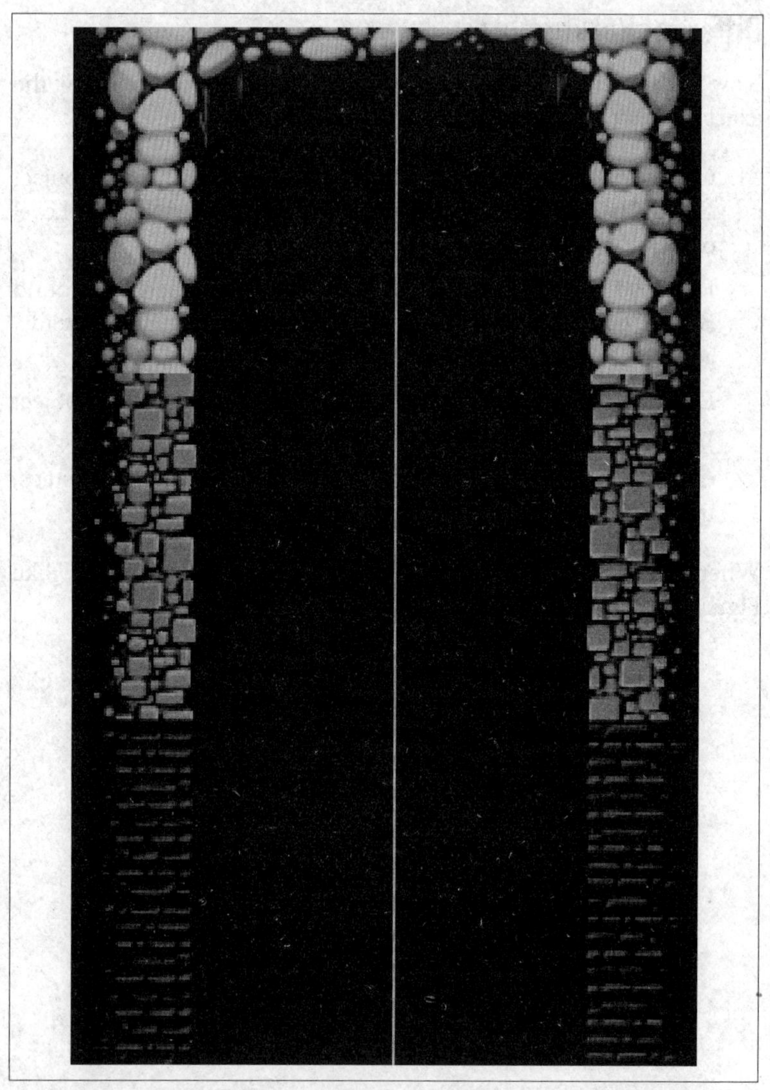

Figure 7-14. The background areas

There's one problem with this setup: the backgrounds tile well when they're touching background objects of the same color, but there are harsh lines where the different colors meet.

To fix this, we'll overlay sprites that cover up this discontinuity. These sprites will be on the "Level Foreground" layer, and will be set up to appear over everything else in the game.

1. *Add the BlueBarrier sprite.* This sprite is designed to mask the line between the Brown and Blue backgrounds. Position it at the point where the Brown and Blue backgrounds meet, and make it a child of the Level object.
2. *Add the RedBarrier sprite.* This one's designed to mask the line between the Blue and Red backgounds; place it at the meeting point between Blue and Red, and make it a child of the Level object.
3. *Update the sorting layers for both sprites.* Select both the BlueBarrier and RedBarrier sprites, and set their Sorting Layer to "Level Foreground."

   Next, set the Order in Layer to 1. This will make the barriers appear over the side walls.

When you're done, the level should look like Figure 7-15.

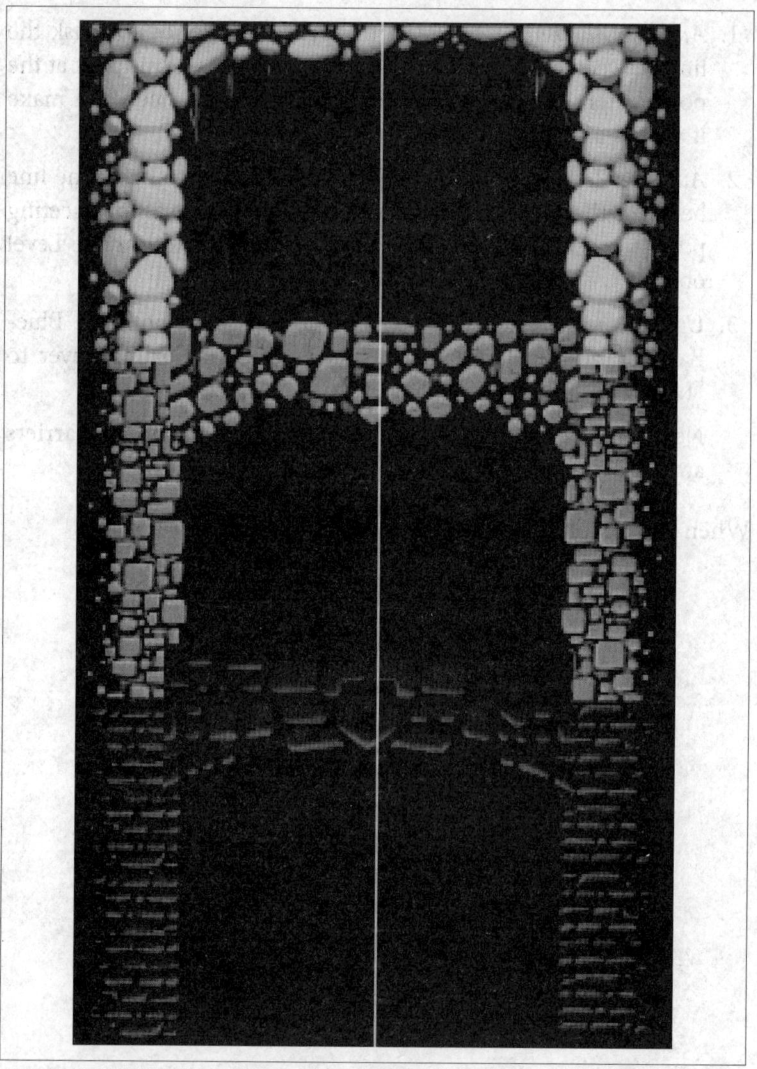

Figure 7-15. The backgrounds, with Barrier sprites

The Bottom of the Well

There's one last thing to add: the well needs a bottom. In this game, the well is dry, and has drifts of sand that cover the very bottom of the well. Some of this sand has blown up against the walls, too. To add this to the scene, follow these steps:

1. *Create a container object for the well bottom sprites.* Create a new empty game object named "Well Bottom". Make it a child of the Level object.
2. *Add the well's bottom sprite.* Drag in the Bottom sprite, and add it as a child of the Well Bottom object.

   Set the sprite's sorting layer to Level Background, and set its Order In Layer to 2. This places it above the "Background" and "Background Side" sprites, but behind everything else in the game.

   Position the sprite at the bottom of the well, and set its X position to 0, which will keep it in line with the rest of the level's sprites.
3. *Add the side decoration sprite to the left of the well.* Drag in the SandySide sprite, and add it as a child of the Well Bottom object.

   Set the Sorting Layer to "Level Foreground." Set the Order in Layer to 1, so that it appears over the walls.

   Next, move the sprite to the left, so that it lines up with the walls (see Figure 7-16 for an example of how it should look).

Figure 7-16. The SandySide sprite, lined up with the bottom of the well

4. *Add the righthand side object.* Duplicate the SandySide sprite. Set its X Scale to -1 to flip it, and then move to the right of the well.
5. *Ensure that the treasure is in the right position.* Reposition the treasure sprite so that it's in the middle of the sand dune.

When you're all done, it should look like Figure 7-17.

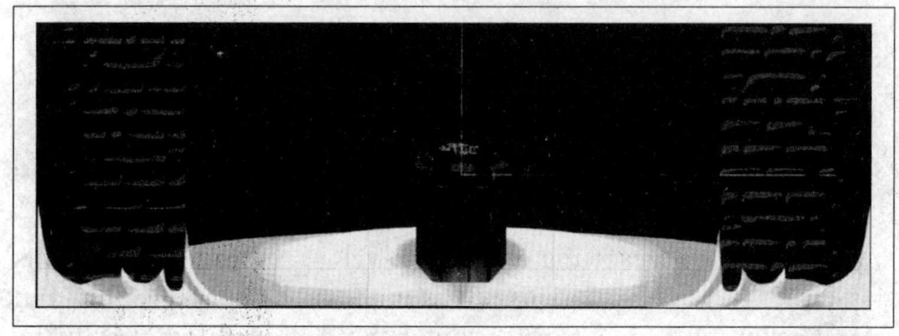

Figure 7-17. The finished well bottom

Updating the Camera

There's now one last thing you need to do to make the new background fit into the game: update the camera. There are two changes that need to be made: first, the camera needs to be updated so that the player can see the entire level, and second, the script that constrains the position of the camera needs to be updated to take into account the updated size of the level. Configure the camera by following these steps:

1. *Update the camera's size.* Select the Main Camera object, and change the camera's Ortho Size to 7. This will give the player a sufficiently wide view on the whole level.

2. *Update the camera's limits.* Because we've changed the amount that the camera can see, we also need to adjust the limits of the Camera. Change the camera's Top Limit to 11.5.

   You'll also need to adjust the Bottom Limit, but the value you choose here will depend on how deep you've made the well.

   The best way to work it out is to lower the gnome as far as you can, and if the camera stops moving before you reach the bottom of the well, lower the Bottom Limit; if the camera goes below the bottom of the well (revealing the blue background), raise the Bottom Limit.

   Take note of the value before you stop the game, because it will reset back to its original value when you end the game; after you stop the game, enter the number you wrote down into the Bottom Limit field.

User Interface

It's time to improve the look and feel of the game's UI. Earlier, when we were setting up the interface, we used the standard Unity-provided buttons. While they're capable, they don't really suit the look and feel of the game at all, and we'll need to replace the button's imagery with better stuff.

Additionally, we need to show a Game Over screen when the gnome reaches the top, as well as a screen that appears when the player pauses the game.

Before you continue, make sure that you've imported the sprites for this section. Import the *Interface* folder of sprites, and put this folder in the *Sprites* folder.

These sprites are designed to be high resolution, so that they can be used in a variety of different situations. In order for them to be useful as buttons in the game, Unity needs to know how large they should be when added to the Canvas. You do this by adjusting the Pixels Per Unit value for the sprites, which controls their scale when added to a UI component or sprite renderer.

Configure the sprites by selecting all of the images in this folder (except for "You-Win"), and changing their Pixels Per Unit to 2500.

We'll start by updating the Up and Down buttons, which currently appear at the bottom right of the window, to use nicer images. To do that, we'll need to remove the label from the button, and also adjust the size and position of each button to suit their new images. Here are the steps you'll need to follow:

1. *Remove the label from the Down button.* Find the Down Button object, and remove the Text object that's attached as a child.
2. *Update the sprite.* Select the Down Button object, and change the Source Image property to the Down sprite (which is in the *Interface* folder).

   Click the Set Native Size button, and the button will adjust its size.

   Finally, adjust the position of the button so that it's still in the bottom-right corner of the screen.
3. *Update the Up button.* Repeat the same process for the Up button. Remove the Text child object, and change the Source Image

to the Up sprite. Next, click Set Native Size, and update the button's position so that it's just above the Down button.

4. *Test the game.* The buttons will still work, but will look much nicer (see Figure 7-18).

Figure 7-18. The updated Up and Down buttons

We'll now group these buttons into a container. This is for two reasons: first, it's good to keep the UI organized, and second, by grouping them into an object you'll be able to enable and disable everything at once. This will be useful very soon, when you implement the Pause menu. Follow these steps to set it up:

1. *Create the parent object for the buttons.* Create a new empty game object, named "Gameplay Menu". Make it a child of the Canvas.
2. *Set the object to fill the entire screen.* Set the Gameplay Menu's anchors to stretch horizontally and vertically. Do this by clicking on the anchors near the top left, and clicking the option at the bottom right of the menu that pops up (see Figure 7-19).

   Once you've done that, set Left, Top, Right, and Bottom to 0. This makes the whole object fill its entire parent (which is the canvas, so the whole thing will fill the entire canvas).

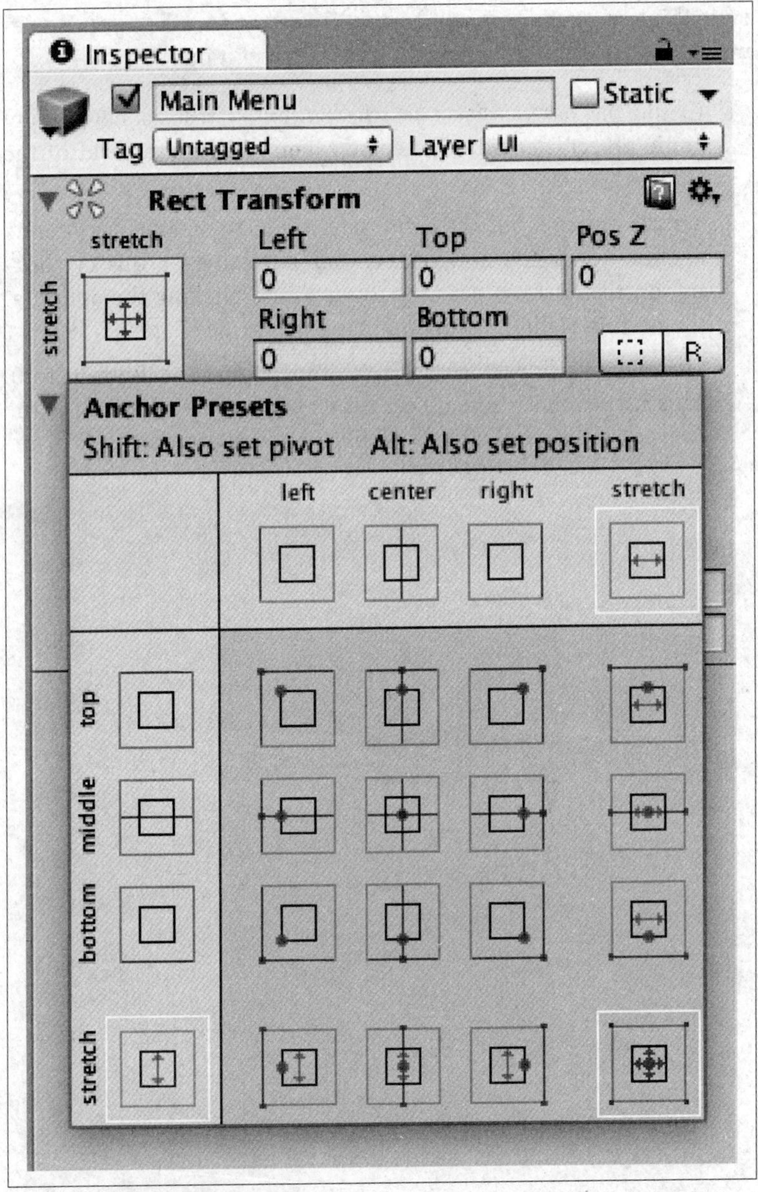

Figure 7-19. Setting the anchors of the object to stretch horizontally and vertically

3. *Move the buttons into the Gameplay Menu object.* Drag both the Up button and the Down button's entries in the hierarchy onto the Gameplay Menu object.

Next, we'll create the "You Win" graphic. This will display an image to the player, as well as a button that lets them play the game again. To prepare it, follow these steps:

1. *Create the container object for the Game Over screen.* Create a new empty game object named "Game Over", and make it a child of the Canvas.

   Make it stretch horizontally and vertically by following the same steps as for the Gameplay Menu object.

2. *Add the Game Over image.* Create a new Image game object by opening the GameObject menu and choosing UI → Image. Make this new Image a child of the Game Over object that you just created.

   Make the new Image object's anchors set to stretch horizontally and vertically. Set the Left and Right margins to 30, and set the Bottom Margin to 60. This will give the image some padding at the sides, and will also ensure that it doesn't cover up the New Game button that you're about to add.

   Set the Image's Source Image property to the You Win sprite, and turn on the Preserve Aspect option to prevent it from stretching.

3. *Add the New Game button.* Add a new Button to the Game Over object by opening the GameObject menu and selecting UI → Button.

   Set the text of the new button's label to read "New Game", and set the button's anchors to the bottom-center.

   Move the button to the bottom-center of the screen. When you're all done, the interface should look like Figure 7-20.

Figure 7-20. The Game Over interface

4. *Connect the New Game button to the Game Manager.* When the button is clicked, we want the Game Manager to reset the game. We can do this by calling the `GameManager` script's `RestartGame` function.

   Click on the + button at the bottom of the Button's inspector, and drag the Game Manager into the slot that appears. Next, change the function to `GameManager` → `RestartGame`.

We now need to connect the Game Manager to these new UI elements. The `GameManager` script is already set up to enable and disable the appropriate user interface elements based on the state of the game: when the game is playing, it will attempt to activate whatever object is in the "Gameplay Menu" variable, and deactivate the other menus. Follow these steps to configure and test it out:

1. *Connect the Game Manager to the menus.* Select the Game Manager, and drag the Gameplay Menu object into the Gameplay Menu slot. Next, drag the Game Over object into the Game Over Menu slot.
2. *Test the game.* Lower the gnome all the way down to the bottom of the well, pick up the treasure, and reach the exit. You'll get a Game Over screen.

There's one last menu that we need to set up: the Pause menu, along with the button that's used to pause the game. The Pause button will appear at the top-right of the screen, and when the player taps it, the game will freeze and display buttons for resuming the game and restarting it.

To set up the pause button, create a new Button object, and name it "Menu Button". Make it a child of the Gameplay Menu object.

- Remove the Text child object, and set the button's Source Image to the Menu sprite.
- Click Set Native Size, and move it to the top right of the canvas. Set anchors to top right.
- When you're done, the new button should look like Figure 7-21.

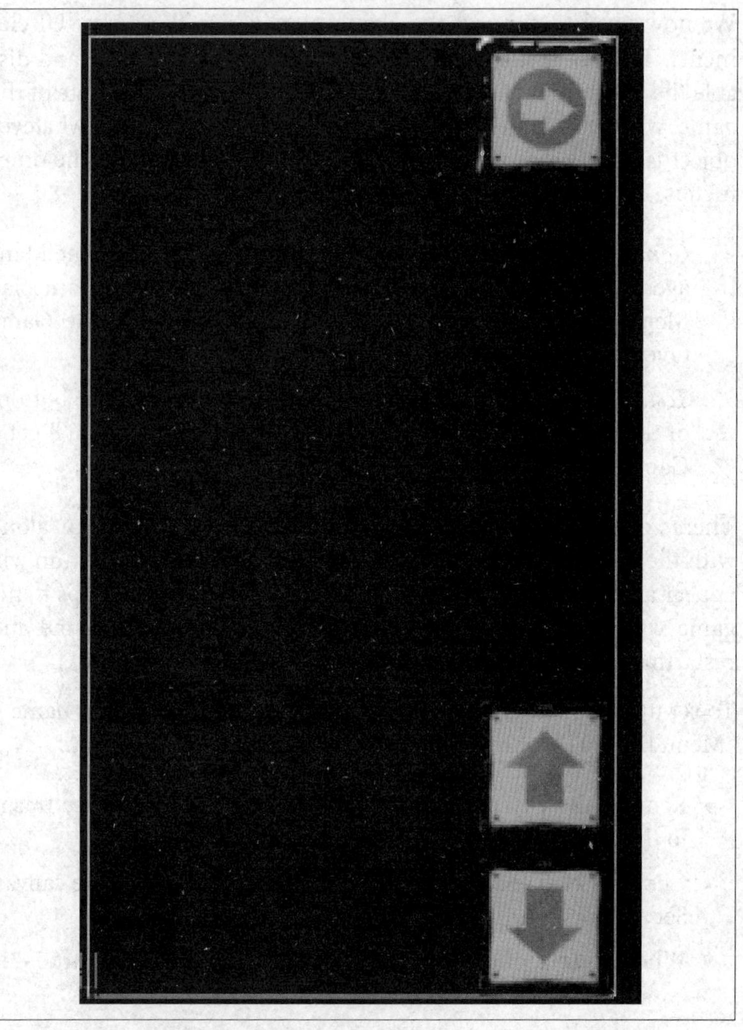
Figure 7-21. The Menu button

Next, we need to connect this button to the Game Manager. When the button is tapped, it will instruct the Game Manager to enter the Paused state. Doing this will show the Pause Menu (which we're about to create), hide the Gameplay Menu, and also freeze the game.

To connect the Menu button to the Game Manager, click the + button at the bottom of the button's inspector, and drag the Game Manager into the slot that appears.

Make the button call `GameManager.SetPaused`. Turn the checkbox on, so that the `SetPaused` button is sent a `true` parameter when the button is tapped.

We can now set up a menu to appear when we pause the game:

1. *Create the Main Menu container.* Make a new empty object named "Main Menu". Make it a child of the Canvas, and set its anchors to stretch horizontally and vertically. Set the Left, Right, Top, and Bottom margins to 0.
2. *Add the buttons to the Main Menu.* Add two buttons, named "Restart" and "Resume". Make both of these buttons children of the Main Menu object that you just created, and update the text of their respective labels to read "Restart Game" and "Resume Game".

   When you're done, the Main Menu should look like Figure 7-22.

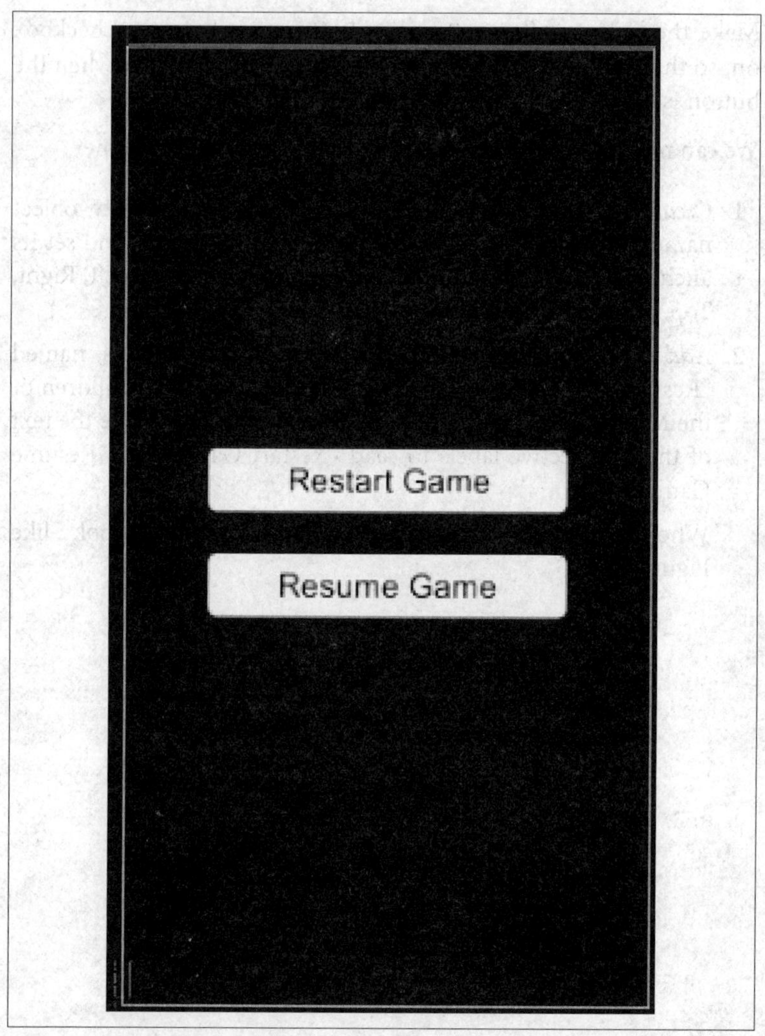

Figure 7-22. The Main Menu

3. *Connect the buttons to the Game Manager.* Select the Restart button, and make it call the Game Manager object's `GameManager.RestartGame` function.

   Next, select the Resume button, and make it call the Game Manager's `GameManager.Reset` function.

4. *Connect the Main Menu to the Game Manager.* The Game Manager needs to know which object should appear when the `SetPaused` function is called. Select the Game Manager, and drag

the Main Menu object into the Game Manager's Main Menu slot.

5. *Test the game.* You can now pause the game and resume it; additionally, you can restart the whole game.

Invincibility Mode

The idea of video game cheat codes actually comes from a very practical requirement. When you're building a game, it can get quite tedious to have to successfully defeat the various traps and puzzles that your game contains in order to reach the specific part that you want to test. In order to speed up development, it's very common to add tools that change the way the game plays: shooting games frequently have codes that make enemies not attack the player, and strategy games can disable the fog of war.

This game is no different: in order to build the game, you shouldn't have to deal with each and every obstacle, every time you run the game. To that end, we'll add a tool for making the gnome invincible.

This will be implemented as a checkbox (sometimes called a *toggle*), which appears at the top-left of the screen. When turned on, the gnome will never die. It will still "receive damage," which means that the various particle effects that you'll be adding in the next chapter will still appear, which is useful for testing.

To keep things organized, this checkbox will be contained inside a container object, just like the other UI components. Let's get started by creating this container:

1. *Create the Debug Menu container.* Create a new empty game object called "Debug Menu", and make it a child of the Canvas. Set its anchors to stretch horizontally and vertically, and make it fill the screen by setting the Left, Right, Top, and Bottom margins to zero.
2. *Add the Invincible toggle.* Create a new Toggle object by opening the GameObject menu and choosing UI → Toggle. Name this object "Invincible".

   Set the new object's anchors to Top Left, and move it to the top-left of the canvas.

3. *Configure the toggle.* Select the Label object, which is a child of the toggle you just added, and set the Text component's color to white. Set the label's text to read "Invincible".

   Set the Toggle object's "Is On" property to off.

   When you're done, the toggle should look like Figure 7-23.

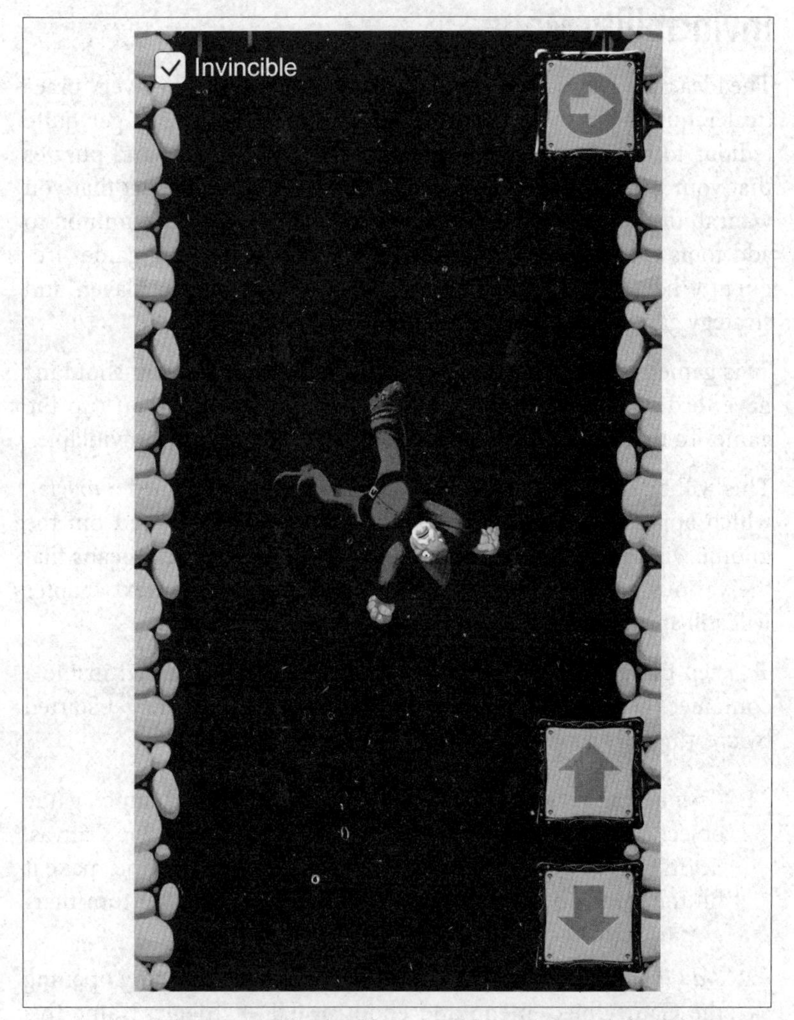

Figure 7-23. The Invincible checkbox, visible at the top-left of the screen

4. *Connect the toggle to the Game Manager.* Add a new entry to the Invincible toggle's Value Changed event by clicking the + button. Drag the Game Manager into the slot that appears, and change the function to `GameManager.gnomeInvincible`. Now, when the toggle changes value, the `gnomeInvincible` property will change.
5. *Test the game.* Play the game, and turn on Invincible. The gnome will now not die when he touches a trap!

Wrapping Up

The game is looking pretty good, now. The core gameplay is present and feeling nice, and you've added some developer tools of your own to make play-testing easier. There's still more that we can do, though. In the next chapter, we'll add more content and polish, and we'll finish up our development of the game by building out the menu structure and audio.

CHAPTER 8

Final Touches on Gnome's Well

More Traps and Level Objects

The game is starting to take shape: the gnome's art has been updated, the UI has been updated, and the backgrounds look nice. Currently, we only have one type of trap: the brown spikes. Our next steps will be to create two more themed versions of these static spikes, to add a bit more variety.

We'll also add a new type of trap: the spinning blade. The spinning blade deals the same kind of damage as the spikes, but is slightly more complicated—it's composed of three sprites, one of which is animated.

Finally, we'll add some non-damage-dealing stuff, in the form of walls and blocks that the player will need to navigate around. These objects, when placed in conjunction with traps, will force the player to think carefully about how they're going to navigate the level.

Spikes

Let's start with the themed spikes. We currently already have a prefab for the existing sprites; all that needs to change is to update the sprites and regenerate the collider. To do so, follow these steps:

1. *Create the new prefabs for the spikes.* Select the SpikesBrown prefab, and create a duplicate of it by pressing Ctrl-D (Command-D on a Mac). Name this new object "SpikesBlue".

   Make another copy, and call it "SpikesRed".

2. *Update the sprite.* Select the SpikesBlue prefab, and change the sprite to the SpikesBlue image.
3. *Update the polygon collider.* Because the polygon collider is on the same object as a Sprite Renderer, the collider uses the sprite to calculate its shape. However, it won't automatically update the shape when the sprite changes; to fix this, you'll need to reset the polygon collider.

   Click the Gear icon at the top right of the Polygon Collider 2D component, and click Reset in the menu that appears.
4. *Update the SpikesRed object.* Now that you're all done with the SpikesBlue object, follow the same steps for the SpikesRed object (and use the SpikesRed image.)

When you're done, you can add a few SpikesBlue and SpikesRed objects to the level.

Spinning Blade

Next up, we'll add the spinning blade. The spinning blade pokes out into the game a little further than the spikes, and contains a nasty-looking circular saw. In terms of the underlying logic, the spinning blade is actually identical to the spikes—when the gnome touches it, he dies. However, adding a variety of different traps to the game helps to break up the flow of the level, and maintains player interest.

Because the spinning blade is animated, we'll build it up using multiple sprites. Additionally, one of these sprites—the circular saw—will be set up to rotate at high speed.

To build the spinning blade, drag the SpinnerArm sprite into the scene, and set its Sorting Layer to "Level Objects."

Drag out the SpinnerBladesClean sprite, and add it as a child to the SpinnerArm object. Set its Sorting Layer to "Level Objects," and set its Order in Layer to 1. Position it at the top of the arm, then set the X position to 0, so that it's exactly centered.

Drag out the SpinnerHubcab sprite, and add it as a child to SpinnerArm as well. Set the Sorting Layer to "Level Objects," and set the Order in Layer to 2. Set the X position to 0 as well.

When you're done, the spinner should look like Figure 8-1.

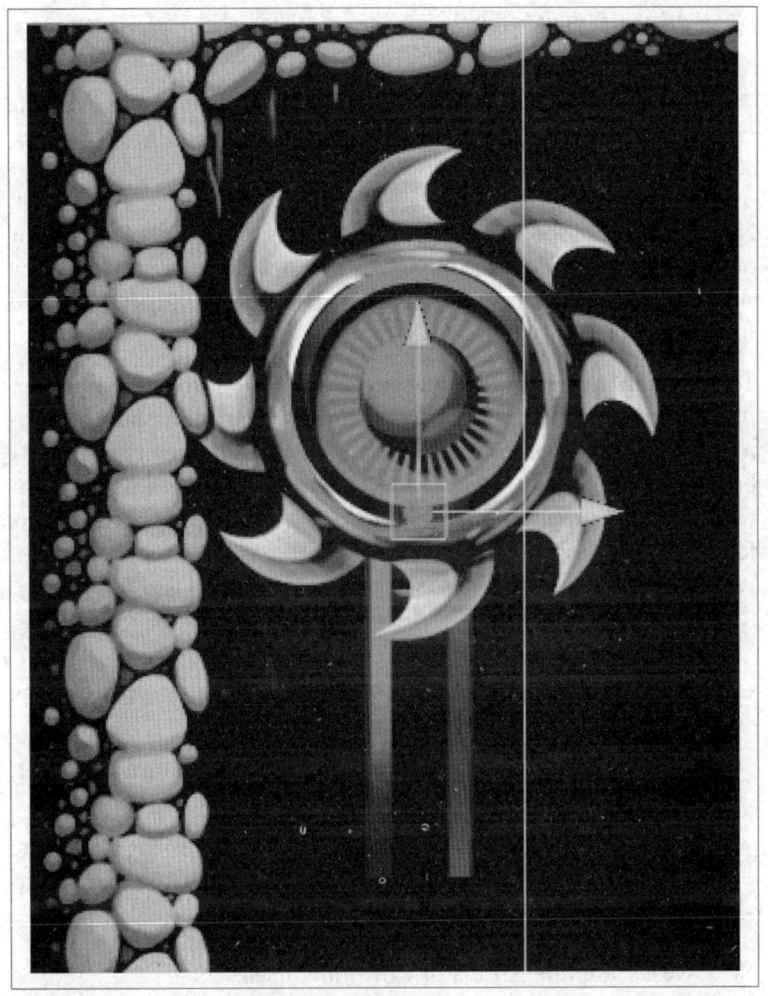

Figure 8-1. The constructed spinner

We'll now add the bit that makes it damage the gnome: a `SignalOn Touch` script. The `SignalOnTouch` script sends a message when the gnome touches the collider that's attached to the object; to make it work, we'll also need to add a collider. Follow these steps to set it all up:

1. *Add a collider to the blades.* Select the SpinnerBladesClean object, and add a Circle Collider 2D. Reduce the radius to 2; this reduces the size of the hitbox, and makes it slightly easier to deal with the spinning blade.

2. *Add the SignalOnTouch component.* Click the Add Component button, and add a `SignalOnTouch` script.

   Click the + button at the bottom of the Inspector, and drag the Game Manager into the slot. Change the function to `GameManager.TrapTouched`.

Next, we'll make the blade rotate. To do this, we'll add an Animator object, and configure it to run an Animation. The Animation is very simple: all that it needs to do is rotate whatever object it's attached to in a full circle.

To set up an Animator, you need to create an Animator Controller. Animator Controllers allow you to define which animation the Animator is currently playing, based on different parameters. We won't be making use of any of the advanced features of the Animator Controller in this game, but it's useful to know that it exists. To set it up, follow these steps:

1. *Add the Animator.* Select the blades and add a new Animator component.
2. *Create the Animator Controller.* In the *Level* folder, create a new Animator Controller asset named "Spinner".

   While you're in the *Level* folder, create a new Animation asset, called "Spinning".
3. *Make the Animator use the new Animator Controller.* Select the blades, and drag the Animator Controller you just created onto the Controller slot.

Next, we'll set up the Animator Controller itself:

1. *Open the Animator.* Double click the Animator Controller and the Animation tab will open.
2. *Add the Spinning animation to the Animator Controller.* Drag the Spinning animation into the Animator pane. The Animator Controller should now have a single animation state in it, as well as the preexisting Entry, Exit, and Any State items (see Figure 8-2).

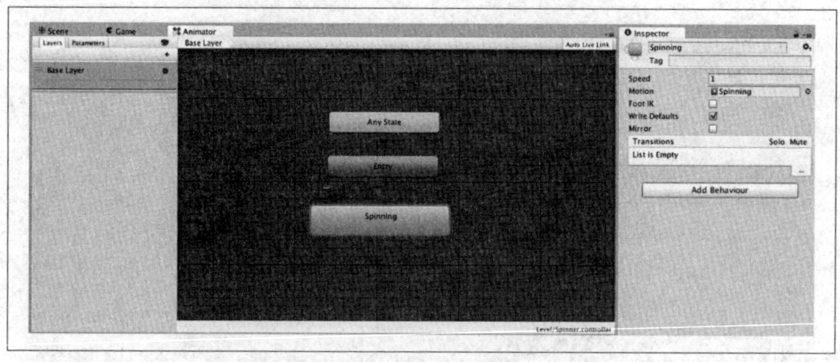

Figure 8-2. The Animator Controller for the Spinner

The Animator is now set up to use the Animator Controller, which itself is set up to start playing the "Spinning" animation. It's time to set up this animation to actually make things spin.

1. *Ensure that the spinner's blades are selected.* Go back to the Scene view, and select the blades again.
2. *Open the Animation pane.* Open the Window menu, and choose Animation. The Animation tab will open; drag the tab to somewhere convenient to you. You can also dock it to another section of Unity by dragging the tab at the top of the pane around the main Unity window.

   Ensure that the Spinning animation is selected in the top left of the Animation pane before you continue.
3. *Add a curve for the spinner's Rotation property.* Click the Add Property button, and a list of animatable components will appear. Navigate to the Transform → Rotation element, and click the + button at the righthand side of the list.

By default, new properties come with two keyframes—one at the start of the animation and one at the end (see Figure 8-3).

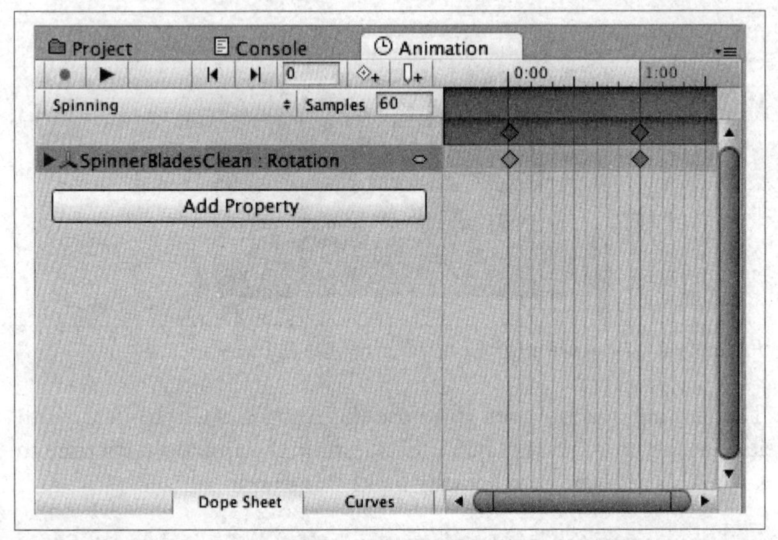

Figure 8-3. The keyframes for the newly created animation

We want the object to rotate 360 degrees. This means that at the start of the animation, the object should be rotated 0 degrees, and at the end of the animation, it should be rotated 360 degrees. To make this change, we'll need to modify the last keyframe in the animation:

1. *Select the rightmost keyframe.*
2. Click on the rightmost diamond in the Animation pane, and the animation will jump to that point in the timeline. Unity will now be in *record mode*, which means that your changes to the spinner will be logged. You'll also notice that the controls at the top of the Unity window will be red, to remind you of this fact.

   When you look at the Transform component in the Inspector, you'll also notice that the Rotation values are red.
3. *Update the rotation.* Change the Z rotation to 360.
4. *Test the animation.* Click the Play button in the Animation tab, and watch the blades spin. If they aren't spinning fast enough, click and drag the final keyframe so that it's closer to the start. This reduces the duration of the animation, and makes the object complete its revolution faster.
5. *Make the animation loop.* Go to the Project pane, and select the Spinning animation asset you created. In the Inspector, make sure the Loop Time checkbox is selected.

6. *Play the game.* The blades of the circular saw will now be rotating.

There's one last thing to do before the spinner is ready for use—it needs to be scaled down, to fit with the rest of the game:

1. *Scale the spinner.* Select the parent SpinnerArm object, and set the X and Y scale values to 0.4.
2. *Make the spinner into a prefab.* Drag the SpinnerArm object into the Project pane. This will create a new prefab called SpinnerArm; rename it "Spinner".

You can now rotate the spinner, and place it in the level; the gnome will die when he touches it.

Blocks

In addition to traps, it's also a good idea to add obstacles that *don't* kill the gnome when he touches them. These blocks serve to slow the player down, and force them to think about how they're going to get around the different traps you've added.

These blocks are among the simplest objects you'll add to the game: all they're made of is a sprite renderer and a collider. Because they're all so simple and similar to each other, you can produce the prefabs for them all at the same time. Here's what you'll need to do to set them up:

1. *Drag out the block sprites.* Add the BlockSquareBlue, BlockSquareRed, and BlockSquareBrown sprites to the scene. Next, add the BlockLongBlue, BlockLongRed, and BlockLongBrown sprites to the scene.
2. *Add the colliders.* Select all six of these objects, and click the Add Component button at the bottom of the Inspector. Add a Box Collider 2D component, and each of the blocks will be given a green box collision shape.
3. *Convert them to prefabs.* Drag each block into the *Level* folder to create prefabs.

You're done, and can now add blocks and walls to the level. That was easy.

Particle Effects

When the gnome dies, having him simply fall apart isn't a terribly satisfying visual effect. To create a more interesting effect, we're going to add particle systems.

In particular, we're going to add a particle effect that appears when the gnome touches a trap (the "blood explosion"), and an effect that appears when one of the gnome's limbs detaches (the "blood fountain").

Defining the Particle Material

Because both of these particle systems will emit the same thing (that is, gnome blood), we'll start by creating a single material that's shared between the two. Follow these steps to create and prepare it for use:

1. *Configure the Blood texture.* Find the Blood texture, and select it. Change its type from Sprite to Default, and ensure that the "Alpha Is Transparency" setting is on (see Figure 8-4).
2. *Create the Blood material.* Create a new Material asset by opening the Asset menu and choosing Create → Material. Name this material "Blood", and change the shader to Unlit → Transparent. Next, drag the Blood texture into the Texture slot. When you're done, the Inspector should look like Figure 8-5.

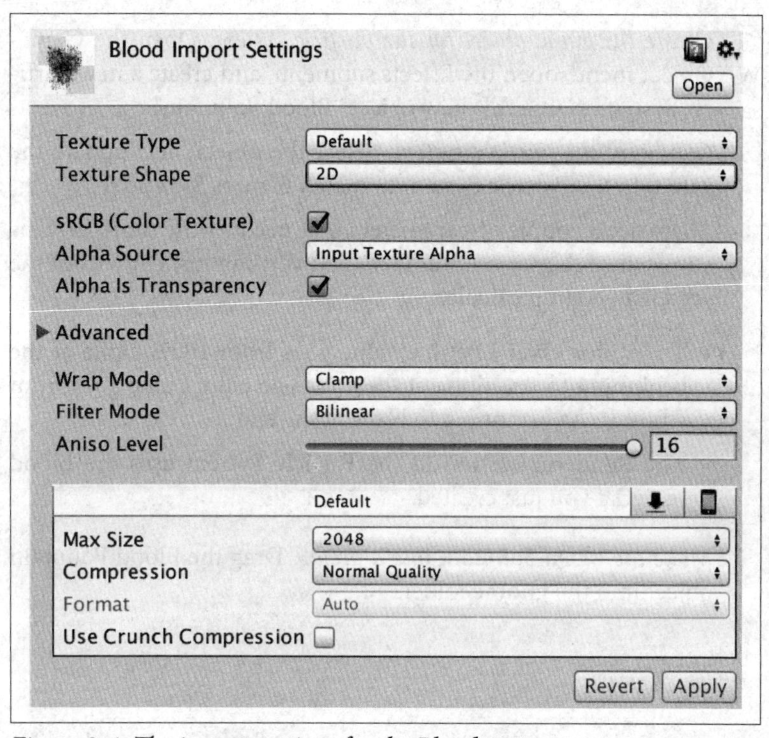

Figure 8-4. The import settings for the Blood texture

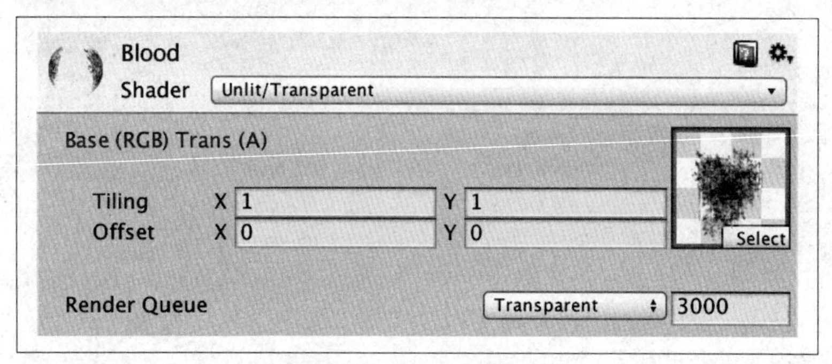

Figure 8-5. The material for the particle effect

The Blood Fountain

The material is now ready for use, so it's time to start building the particle effects. We'll start with the Blood Fountain effect, which creates a stream of particles that shoot in a specific direction and eventually fade out. Here's how to set it up:

1. *Create the game object for the particle system.* Open the GameObject menu, open the Effects submenu, and create a new Particle System. Name this new object "Blood Fountain".
2. *Configure the particle system.* Select the object, and update the values in the Particle System to match Figures 8-6 and 8-7.

   There are a couple of parameters that need a little more explanation, since they're not numbers that you can just copy from the screenshots. In particular:

   - The Color Over Lifetime value goes from 100% alpha at the beginning to 0% alpha at the end. The color value goes from white at the beginning to black at the end.
   - The Renderer section of the Particle System uses the Blood material you just created.
3. *Make the Blood Fountain into a prefab.* Drag the Blood Fountain object into the gnome folder.

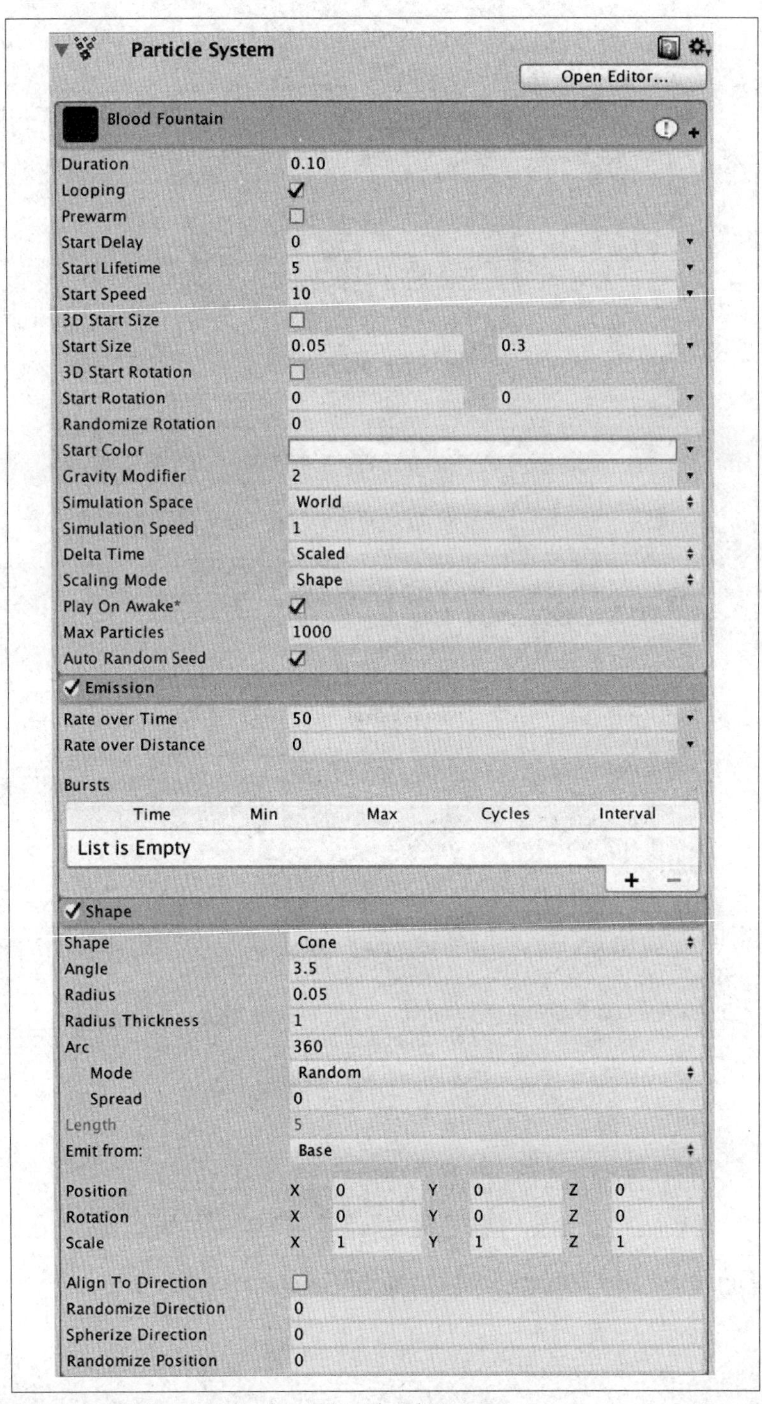

Figure 8-6. The settings from the Blood Fountain

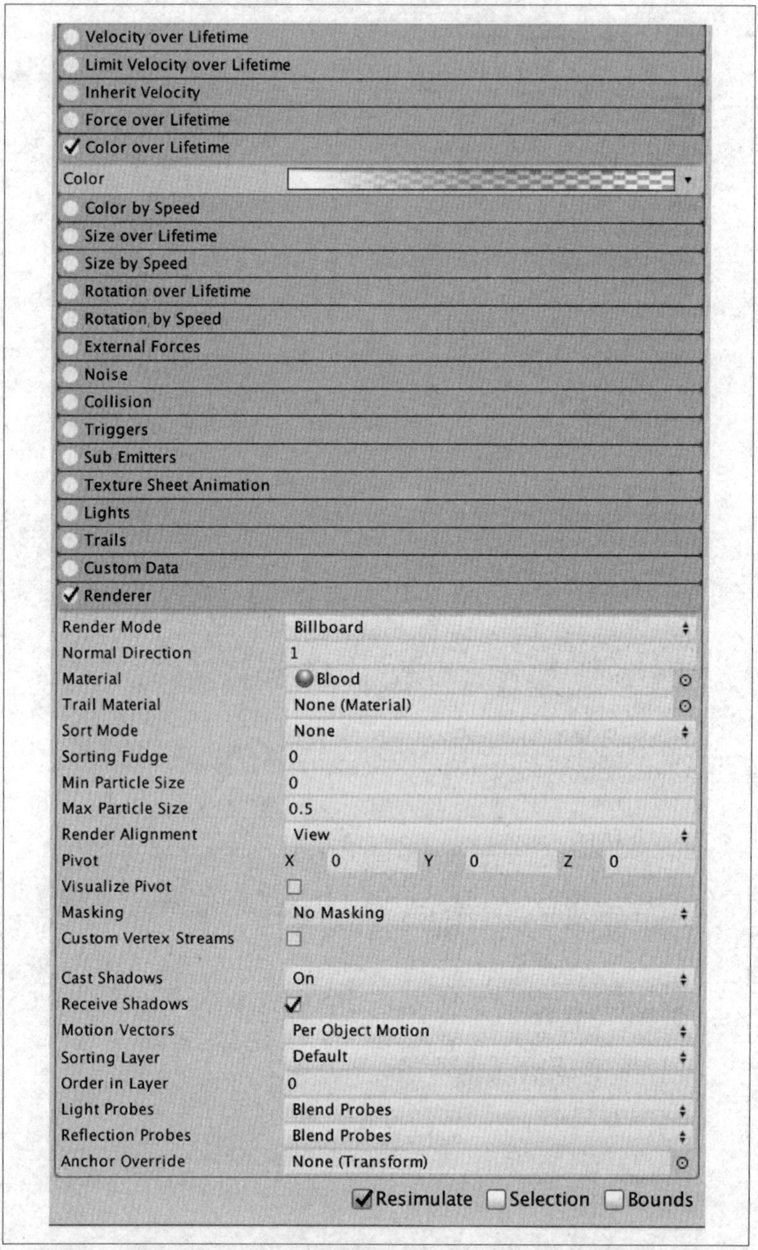

- Velocity over Lifetime
- Limit Velocity over Lifetime
- Inherit Velocity
- Force over Lifetime
- ✓ Color over Lifetime

Color	

- Color by Speed
- Size over Lifetime
- Size by Speed
- Rotation over Lifetime
- Rotation by Speed
- External Forces
- Noise
- Collision
- Triggers
- Sub Emitters
- Texture Sheet Animation
- Lights
- Trails
- Custom Data
- ✓ Renderer

Render Mode	Billboard
Normal Direction	1
Material	Blood
Trail Material	None (Material)
Sort Mode	None
Sorting Fudge	0
Min Particle Size	0
Max Particle Size	0.5
Render Alignment	View
Pivot	X 0 Y 0 Z 0
Visualize Pivot	☐
Masking	No Masking
Custom Vertex Streams	☐
Cast Shadows	On
Receive Shadows	✓
Motion Vectors	Per Object Motion
Sorting Layer	Default
Order in Layer	0
Light Probes	Blend Probes
Reflection Probes	Blend Probes
Anchor Override	None (Transform)

✓ Resimulate ☐ Selection ☐ Bounds

Figure 8-7. The settings from the Blood Fountain (cont.)

The Blood Explosion

Next, we'll create the Blood Explosion prefab, which emits a single burst of particles, rather than creating a continuous stream of them.

1. *Create the particle system object.* Create another Particle System game object, and name it "Blood Explosion".
2. *Configure the particle system.* Update the values in the Inspector to match those in Figure 8-8.

   This particle system uses the same material and color over lifetime settings as the Blood Fountain effect; the only major differences are the fact that it uses a circle emitter, and the emission rate is set up to emit all of its particles in a single burst.
3. *Add a `RemoveAfterDelay` script.* To keep the scene tidy, the Blood Explosion should remove itself after a certain amount of time.

   Add a `RemoveAfterDelay` component to the object, and set the Delay property to 2.
4. *Make the Blood Explosion into a prefab.*

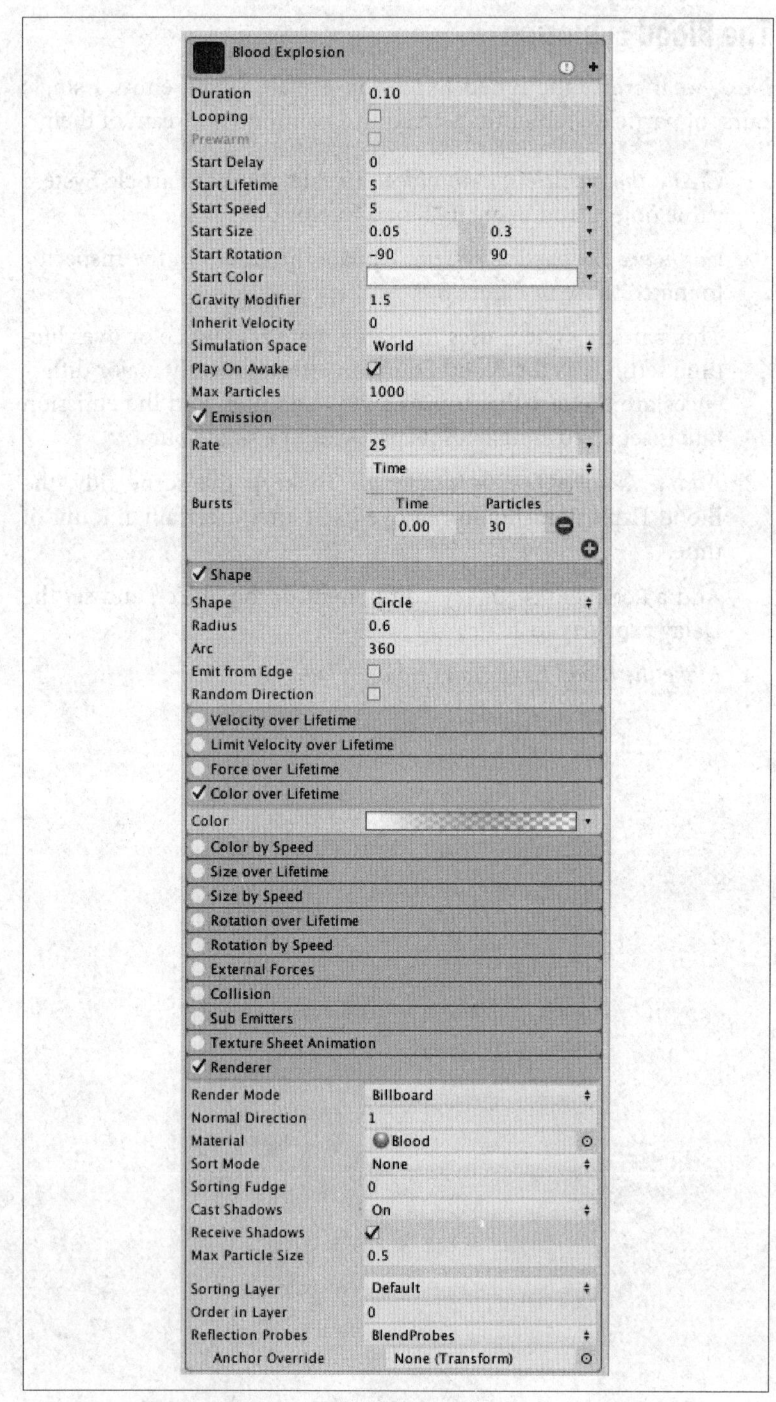

Figure 8-8. The Blood Explosion settings

You're just about ready to start using it in the game.

Using the Particle Systems

To make the game use these particle systems, you need to connect them to the Gnome prefab. Here's how to set it up:

1. *Select the Gnome prefab.* Don't forget to select the correct one—you want the new Gnome prefab, not the old Prototype Gnome prefab.
2. *Connect the particle systems to the Gnome.* Drag the Blood Explosion prefab into the Death Prefab slot, and drag the Blood Fountain prefab into the Blood Fountain slot.
3. *Test the game.* Make the gnome touch a trap, and you'll see blood.

Main Menu

The core of the game is now complete and polished. It's now time to work on some of the features that all games need, as opposed to the features that are specific to *Gnome's Well.* To wit: you need a title screen, and a way to get from the title screen into the game.

This will be implemented as a separate scene, to keep the game separate. Because the menu is a simpler scene than the full game, the menu will load faster than the game, and the player will be looking at stuff sooner. Additionally, the menu will start loading the full game in the background; when the player taps the New Game button, the game will finish up its loading, and switch scenes. The net result will be that the game appears to start up much more quickly. To set it up, follow these steps:

1. *Create a new scene.* Open the File menu, and choose New Scene. Immediately save this new scene by opening the File menu again and choosing Save Scene. Name this scene "Menu".
2. *Add the background image.* Open the GameObject menu, and choose UI → Image.

   Set the image's Source Image to the Main Menu Background sprite.

Set the image's anchors to stretch vertically, and to be centered horizontally. Set the X position to 0, the Top margin 0, the Bottom margin to 0, and the width to 800.

Turn on Preserve Aspect on the image, to prevent it from stretching.

The Inspector should now look like Figure 8-9, and the image itself should look like Figure 8-10.

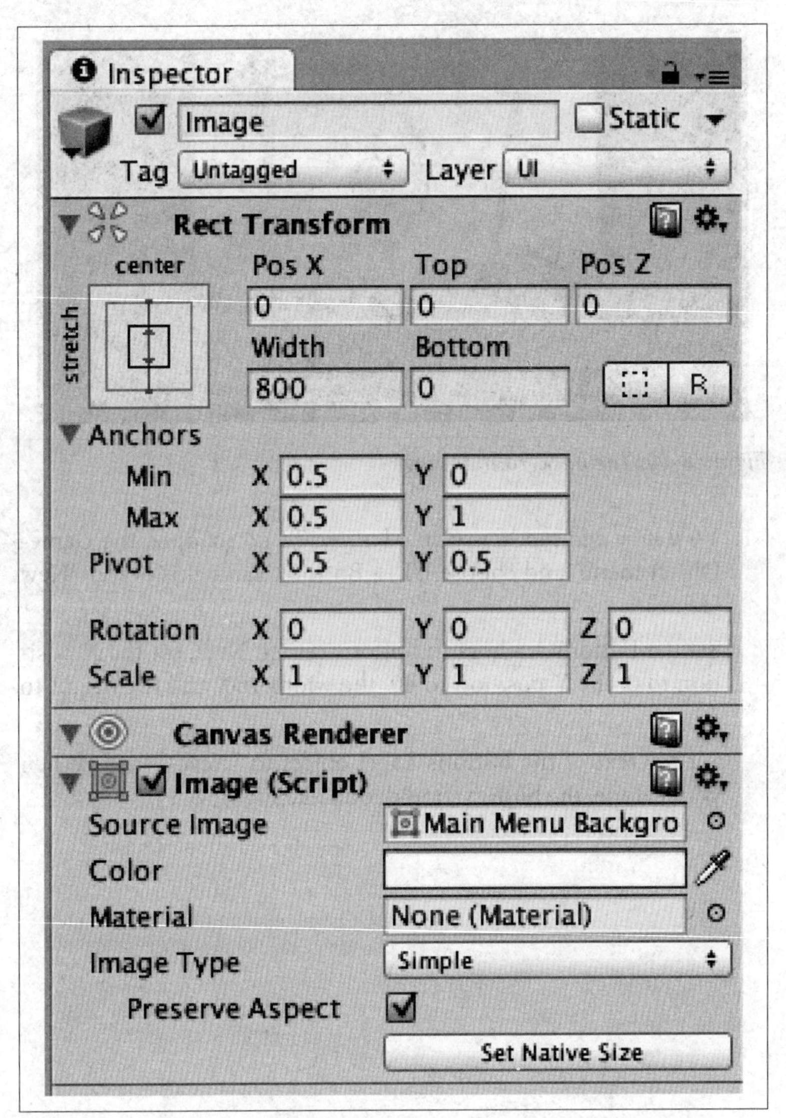

Figure 8-9. The Inspector for the main menu's background image

Figure 8-10. The background image

1. *We'll now add the New Game button.* To do so, open the GameObject menu, and choose UI → Button. Name this object "New Game".

   Set the button's anchors to bottom-center. Next, set the X position to 0, the Y position to 40, the width 160, and the height to 30.

   Set the text of the button's Label object to "New Game". When you're done, the button should look like Figure 8-11.

Figure 8-11. The menu, with the button added

Scene Loading

When the player taps the New Game button, we want an overlay to appear that tells the player that the game is loading. Let's get that set up by doing the following:

1. *Create the overlay object.* Create a new empty game object named "Loading Overlay". Make it a child of the Canvas object.

Make the overlay's anchors stretch vertically and horizontally, and set the Top, Bottom, Left, and Right margins to zero. This will make it fill the screen.

2. *Add an Image component.* With the Loading Overlay object still selected, click the Add Component button, and add an Image component. The canvas will fill with white.

   Change the Color property to black, with a transparency value of 50%. The overlay will now be a translucent black cover.

3. *Add a label.* Add a Text object, and make it a child of the Loading Overlay.

   Set the label's anchors to center horizontally and vertically. Set the Left, Top, Right, and Bottom positions to 0.

   Next, increase the Text component's font size, and make the text vertically and horizontally centered. Set color to white, and set the text to "Loading..."

Now that the overlay has been set up, we'll add the code that actually loads the full game, and switches scenes when the New Game button is tapped. For convenience, we'll add this to the Main Camera, though you can add it to a new empty game object, if you prefer. Follow these steps to get it done:

1. *Add the `MainMenu` code to the Main Camera.* Select the Main Menu and add a new C# script called `MainMenu`.

   Add the following code to *MainMenu.cs*:

```
using UnityEngine.SceneManagement;

// Manages the main menu.
public class MainMenu : MonoBehaviour {

  // The name of the scene that contains the game itself.
  public string sceneToLoad;

  // The UI component that contains the "Loading..." text.
  public RectTransform loadingOverlay;

  // Represents the scene background loading. This is
  // used to control when the scene should switch over.
  AsyncOperation sceneLoadingOperation;

  // On start, begin loading the game.
  public void Start() {
```

```
        // Ensure the 'loading' overlay is invisible
        loadingOverlay.gameObject.SetActive(false);

        // Begin loading in the scene in the background...
        sceneLoadingOperation =
          SceneManager.LoadSceneAsync(sceneToLoad);

        // ...but don't actually switch to the new scene until
        // we're ready.
        sceneLoadingOperation.allowSceneActivation = false;

    }

    // Called when the New Game button is tapped.
    public void LoadScene() {

        // Make the 'Loading' overlay visible
        loadingOverlay.gameObject.SetActive(true);

        // Tell the scene loading operation to switch scenes
        // when it's done loading.
        sceneLoadingOperation.allowSceneActivation = true;

    }

}
```

The Main Menu script is responsible for two things: loading the game scene in the background, and responding to the player tapping the New Game button. In the `Start` method, the `SceneManager` is asked to begin loading the scene in the background. This is returned as an `AsyncOperation` object called `sceneLoadingOperation`, which gives us control over how the loading is performed. In this case, we tell the `sceneLoadingOperation` that the new scene should not be activated when loading is complete. Doing this means that after loading is complete, the loading operation will wait until the user is ready to proceed to the next menu.

This is done in the `LoadScene` method, which is called when the user taps on the New Game button. First, the "loading" overlay that you just set up is made to appear; next, the scene loading operation is told that it is allowed to activate the scene after loading completes. Doing this means that if the scene has finished loading, it will immediately appear; if the scene hasn't finished loading yet, it will appear as soon as loading finishes.

Structuring the main menu this way means that the entire game will appear to load faster. Because the main menu requires fewer resources that need to be loaded than the main game does, it will appear faster; when it appears, the user will take a moment to click on the New Game button. This is time that the game will spend loading the new scene; however, because the user wasn't stuck staring at a "please wait" screen, the whole thing will feel faster than if the game had launched directly to the game itself.

Follow these steps:

1. *Configure the Main Menu component.* Set the Scene to Load variable to *Main* (that is, the name of the game's main scene). Set the Loading Overlay variable to the Loading Overlay that you just made.
2. *Make the button load the scene.* Select the New Game button, and make it run Main Camera's `MainMenu.LoadScene` function.

Finally, we need to set up the list of scenes to include in the build. `Application.LoadLevel` and its related functions are only able to load scenes that appear in the list of scenes that are included in the build, which means that we need to ensure that both the Main and Menu scenes are present. Here's how to do it:

1. *Open the Build Settings window.* Do this by opening the File menu, and choosing File → Build Settings.
2. *Add the scenes to the Scenes In Build list.* Drag both the Main and Menu scene files from the *Assets* folder into the Scenes In Build list. Make sure that Menu is first in the list, since that's the scene that should appear when the game starts.
3. *Test the game.* Run the game, and click the New Game button. You'll end up in the game!

Audio

There's one last piece of polish to add: sound effects. Without sound, the game is merely a horrifying vignette of gnome death, and we need to fix that.

Fortunately, the code that you've already added to the game is set up to make adding sound easy to do. The Signal On Touch script will make sound play when the gnome touches the corresponding collider, but only if there's an audio source attached. To make it happen, you'll need to add Audio Source components to the various prefabs.

Additionally, the Game Manager script plays sounds when the Gnome dies, and also when the gnome successfully reaches the exit while holding the treasure. Again, you'll need to add Audio Source components to the Game Manager. To do that, follow these steps:

1. *Add Audio Source components to the spikes.* Find the SpikesBrown prefab, and add a new Audio Source component.

   Attach the "Death By Static Object" sound to the new Audio Source. Ensure that Loop and Play On Awake are both turned off.

   Repeat this for the SpikesRed and SpikesBlue prefabs.

2. *Add an Audio Source component to the Spinner.* Find the Spinner prefab, and add a new Audio Source component. Attach the "Death by Moving Object" sound to the Audio Source. Again, make sure that Loop and Play On Awake are both off.

3. *Add an Audio Source component to the Treasure.* Find the Treasure at the bottom of the well, and add a new Audio Source component. Attach the "Treasure Collected" sound to the Audio Source. Once more, make sure that Loop and Play On Awake are both turned turned off.

4. *Add an Audio Source component to the Game Manager.* Finally, select the Game Manager object, and add an Audio Source component. Leave the Audio Clip property empty; instead, attach the "Game Over" sound to the Gnome Died Sound slot, and the "You Win" sound to the Game Over Sound slot.

5. *Test the game.* You'll now hear sound effects when the gnome dies, picks up the treasure, and wins the game.

Wrapping Up and Challenges

You're now all done building *Gnome's Well That Ends Well*, and you should be looking at a game that looks a little something like Figure 8-12. Congratulations!

Figure 8-12. The final game

From this point, there are a number of additional things you can do to explore the possiblities of this game:

Add the ghost
: In “Setting Up the Gnome’s Code” on page 96, we set it up so that when the gnome dies, it creates an object. As your next

step, create a prefab that displays a ghost sprite (we've included one in the resources) that travels upward. Consider also using a particle effect to make it leave an ethereal trail.

Add more traps
: We've included assets for two additional traps: the *swinging blade* and the *flamethrower*. The swinging blade is a big blade, attached to a chain, and is designed to swing from left to right. You'll need to use an Animator to make it move. The flamethrower is an object that's designed to shoot fireballs at the gnome; when they hit the gnome, they should call the Game Manager's `FireTrapTouched` function. Don't forget to investigate the burned skeleton sprites for the gnome!

Build more levels
: The game is designed to only have a single level, but there's no reason not to add more.

Add more effects
: Make particles appear around the treasure (use the Shiny1 and Shiny2 images.) Make particles come off the walls when the player hits them.

PART III

Building a 3D Game: Space Shooter

In this part, we'll build a second game from scratch. Unlike the game we built in Part II, this game will take place in 3D. You'll build a space combat simulator, in which the player has to defend a space station from incoming asteroids. As part of this, we'll be exploring systems that frequently appear in other games, such as projectile shooting, respawning objects, and managing the appearance of 3D models. It's going to be a blast. (Our editors let us keep that pun in.)

CHAPTER 9

Building a Space Shooter

In addition to being an excellent platform for building 2D games, Unity is also great for creating 3D content. Unity was designed as a 3D engine long before its 2D features came on the scene, and as such, Unity's features were first built for 3D games.

In this chapter, you'll learn how to use Unity to build *Rockfall*, a 3D space-simulator game. This style of game was originally popular in the mid-1990s, when games like *Star Wars: X-Wing* (1993) and *Descent: Freespace* (1998) gave players the freedom to fly around in open space, shooting at bad guys and blowing space up. These kinds of games are closely related to flight simulators, but because they're not expected to be realistic implementations of flight physics, game developers can get away with more fun-oriented mechanics.

That's not to say that arcade-style flight simulators don't exist, but it's more common to find an arcade-style spaceflight simulator than a realistic one. The biggest exception, in recent years, has been *Kerbal Space Program*, which is so realistic in its spaceflight physics simulation that it's about as far removed from the type of game covered in this chapter as you can get. If you really want to learn about orbital mechanics, and what happens when you thrust prograde at the apoapsis, this is the kind of game for you.

It's therefore fairly reasonable to say that the term "space simulator," though more common for the type of game in this chapter, might be better off as "space combat simulator."

Enough waffling about labels. Let's start shooting laser cannons.

At the end of these next few chapters, you'll have a game that looks like Figure 9-1.

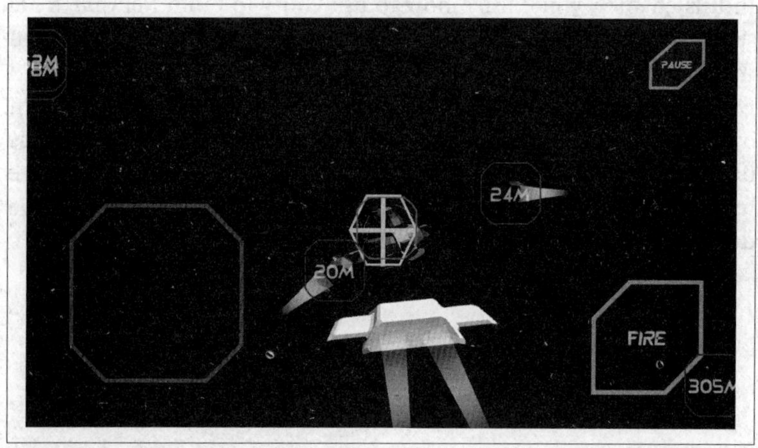

Figure 9-1. The finished game

Designing the Game

When we began to design the game, we decided on some key constraints:

- Gameplay sessions should be a couple of minutes, at most.
- The controls should be very simple, and try to keep to a minimum of "move" and "shoot."
- The game should focus on multiple short-term challenges, instead of a single one. That is, lots of little enemies, rather than a single boss fight. (This is the opposite of *Gnome's Well*, which is the 2D game we discussed in Part II.)
- The game should primarily be about shooting laser beams in space. There aren't enough video games about shooting laser beams in space. There never *can* be.

It's almost always a good idea to start thinking about high-level concepts on paper. Thinking on paper gives you an unstructured approach that's useful for discovering new ideas that fit in nicely with your overall plan. To that end, we sat down and sketched up a quick idea of the game (seen in Figure 9-2).

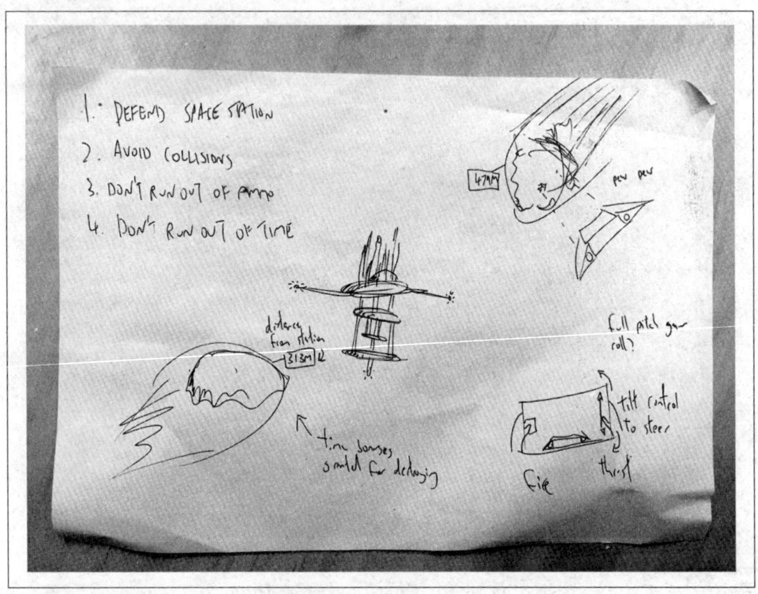

Figure 9-2. The original sketched idea for the game

The whole sketch is deliberately rough, and drawn very quickly, but you can see a number of additional items popping up: asteroids are moving toward a space station, the player uses an on-screen joystick to pilot the spaceship, and taps a fire button to shoot lasers. Some more specific details are visible as well, which were the result of thinking about how to represent this kind of scene: things like a label that shows how far away the asteroids are from the space station, and thoughts on how the player would hold the device.

Once we had this rough sketch, we grabbed an artist friend, Rex Smeal (*https://twitter.com/RexSmeal*), and asked him to turn Jon's messy sketch into something a little more fleshed out. While doing this wasn't an absolutely critical part of the game design process, it helped us to figure out the overall feel of the game. In particular, we realized that the central space station that the player is defending needs to have a a significant amount of attention paid to it, since it needs to look like it's something worth saving. After grabbing our local artist and describing the game to him, he came up with the design shown in Figure 9-3; once we'd locked down the design together, Rex refined the design into something that could be modeled (Figure 9-4).

Figure 9-3. Rex's initial concept for the game's look

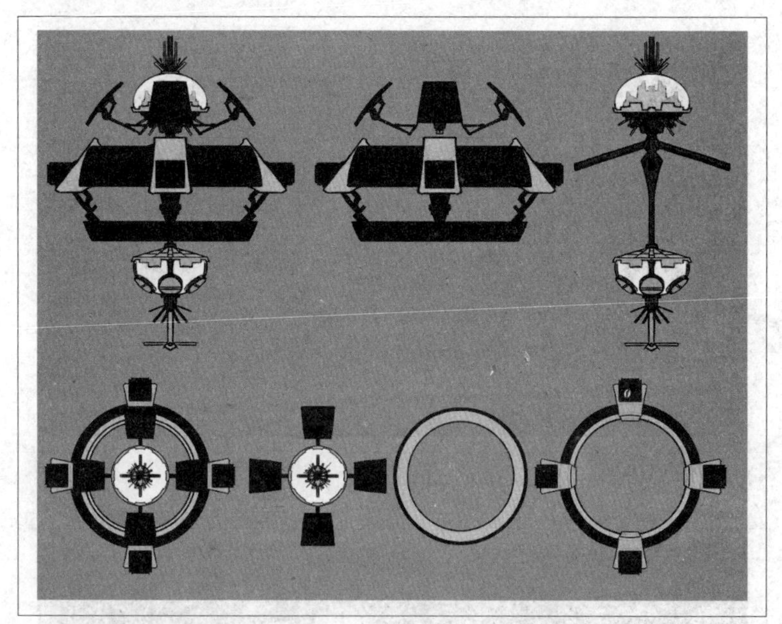

Figure 9-4. The refined concept art for the space station, ready for modeling

Following this design, we modeled it up in Blender. During the development of the station, we decided that a low-poly approach to the art (inspired by artists like Heather Penn (*https://twitter.com/heatpenn*) and Timothy Reynolds (*http://turnislefthome.com*)) would work well, due to its simple style. (That's not to say that low-poly art is simple or easy to make; just that it's easier to work with the style for the same reason that drawing with a pencil can be simpler than painting with oils.)

You can see the station in Figure 9-5. Additionally, we modeled the spaceship and an asteroid in Blender as well, which you can see in Figures 9-6 and 9-7.

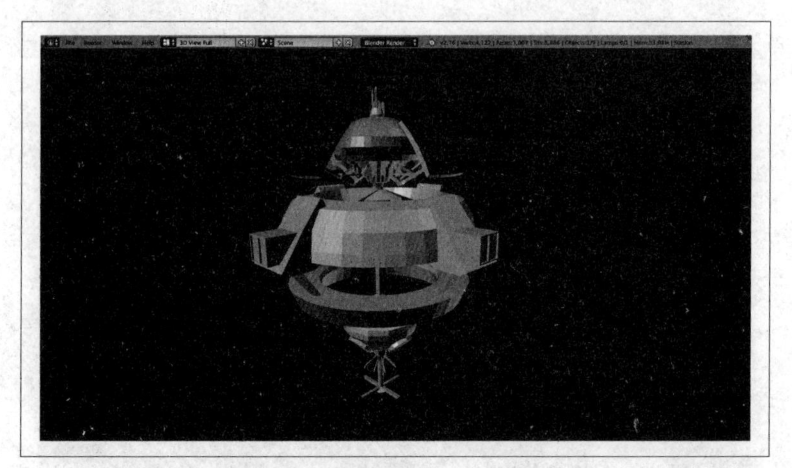

Figure 9-5. The modeled space station

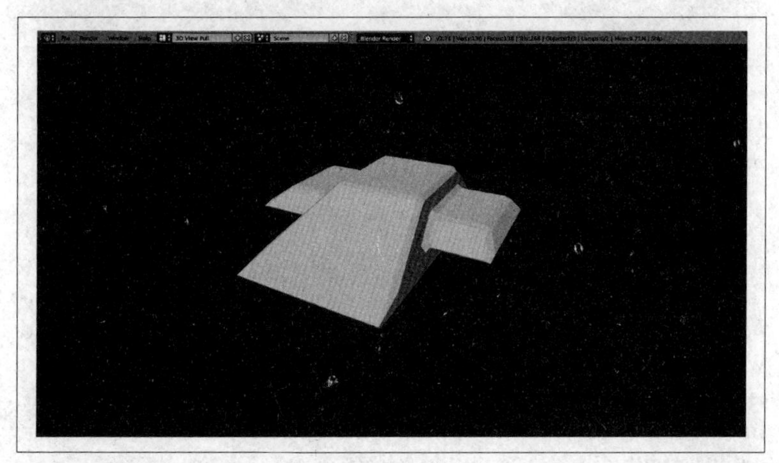

Figure 9-6. The modeled spaceship

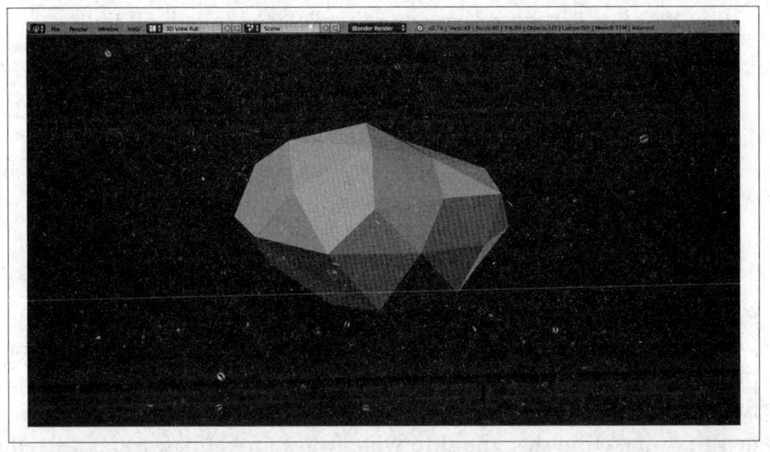

Figure 9-7. The modeled asteroid

Getting the Assets

To build this game, you'll be working with several resources, including sound effects, models, and textures that we've packaged for you. You'll need to download these resources first. The files are organized into a folder structure to make it easy for you to find everything.

To download the assets, grab them from the project's GitHub page (*http://bit.ly/rockfall-assets*).

Architecture

The game's architecture is, at its core, very similar to the one used in *Gnome's Well That Ends Well*. A central game manager is in charge of instantiating critical game objects, like the player-controlled spaceship and the space station; this manager is also notified about when the game ends, which happens when the player dies.

The user interface in the game is slightly more complex than the one we created previously. In *Gnome's Well*, the in-game controls were two buttons, plus tilt controls; in a 3D game where the player can conceivably move in any direction, tilt controls tend not to work very well. Instead, the game will use an on-screen "joystick"—a region on the screen that detects touches, and lets the user drag a finger to indicate direction. This will feed information to a shared input manager, which the spaceship uses to adjust its flight.

Tilt controls are more challenging to pull off in 3D games, but that doesn't mean that it's impossible to do them well. NOVA 3 (*http://bit.ly/nova-3*) is an FPS that uses tilt controls that allow the player to turn their character and to aim quite precisely, and it's worth playing this game to get a feel for how they did their input.

The flight model used in the game will be deliberately slightly unrealistic. The simplest, and most realistic approach, would be to simply model a physics object that applies a forward thrust, and uses physical forces to rotate the craft. However, this would be difficult to fly, and would lead the player to getting lost easily. Instead, we decided to do fake physics: the ship will always move forward at a fixed speed, and has no momentum. Additionally, the player can't roll the ship, and any roll will be corrected (that is, unlike the real outer space, space in this game has an "up").

Design and Direction

Every single one of the design decisions made for the games in this book are entirely arbitrary. Even though we decided that we didn't want to do a physical flight model in this game, it doesn't mean that physical flight models are something to avoid in arcade-style flight simulations. Play around, and see what you come up with. Don't decide that games are one certain way based on what some book author told you. They could just be making it all up as they go.

The asteroids will be prefabs that are created by a dedicated "asteroid spawner" object. This object will instantiate asteroids every so often, and aim them toward the space station. When the asteroids collide with the space station, they'll decrease the station's hit points; when the space station has no more hit points, it's destroyed, and the game is over.

Creating the Scene

Let's get started by setting up the scene. We'll create a new Unity project, and then we'll create the spaceship, which we'll fly around the scene. Follow these steps to get started:

1. *Create the project.* Create a new Unity project called Rockfall, and set its mode to 3D (see Figure 9-8).

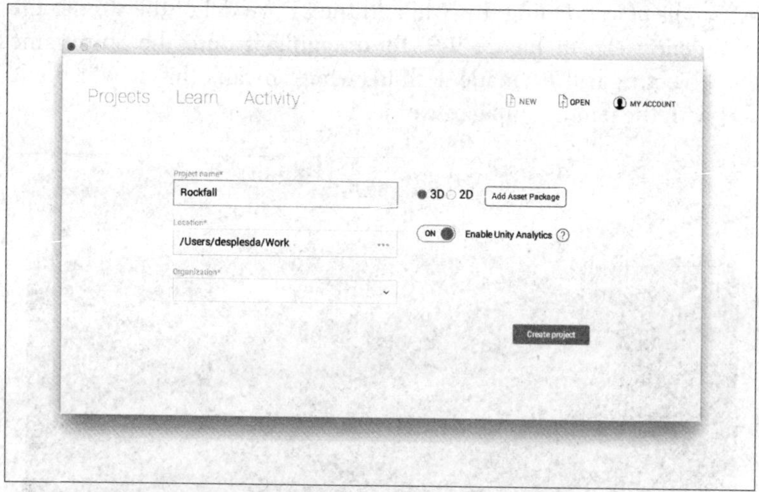

Figure 9-8. Creating the project

2. *Save the new scene.* Once Unity has created your project and shows the empty scene, save it by opening the File menu and choosing Save. Save the scene as *Main.scene* in the *Assets* folder.
3. *Import the downloaded assets.* Double-click on the *.unitypackage* file that you downloaded in "Getting the Assets" on page 209. Import all of the assets into the project.

You're now ready to start building a spaceship.

Ship

We'll start the game with the spaceship, using the spaceship model that you downloaded in "Getting the Assets" on page 209.

The Ship object itself will be an invisible object that contains only scripts; multiple child objects will be attached to it, which handle the specific tasks of appearing on screen.

1. *Create the Ship object.* Open the GameObject menu, and choose Create Empty. A new `GameObject` will appear in the scene; rename it to "Ship".

We'll now add the ship's model.

2. *Open the Models folder, and drag the Ship model onto the `Ship` game object.* Doing this will add the 3D model of the ship to the scene, seen in Figure 9-9. By dragging it onto the Ship game object, it will be made a child, which means that it will move with the parent Ship game object.

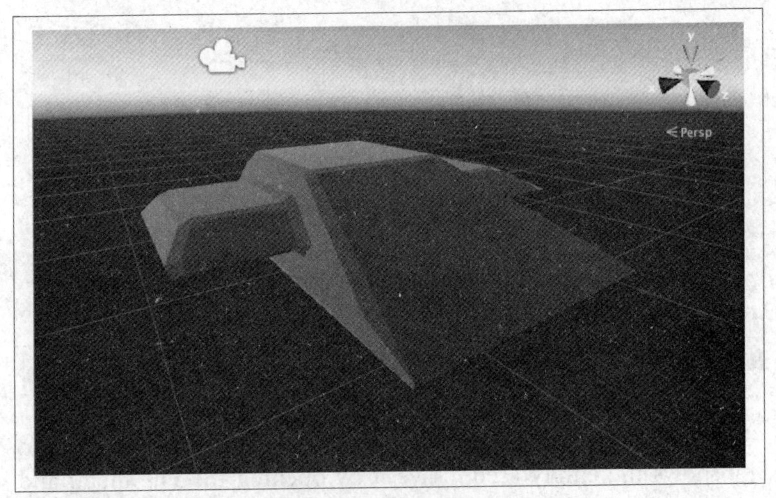

Figure 9-9. The ship model in the scene

3. *Rename the model object to "Graphics".*

Next, we need to make the Graphics object be at the same location as the Ship object.

4. *Select the Graphics object.* Click on the gear icon at the top-right of the Transform component and select Reset Position (see Figure 9-10).

Leave the rotation at (-90, 0, 0). This is needed because the ship was modeled in Blender, which uses a different coordinate system to Unity; specifically, Blender's "up" direction is the Z axis, while Unity's "up" is the Y axis. To fix this, Unity automatically rotates Blender models to compensate.

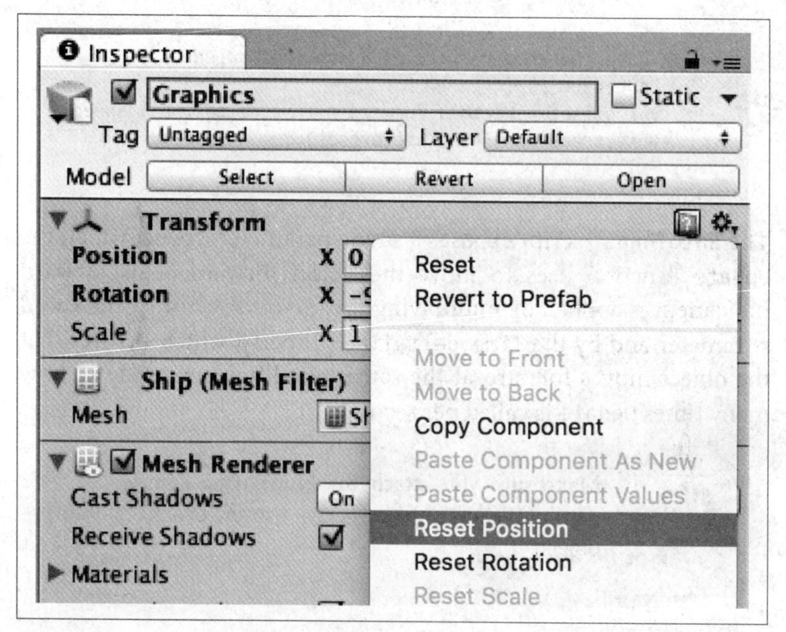

Figure 9-10. Resetting the position of the Graphics object

We want the ship to collide with objects. To do that, we'll add a Collider.

5. *Add a Box Collider to the ship.* Select the Ship object (that is, the parent of the Graphics object), and click the Add Component button at the bottom of the Inspector. Choose Physics → Box Collider.

   Once the collider has been added, turn Is Trigger on, and set the box's Size to (2, 1.2, 3). This will create a box surrounding the player.

The ship needs to travel forward at a constant speed. To do this, we'll add a script that translates whatever object that it's attached to.

6. *Add the `ShipThrust` script.* With the Ship object still selected, click the Add Component button at the bottom of the Inspector. Create a new C# script called *ShipThrust.cs*.

   Once it's added, open *ShipThrust.cs*, and add the following code:

```
public class ShipThrust : MonoBehaviour {

  public float speed = 5.0f;
```

```
    // Move the ship forward at a constant speed
    void Update () {
      var offset = Vector3.forward * Time.deltaTime * speed;
      this.transform.Translate(offset);
    }
}
```

The ShipThrust script exposes a single parameter, speed, which the Update function uses to move the object forward. This forward movement is applied by multiplying the forward vector by the speed parameter and by the Time.deltaTime property, which ensures that the object moves forward at the same speed independently of how many times Update is called per second.

Make sure you attach the ShipThrust component to the Ship object, and not to the Graphics object.

7. *Test the game.* Hit the Play button, and watch as the ship starts moving forward.

Camera follow

The next step is to make the camera follow the spaceship as it moves. There are several options for how you can do this: the most primitive is to put the camera inside the Ship object, so that it moves with it. However, this tends to look kind of bad, since it means that the ship will never appear to be rotating relative to the camera.

A better solution is to keep the camera as a separate object, and add a script that makes the camera slowly move into the right position over time. This means that when the ship performs a sharp turn, it will take a moment for the camera to compensate—which is exactly how it would look if a real camera operator was trying to follow an object around.

1. *Add the `SmoothFollow` script to the main camera.* Select the Main Camera, and click the Add Component button. Add a new C# script called *SmoothFollow.cs*.

   Open the file, and add the following code:

```
public class SmoothFollow : MonoBehaviour
{

  // The target we are following
  public Transform target;

  // The height we want the camera to be above the target
  public float height = 5.0f;

  // The distance to the target, not counting height
  public float distance = 10.0f;

  // How much we slow down changes in rotation and height
  public float rotationDamping;
  public float heightDamping;

  // Update is called once per frame
  void LateUpdate()
  {
    // Bail out if we don't have a target
    if (!target)
      return;

    // Calculate the current rotation angles
    var wantedRotationAngle = target.eulerAngles.y;
    var wantedHeight = target.position.y + height;

    // Note where we're currently positioned and looking
    var currentRotationAngle = transform.eulerAngles.y;
    var currentHeight = transform.position.y;

    // Damp the rotation around the y-axis
    currentRotationAngle
      = Mathf.LerpAngle(currentRotationAngle,
        wantedRotationAngle,
          rotationDamping * Time.deltaTime);

    // Damp the height
    currentHeight = Mathf.Lerp(currentHeight,
      wantedHeight, heightDamping * Time.deltaTime);

    // Convert the angle into a rotation
    var currentRotation
      = Quaternion.Euler(0, currentRotationAngle, 0);

    // Set the position of the camera on the x-z plane to:
    // "distance" meters behind the target
    transform.position = target.position;
    transform.position -=
      currentRotation * Vector3.forward * distance;
```

```
        // Set the position of the camera using our new height
        transform.position = new Vector3(transform.position.x,
          currentHeight, transform.position.z);

        // Finally, look at where the target is looking
        transform.rotation = Quaternion.Lerp(transform.rotation,
          target.rotation,
            rotationDamping * Time.deltaTime);

    }
}
```

The *SmoothFollow.cs* script that appears in this book is based on code provided by Unity. We've adapted it slightly so that it's better suited for the flight simulator. If you want to see the original version of this code, you can find it in the Utility package, which you can import by opening the Assets menu and choosing Import Package → Utility. After importing, you'll find the original *SmoothFollow.cs* file in Standardard Assets → Utility.

`SmoothFollow` works by calculating a location in 3D space for where the camera should be, and then calculating a point *between* that location and where the camera is right now. When applied over multiple frames, this has the effect of making the camera gradually approach that point, but also to slow to a stop as it gets closer. Additionally, because the location of where the camera should be is changing every frame, the camera will always be slightly lagging behind—which is exactly what you want.

2. *Configure the SmoothFollow component.* Drag the Ship object into the Target field.
3. *Test the game.* Hit the Play button. When the game starts, the Game panel will no longer show the ship moving; instead, the camera will be following along. You'll be able to see this in action in the Scene panel.

Space Station

The space station, which is under threat from the incoming asteroids, will follow the same development pattern as the ship: we'll create an empty game object, and attach the model to it. The space station is also simpler than the spaceship, since it's entirely passive: it just sits there, and has rocks thrown at it. Follow these steps to set it up:

1. *Create a container for the space station.* Create a new empty game object, and name it "Space Station".
2. *Add the model as a child object.* Open the *Models* folder, and drag the Station model onto the Space Station game object.
3. *Reset the position of the Station model object.* Select the Station object you just added, and right-click the Transform component. Choose Reset Position, just like you did for the Ship's model.

When you're done, the space station should look like Figure 9-11.

Figure 9-11. The space station

Now that the model has been added, it's worth taking a quick look at the structure of the model, and to ensure that it's got colliders. The station's colliders are important, because the asteroids (which we'll add eventually) will need something to collide with.

Select the model object, and expand it to show its children. The station model is composed of multiple submeshes; the main one is named Station. Select it.

Look at the Inspector. In addition to the Mesh Filter and Mesh Renderer, you'll also see a Mesh Collider (see Figure 9-12). If you *don't* see this, see Models and Colliders.

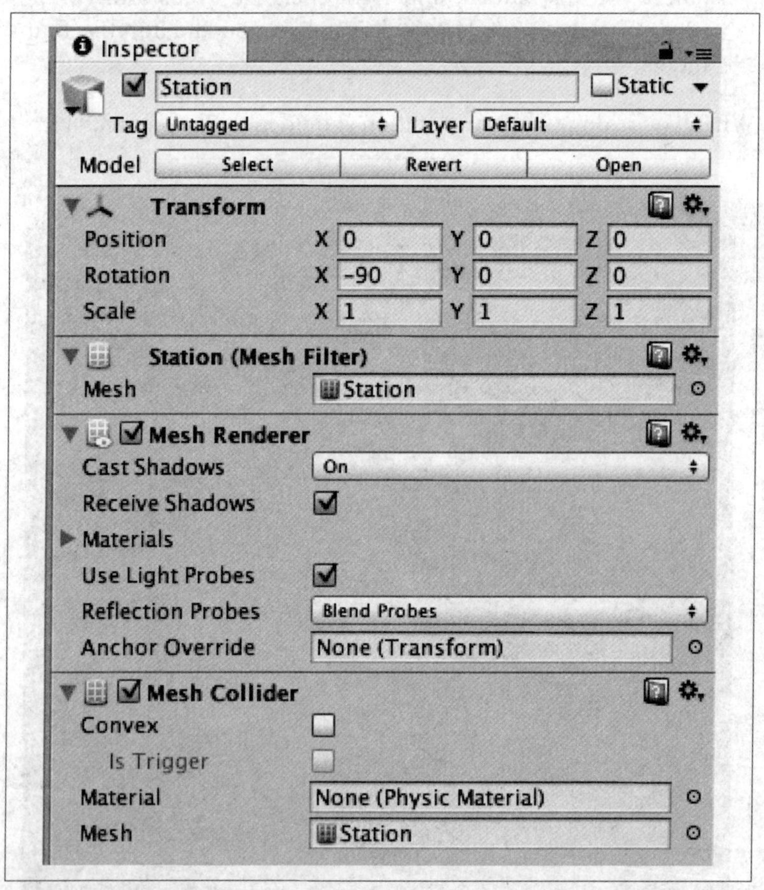

Figure 9-12. The space station's collider

Models and Colliders

When you import a model, Unity can create a collider for it as well. When you imported the model from the *Asset* package, you also imported the settings that we created for it, which include the setting to create a collider for the station. (We did the same thing for the Ship and Asteroid models, too.)

If you don't see it, or if you're importing your own model and want to know how to set it up, you can view and change these settings by selecting the model itself (that is, the file in the *Models* folder), and looking at the settings (Figure 9-13). In particular, note that the Generate Colliders option is selected.

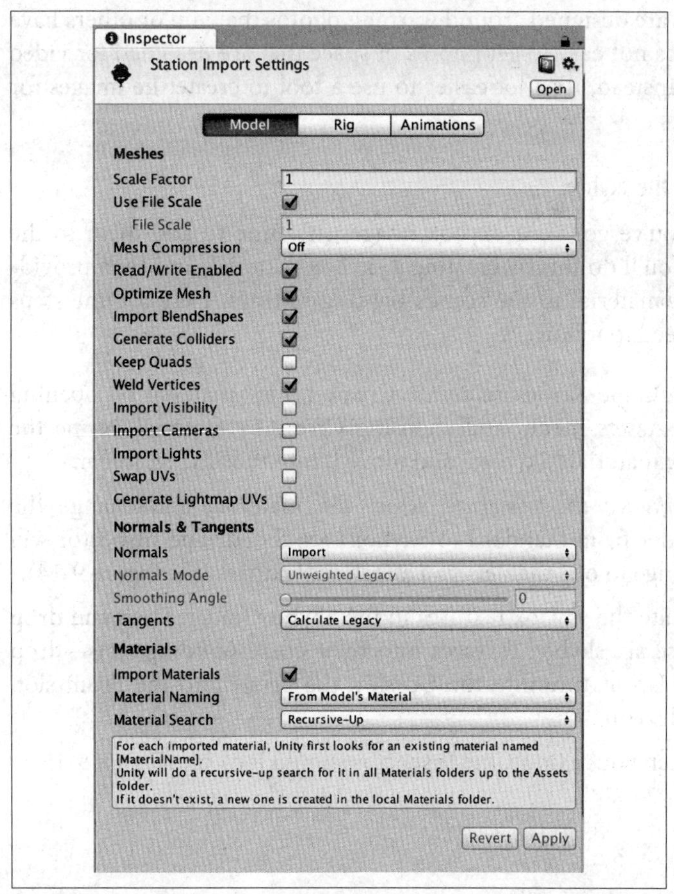

Figure 9-13. The import settings for the space station model

Skybox

Currently, the skybox is the Unity default, which is designed to suit a game set on the surface of a planet. Changing that skybox to one designed to look like you're floating in space will go a long way toward making the game feel correct.

Skyboxes work by creating a virtual cube that's always drawn underneath everything else in the scene, and never moves relative to the camera. This creates the impression that the textures on that cube are infinitely far away—hence, "skybox."

In order to create the illusion that you're inside a sphere instead of a boxy cube, the textures on the skybox need to be distorted so that there aren't visible seams at the edges. There are multiple ways to do this, including several plug-ins for Adobe Photoshop; however, most of these are designed around warping photos that you or others have taken. It's not easy to get photos of space that are designed for video games; instead, it's a lot easier to use a tool to create the images for you.

Creating the skybox

Once you've got your skybox images, it's time to add them to the game. You'll do this by creating a skybox material, and then providing that material to the scene's lighting settings. Here are the steps you'll need to follow:

1. *Create the Skybox material.* Create a new material by opening the Assets menu, and choosing Create → Material. Name the new material "Skybox", and move it into the Skybox folder.
2. *Configure the material.* Select the material, and change the shader from Standard to Skybox → 6 Sided. The Inspector will change to one that lets you attach six textures (see Figure 9-14).

   Locate the skybox textures in the Skybox folder. Drag and drop these six skybox textures into their corresponding slots—drop the Up texture into the Up slot, the Front into the Front slot, and so on.

   When you're done, the Inspector should look like Figure 9-15.

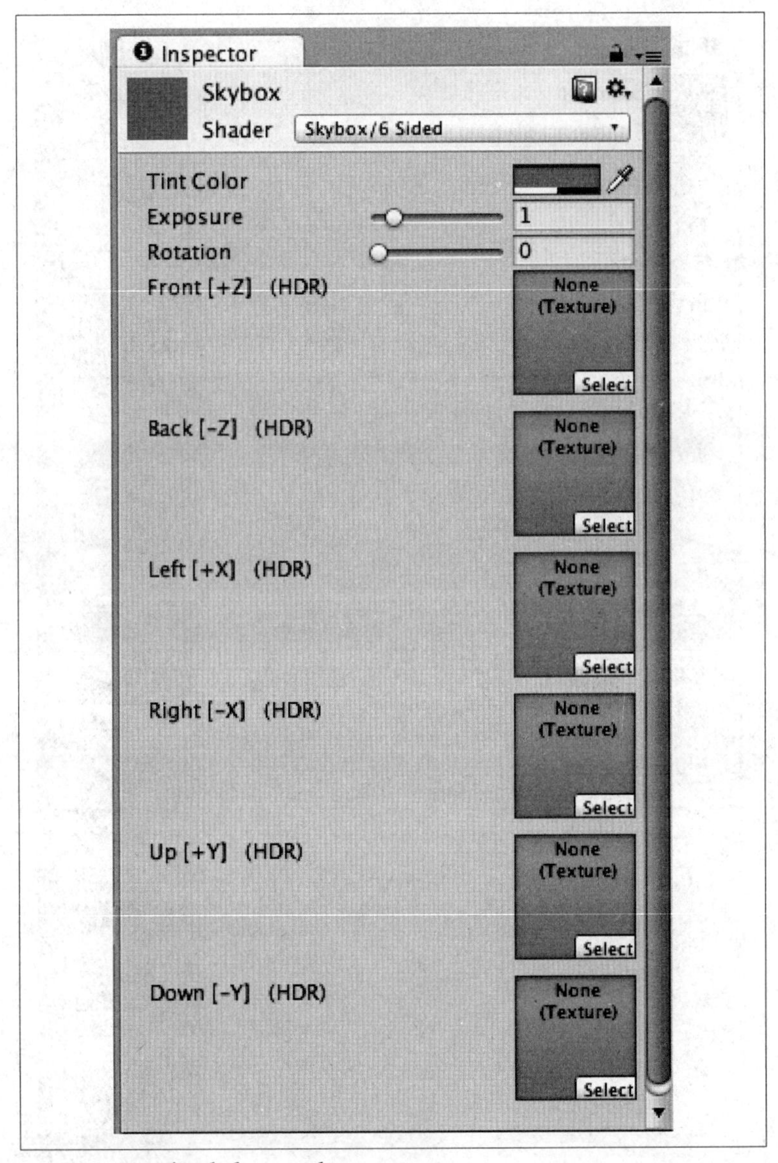

Figure 9-14. The skybox, with no textures

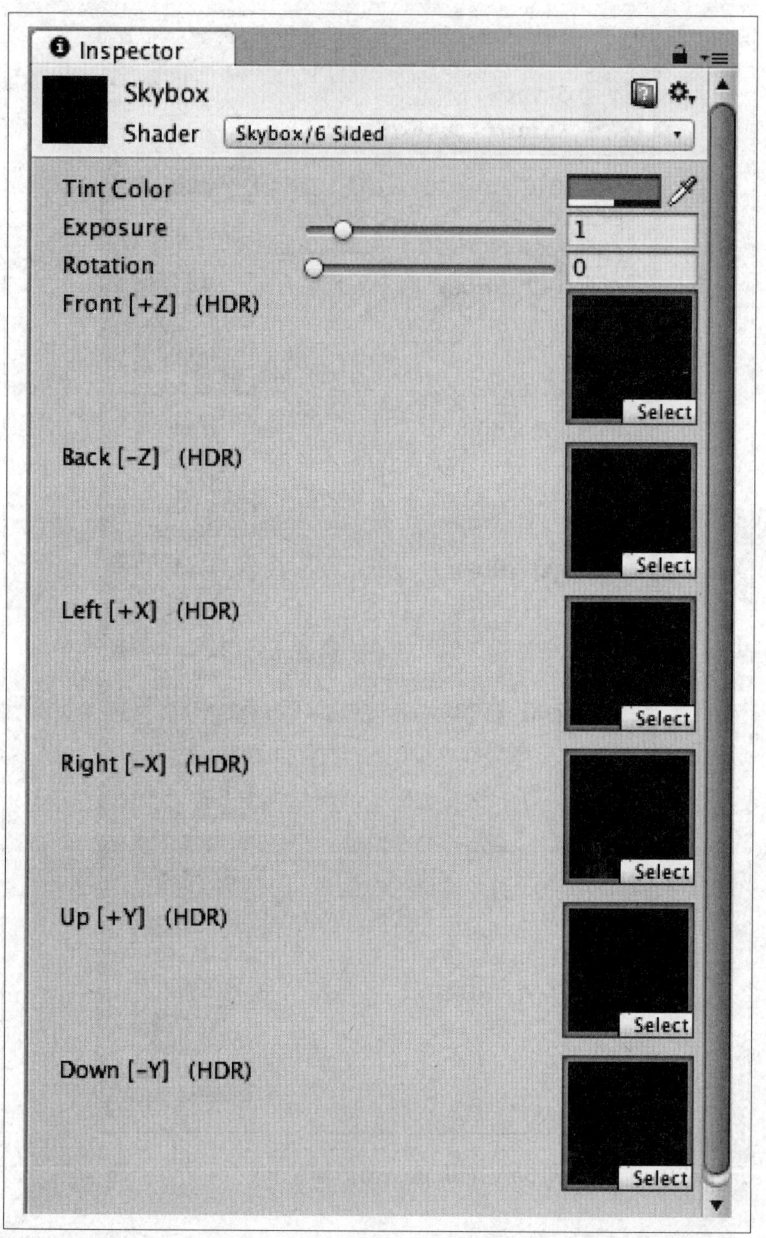

Figure 9-15. The skybox, with textures attached

3. *Connect the skybox to the lighting settings.* Open the Window menu, and choose Lighting → Settings. The Lighting panel will appear; near the top of the panel, you'll see a slot labeled "Skybox." Drop the Skybox material that you just attached into the slot (see Figure 9-16).

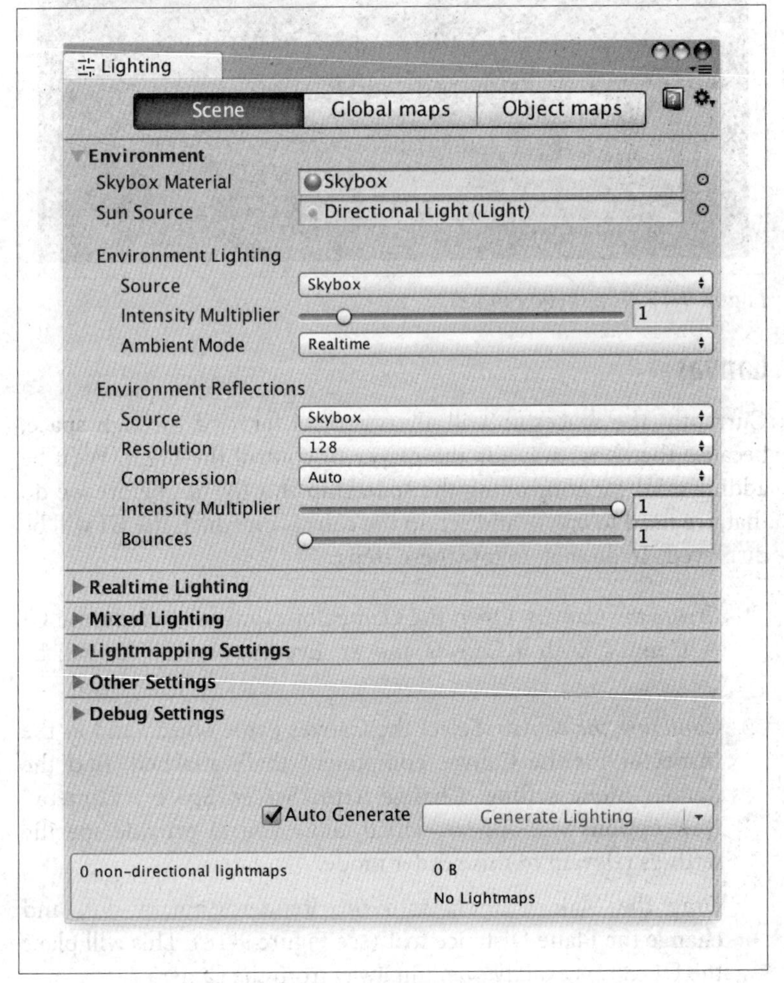

Figure 9-16. Creating the lighting settings

When you're done, the sky will be replaced with the space images (as seen in Figure 9-17). Additionally, Unity's lighting system uses information from the skybox to influence how objects are lit; if you look closely, you'll notice that the spaceship and the space station are both slightly tinted green, because the skybox images are green.

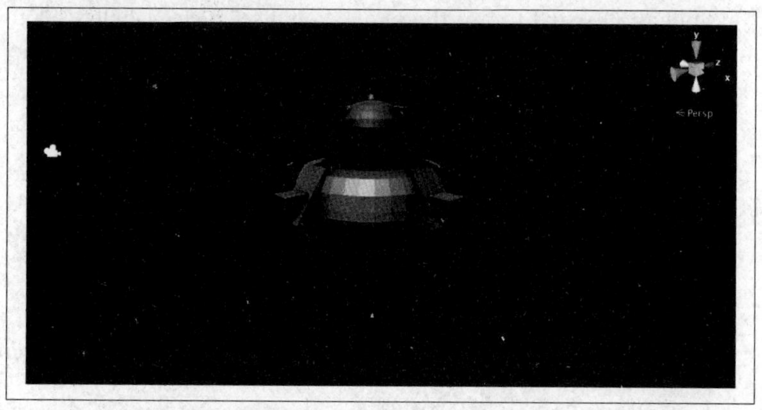

Figure 9-17. The skybox in use

Canvas

Currently, the spaceship will always travel forward through space, because there's no way for the player to control the flight. We'll be adding a UI for controlling the spaceship shortly, but before we do that, we need to create and set up the canvas on which the UI will be displayed. To do that, follow these steps:

1. *Create the Canvas.* Open the GameObject menu, and choose UI → Canvas. Both a Canvas and an EventSystem object will be created.
2. *Configure the canvas.* Select the Canvas game object, and in the Inspector for the Canvas component that's attached, find the Render Mode setting. Change it to "Screen Space - Camera." New options will appear, which allow you to provide specific settings relevant to this render mode.

   Drag the Main Camera into the Render Camera slot, and change the Plane Distance to 1 (see Figure 9-18). This will place the UI canvas exactly one unit away from the camera.

   Change the "UI Scale Mode" setting for the Canvas Scaler to "Scale with Screen Size," and set the reference resolution to 1024 x 768, which is the right shape for an iPad.

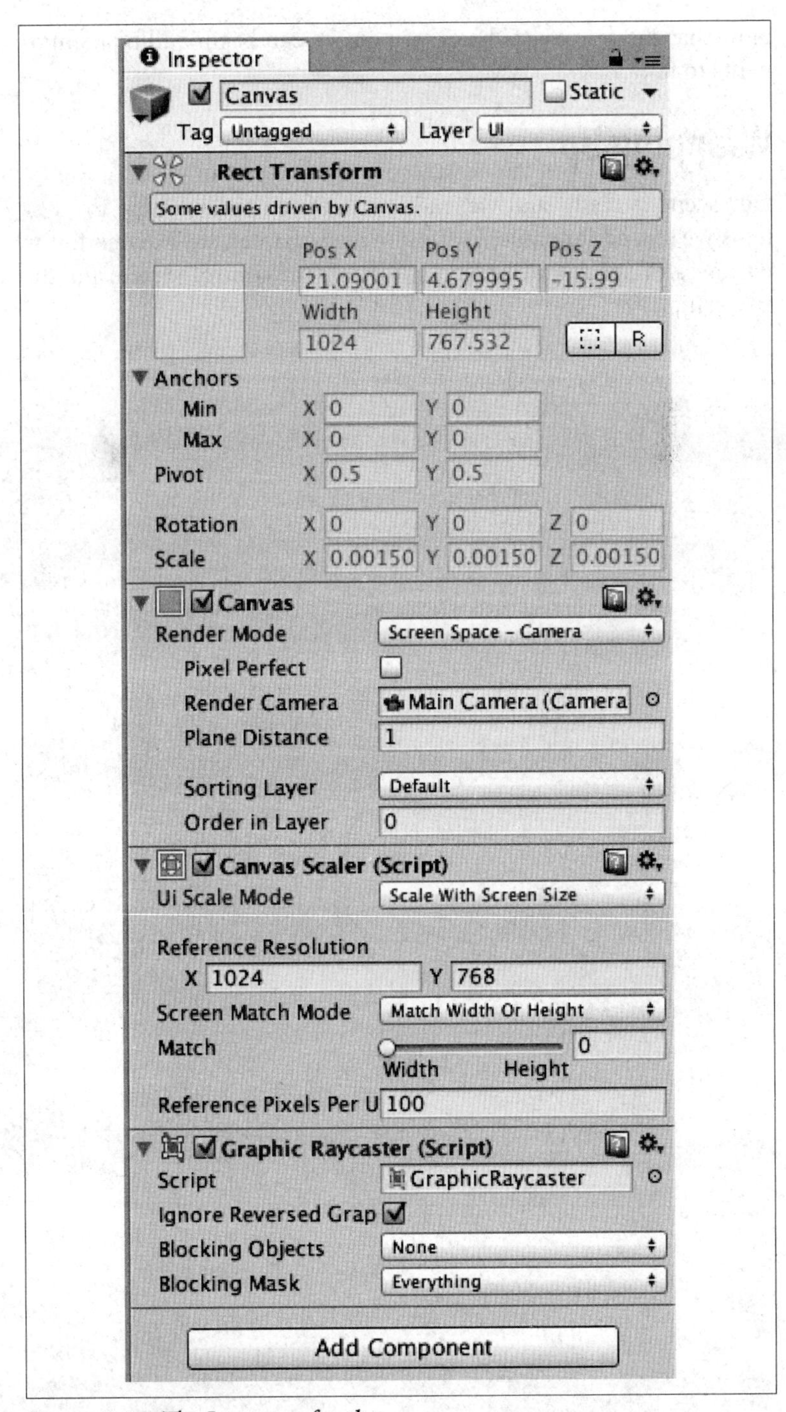

Figure 9-18. The Inspector for the canvas

Now that the canvas has been created, we can begin adding components to it.

Wrapping Up

Our scene is ready, and we're all set to start implementing the systems we'll need for gameplay. In the next chapter, we'll dive into the darkness of space and implement the flight control system for the spaceship.

CHAPTER 10

Input and Flight Control

Once you've got the scene largely laid out, it's time to start adding the basics of gameplay. In this chapter, we'll start building the systems that let you get your ship around in space.

Input

There are two different kinds of input used in the game: a virtual joystick, which lets the player provide directional input used for flight, and a button that signals whether the player wants to be firing the ship's lasers or not.

Don't forget that the only way to properly test a touchscreen game's input is by testing it on a touchscreen. To test your game without building to the device, use the Unity Remote app (see "Unity Remote" on page 79).

Adding the Joystick

We'll begin by creating the joystick. The joystick is composed of two visible components: a large square "pad" in the lower-left corner of the canvas, and a smaller "thumb" in the center of that square. When the user places a finger inside the pad, the joystick will reposition itself so that the thumb is directly under the finger, and is still centered. When the finger moves, the thumb will move with it. <<< Follow these steps to start building the input system:

1. *Create the pad.* Open the GameObject menu, and choose UI → Panel. Name the new panel "Joystick".

   We'll start by making it square, and placing it in the lower-left corner of the screen. Set the anchors to Lower Left. Next, set both the width and height of the panel to 250.

2. *Add the imagery to the pad.* Change the Source Image setting of the Image component to the `Pad` sprite.

When you're done, the panel should look like Figure 10-1.

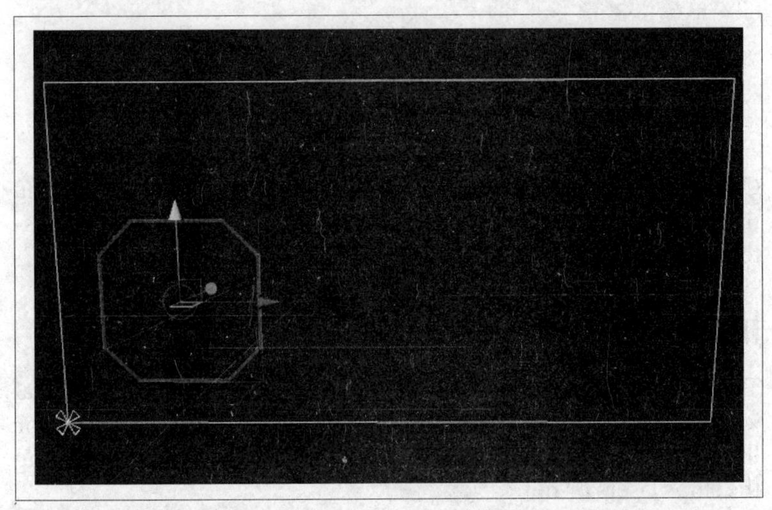

Figure 10-1. The joystick pad

3. *Create the thumb.* Create a second Panel UI object, and name it "Thumb".

   Make the thumb be a child of the Joystick. Set its anchors to Middle Center, and set its width and height to 80. Set the Pos X and Pos Y to 0. This will center the thumb in the middle of the pad. Finally, set the Source Image to the `Thumb` sprite.

4. *Add the `VirtualJoystick` script.* Select the Joystick, and add a new C# script called *VirtualJoystick.cs*. Open the file, and add the following code:

   ```
   // Get access to the Event interfaces
   using UnityEngine.EventSystems;

   // Get access to UI elements
   ```

```
using UnityEngine.UI;

public class VirtualJoystick : MonoBehaviour,
  IBeginDragHandler, IDragHandler, IEndDragHandler {

  // The sprite that gets dragged around
  public RectTransform thumb;

  // The locations of the thumb and joystick when no dragging
  // is happening
  private Vector2 originalPosition;
  private Vector2 originalThumbPosition;

  // The distance that the thumb has been dragged away from
  // its original position
  public Vector2 delta;

  void Start () {
    // When the joystick starts up, record the original
    // positions
    originalPosition
      = this.GetComponent<RectTransform>().localPosition;
    originalThumbPosition = thumb.localPosition;

    // Disable the thumb so that it's not visible
    thumb.gameObject.SetActive(false);

    // Reset the delta to zero
    delta = Vector2.zero;
  }

  // Called when dragging starts
  public void OnBeginDrag (PointerEventData eventData) {

    // Make the thumb visible
    thumb.gameObject.SetActive(true);

    // Figure out where in world-space the drag started
    Vector3 worldPoint = new Vector3();
    RectTransformUtility.ScreenPointToWorldPointInRectangle(
      this.transform as RectTransform,
      eventData.position,
      eventData.enterEventCamera,
      out worldPoint);

    // Place the joystick at that point
    this.GetComponent<RectTransform>().position
      = worldPoint;

    // Ensure that the thumb is in its original location,
    // relative to the joystick
```

```
        thumb.localPosition = originalThumbPosition;
    }

    // Called when the drag moves
    public void OnDrag (PointerEventData eventData) {

        // Work out where the drag is in world space now
        Vector3 worldPoint = new Vector3();
        RectTransformUtility.ScreenPointToWorldPointInRectangle(
            this.transform as RectTransform,
            eventData.position,
            eventData.enterEventCamera,
            out worldPoint);

        // Place the thumb at that point
        thumb.position = worldPoint;

        // Calculate distance from original position
        var size = GetComponent<RectTransform>().rect.size;

        delta = thumb.localPosition;

        delta.x /= size.x / 2.0f;
        delta.y /= size.y / 2.0f;

        delta.x = Mathf.Clamp(delta.x, -1.0f, 1.0f);
        delta.y = Mathf.Clamp(delta.y, -1.0f, 1.0f);

    }

    // Called when dragging ends
    public void OnEndDrag (PointerEventData eventData) {
        // Reset the position of the joystick
        this.GetComponent<RectTransform>().localPosition
            = originalPosition;

        // Reset the distance to zero
        delta = Vector2.zero;

        // Hide the thumb
        thumb.gameObject.SetActive(false);
    }
}
```

The `VirtualJoystick` class implements three key C# interfaces: `IBeginDragHandler`, `IDragHandler`, and `IEndDragHandler`. When the player begins dragging, continues a drag, or finishes dragging anywhere in the Joystick, the script will receive an `OnBeginDrag`, `OnDrag`, and `OnEndDrag` method call, respectively. These methods receive a single parameter: a `PointerEventData` object that contains information about where the finger is on the screen, among lots of other data.

- When a drag *begins*, the pad repositions itself so that its center point is under the finger.
- When a drag *continues*, the thumb is moved to stay under the finger, and the distance from the center of the pad to the thumb is measured and stored in the `delta` property.
- When a drag *ends* (that is, when the touch is lifted from the screen) the pad and thumb reset to their earlier position. The `delta` property is reset to zero.

To finish building the input system:

5. *Configure the joystick.* Select the Joystick object, and drag the Thumb object into the Thumb slot.
6. *Test the joystick.* Run the game, and click and drag inside the joystick pad. The pad will move when the drag begins, and the thumb will move as you continue dragging. Pay attention to the value of the Joystick object's Delta—it should change as you move the thumb around.

The Input Manager

Now that the joystick has been set up, we need a way for the spaceship to get information from it, so that it can be used for steering.

We *could* directly connect the ship to the joystick, but doing that has a problem. During the game, the ship will be destroyed, and new ships will be created. In order to make this possible, the ship needs to be made into a prefab, so that the game manager can make multiple copies of it. However, prefabs aren't allowed to refer to objects in the scene, which means that a freshly created ship object would have no reference to the joystick.

A better way to do it is to create an Input Manager singleton that's always in the scene, and which *does* have a reference to the joystick. Because it's not instantiated from a prefab, we don't need to worry about losing its reference when it's created. When the ship is created, it will use the Input Manager singleton (which it accesses through code) to reach the joystick, and to get at the joystick's value.

1. *Create the Singleton code.* Create a new C# script in the Assets folder named *Singleton.cs*. Open this file, and put the following code in it:

```
// This class allows other objects to refer to a single
// shared object. The GameManager and InputManager classes
// use this.

// To use this, subclass like so:
// public class MyManager : Singleton<MyManager>  {  }

// You can then access the single shared instance of the
// class like so:
// MyManager.instance.DoSomething();

public class Singleton<T> : MonoBehaviour
  where T : MonoBehaviour {

  // The single instance of this class.
  private static T _instance;

  // The accessor. The first time this is called, _instance
  // will be set up. If an appropriate object can't be found,
  // an error will be logged.
  public static T instance {
    get {
      // If we haven't already set up _instance..
      if (_instance == null)
      {
        // Try to find the object.
        _instance = FindObjectOfType<T>();

        // Log if we can't find it.
        if (_instance == null) {
          Debug.LogError("Can't find "
            + typeof(T) + "!");
        }
      }

      // Return the instance so that it can be used!
      return _instance;
    }
  }

}
```

This Singleton code is identical to the Singleton used in *Gnome's Well*. For a description of what it does, see "Creating a Singleton class" on page 81.

2. *Create the Input Manager.* Create a new empty game object called "Input Manager". Add a new C# script to it, called *InputManager.cs*. Open the file, and add the following code:

```
public class InputManager : Singleton<InputManager> {

    // The joystick used to steer the ship.
    public VirtualJoystick steering;

}
```

Currently, `InputManager` serves as a simple data object: it just stores a reference to the `VirtualJoystick`. Later on, we'll be adding some more logic to it, to support things like firing the current spaceship's weapons.

3. *Configure the Input Manager.* Drag the Joystick into the "Steering" slot.

Now that the joystick is set up, we're ready to start using it to control the spaceship's flight.

Flight Control

At the moment, the ship simply moves forward. To control the ship's flight, we just need to change the ship's "forward" direction. We'll be doing this by taking information from the virtual joystick, and using it to update the ship's orientation in space.

Every frame, the ship uses the direction indicated by the joystick, along with a value that controls how quickly the ship should rotate, to generate a new rotation. This is then combined with the ship's *current* orientation, resulting in the new direction the ship is facing.

However, we want the player to not roll over and get confused about where important objects like the space station are. To address that, the steering script also applies an *additional* rotation, which slowly rolls the ship back to a level surface. This makes the ship behave a little more like an aircraft that's flying through an atmosphere, which is more intuitively understandable (but is also less realistic).

1. *Add the `ShipSteering` script.* Select the Ship, and add a new C# script called *ShipSteering.cs*. Open the file, and add the following code:

```
public class ShipSteering : MonoBehaviour {

  // The rate at which the ship turns
  public float turnRate = 6.0f;

  // The strength with which the ship levels out
  public float levelDamping = 1.0f;

  void Update () {

    // Create a new rotation by multiplying the joystick's
    // direction by turnRate, and clamping that to 90% of
    // half a circle.

    // First, get the user's input.
    var steeringInput
      = InputManager.instance.steering.delta;

    // Now, create a rotation amount, as a vector.
    var rotation = new Vector2();

    rotation.y = steeringInput.x;
    rotation.x = steeringInput.y;

    // Multiply by turnRate to get the amount we want to
    // steer by.
    rotation *= turnRate;

    // Turn this into radians by multiplying by 90% of a
    // half-circle
    rotation.x = Mathf.Clamp(
      rotation.x, -Mathf.PI * 0.9f, Mathf.PI * 0.9f);

    // And turn those radians into a rotation quaternion!
    var newOrientation = Quaternion.Euler(rotation);

    // Combine this turn with our current orientation
    transform.rotation *= newOrientation;

    // Next, try to minimize roll!

    // Start by working out what our orientation would be
    // if we weren't rolled around the Z axis at all
```

```
        var levelAngles = transform.eulerAngles;
        levelAngles.z = 0.0f;
        var levelOrientation = Quaternion.Euler(levelAngles);

        // Combine our current orientation with a small amount
        // of this "zero-roll" orientation; when this happens
        // over multiple frames, the object will slowly level
        // out to zero roll
        transform.rotation = Quaternion.Slerp(
          transform.rotation, levelOrientation,
          levelDamping * Time.deltaTime);

    }
}
```

The `ShipSteering` script uses the joystick input to calculate a new, smoothed-out rotation, and applies that to the ship. Having done that, it then applies an additional slight rotation that causes the ship to level out.

2. *Test the steering.* Start the game. The ship will start flying forward; when you click and drag inside the joystick area, the ship will change direction. Using this, you can fly around. Note that if the ship becomes rolled (e.g., if you pull up and then turn to the side), the ship will attempt to roll back to a flat attitude.

Indicators

Because this game involves flying around in 3D space, it's very easy to lose track of the various objects in the game. The space station will (eventually) be under threat from asteroids, and the player will want to know where the station is, and where the asteroids are.

To deal with this, we'll implement a system for showing indicators on the screen that highlight the position of important objects. If the camera can see the objects, then they'll have a circle around them. If the objects are off-screen, then their indicators will appear on the edges of the screen, indicating the direction that you should turn to look at them.

Creating the UI elements

To get started, create an object inside the canvas that will act as the container for all of the indicators. Once that's done, you'll need to build an indicator, which will then be turned into a prefab for reuse. To set them up, follow these steps:

1. *Create the Indicator container.* Select the Canvas, and create a new empty child object by opening the GameObject menu, and choosing Create Empty Child. This will create a new object that has a Rect Transform (which is used for 2D objects like canvas elements), as opposed to an object with a regular Transform (which is used for 3D elements). Set the anchors of the container to stretch horizontally and vertically.

   Name the new object "Indicators".

2. *Create the prototype Indicator.* Create a new Image by opening the GameObject menu and choosing UI → Image.

   Name the new object "Position Indicator". Make it a child of the Indicators object you created in the previous step.

   Drag the Indicator sprite into the sprite's Source Image slot. You'll find it in the UI folder.

3. *Create the text label.* Create a new Text object (again, via the GameObject menu, in the UI submenu). Make this Text object be the child of the Position Indicator sprite object.

   Change the text's color to white, and set alignment to horizontally and vertically centered.

   Change the text's Text to "50m". (The text will change during gameplay, but doing this will mean you have a better idea of how the indicator looks.)

   Set the anchor of the Text to "center middle", and set its X and Y position to zero. This will center the text in the middle of the sprite.

   Finally, we'll use a custom font for the indicator. Locate the CRYSTAL-Regular font, which is in the *Fonts* folder, and drag it into the Text's Font slot. Next, change the Font Size to 28.

   When you're done, the Text component's Inspector should look like Figure 10-2, and the indicator object itself should look like Figure 10-3.

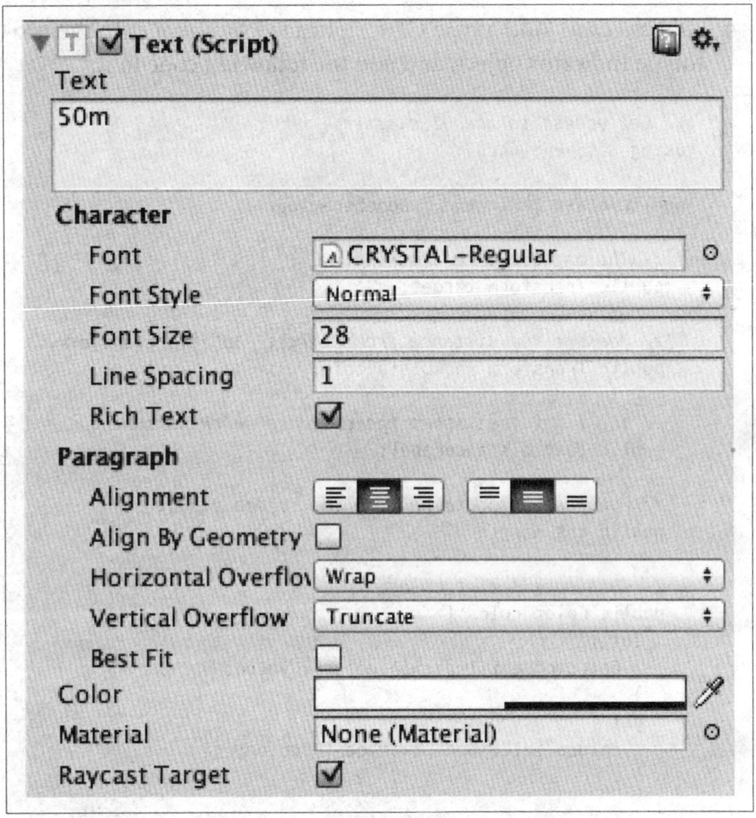

Text (Script)
Text
50m
Character
Font CRYSTAL-Regular
Font Style Normal
Font Size 28
Line Spacing 1
Rich Text
Paragraph
Alignment
Align By Geometry
Horizontal Overflow Wrap
Vertical Overflow Truncate
Best Fit
Color
Material None (Material)
Raycast Target

Figure 10-2. The Inspector for the indicator text labels

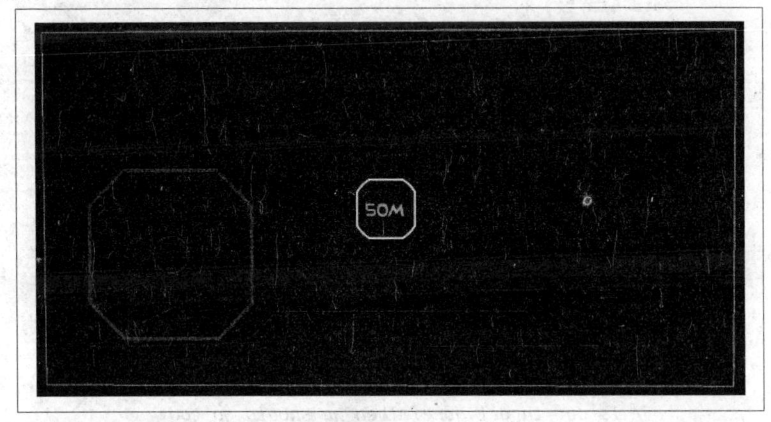

Figure 10-3. The prototype indicator

4. *Add the code.* Add a new C# script called *Indicator.cs* to the prototype Indicator object, and add the following code to it:

```
// Get access to the UI classes
using UnityEngine.UI;

public class Indicator : MonoBehaviour {

  // The object we're tracking.
  public Transform target;

  // Measure the distance from 'target' to this transform.
  public Transform showDistanceTo;

  // The label that shows the distance we're measuring.
  public Text distanceLabel;

  // How far we should be from the screen edges.
  public int margin = 50;

  // Our image's tint color.
  public Color color {
    set {
      GetComponent<Image>().color = value;
    }
    get {
      return GetComponent<Image>().color;
    }
  }

  // Set up the indicator
  void Start() {
    // Hide the label; it will be re-enabled in
    // Update if we have a target
    distanceLabel.enabled = false;

    // On start, wait a frame before appearing to prevent
    // visual glitches
    GetComponent<Image>().enabled = false;

  }

  // Update the indicator's position every frame
  void Update()
  {

    // Is our target gone? Then we should go too
    if (target == null) {
      Destroy (gameObject);
      return;
```

```
    }

    // If we have a target for calculating distance, then
    // calculate it and display it in the distanceLabel
    if (showDistanceTo != null) {

      // Show the label
      distanceLabel.enabled = true;

      // Calculate the distance
      var distance = (int)Vector3.Magnitude(
        showDistanceTo.position - target.position);

      // Show the distance in the label
      distanceLabel.text = distance.ToString() + "m";
    } else {
      // Don't show the label
      distanceLabel.enabled = false;
    }

    GetComponent<Image>().enabled = true;

    // Work out where in screen-space the object is
    var viewportPoint =
      Camera.main.WorldToViewportPoint(target.position);

    // Is the point behind us?
    if (viewportPoint.z < 0) {
      // Push it to the edges of the screen
      viewportPoint.z = 0;
      viewportPoint = viewportPoint.normalized;
      viewportPoint.x *= -Mathf.Infinity;
    }

    // Work out where in view-space we should be
    var screenPoint =
      Camera.main.ViewportToScreenPoint(viewportPoint);

    // Clamp to screen edges
    screenPoint.x = Mathf.Clamp(
      screenPoint.x,
      margin,
      Screen.width - margin * 2);

    screenPoint.y = Mathf.Clamp(
      screenPoint.y,
      margin,
      Screen.height - margin * 2);

    // Work out where in the canvas-space the view-space
    // coordinate is
```

```
        var localPosition = new Vector2();
        RectTransformUtility.ScreenPointToLocalPointInRectangle(
          transform.parent.GetComponent<RectTransform>(),
          screenPoint,
          Camera.main,
          out localPosition);

        // Update our position
        var rectTransform = GetComponent<RectTransform>();
        rectTransform.localPosition = localPosition;

    }
}
```

The indicator code works like this:

- Every frame, in the `Update` method, the 3D coordinates of the object that the indicator is tracking are converted to *viewport space.*

  In viewport space, coordinates represent positions on the screen, where (0,0,0) is the bottom-left of the screen, and (1,1,0) is the top-right. The Z component of a viewport-space coordinate represents distance from the camera, in world units.

  This means that you can very easily tell if something is on screen or not, and if it's behind you or not. If an object's view-space coordinate's X and Y components are not within (0,0) to (1,1), then it's off to the side; if the Z coordinate is less than 0, it's behind you.

- If the object is behind you (that is, the Z coordinate is less than zero), we need to push the marker off to the side. If we don't, then the indicator for something directly behind you would appear in the center of the screen, and the player would be led to believe that the object that the indicator is tracking is in front of you.

  To push the markers to the side, the X component of the viewport is multiplied by negative infinity; by multiplying by infinity, the indicator will always be at the far left or far right of the screen. We multiply by *negative* infinity to compensate for the fact that it's behind you.

- Next, the view-space coordinate is converted to screen-space, and then clamped so that it's never outside of the window. An additional margin parameter is used to bring the indicators in a little bit, to ensure that the text label that shows distance is always readable.

- Finally, this screen-space coordinate is converted into the coordinate space of the indicator container, and is then used to update the position of the indicator. Once this is done, the indicator is now in the right position.

The indicators also do their own cleanup: every frame, they check to see whether their `target` is `null`; if it is, they're destroyed.

There's one final step left to set up the indicator. Once that's done, you can convert this prototype indicator into a prefab.

4. *Connect the distance label.* Drag the Text child object into the Distance Label slot.
5. *Turn the prototype into a prefab.* Drag the Position Indicator object into the Project pane. A new prefab object will be created, which will allow you to create multiple Indicators at runtime.

   Once you've created this prefab, delete the prototype from the scene.

Indicator Manager

The Indicator Manager is a singleton object that manages the process of creating indicators. This object will be used by any other object that needs to add an indicator to the screen—in particular, the space station and the asteroids.

By making this object a singleton, we can create and set up the object in the scene without having to do anything tricky to make objects that are loaded from prefabs aware of the manager.

1. *Create the Indicator Manager.* Create a new empty object, and name it "Indicator Manager."
2. *Add the `IndicatorManager` script.* Add a new C# script to the object, called *IndicatorManager.cs*, and add the following code to it:

```
using UnityEngine.UI;

public class IndicatorManager : Singleton<IndicatorManager> {
```

```
    // The object that all indicators will be children of
    public RectTransform labelContainer;

    // The prefab we'll instantiate for each indicator
    public Indicator indicatorPrefab;

    // This method will be called by other objects
    public Indicator AddIndicator(GameObject target,
      Color color, Sprite sprite = null) {

      // Create the label object
      var newIndicator = Instantiate(indicatorPrefab);

      // Make it track the target
      newIndicator.target = target.transform;

      // Update its color
      newIndicator.color = color;

      // If we received a sprite, set the indicator's sprite
      // to that
      if (sprite != null) {
        newIndicator
          .GetComponent<Image>().sprite = sprite;
      }

      // Add it to the container.
      newIndicator.transform.SetParent(labelContainer, false);

      return newIndicator;
    }

}
```

The Indicator Manager provides a single method, `AddIndicator`, which instantiates an Indicator prefab, configures it with a target object to track and a color to tint the sprite with, and adds it to the indicator container. You can also optionally provide your own `Sprite` to this method, if you want to create a special indicator. (You'll be doing this later, when you add the ship's target reticle.)

Once you've written the `IndicatorManager` source code, you now need to configure it. The manager needs to know two things: which prefab should be instantiated for the indicators, and which object should be their parent.

3. *Set up the Indicator Manager.* Drag the Indicators container object into the Label Container slot, and the Position Indicator prefab into the Indicator Prefab slot.

Next, we'll make the space station run code that adds an indicator when it starts.

4. *Select the space station.*
5. *Add the `SpaceStation` script it.* Add a new C# script to the object called *SpaceStation.cs*, and add the following code to it:

```
public class SpaceStation : MonoBehaviour {

  void Start () {
    IndicatorManager.instance.AddIndicator(
      gameObject,
      Color.green
    );
  }

}
```

This code simply asks the `IndicatorManager` singleton to add a new indicator that tracks this object, and for that indicator to be green.

6. *Run the game.* The station will now have an indicator attached to it.

The distance display won't appear, because the space station doesn't set up the `showDistanceTo` variable. That's on purpose—we'll be setting that up for the Asteroids, but not the station. Having too many numbers on screen can get confusing.

Wrapping Up

Congratulations! You started from next to nothing, and now you've already got spaceflight. In the next chapter, we'll be extending this game and adding actual gameplay.

CHAPTER 11

Adding Weapons and Targeting

Now that you've got a spaceship flying around, it's time to add more gameplay to the whole thing. First, you'll be adding weapons to your spaceship; once that's done, you'll need a target to shoot at.

Weapons

Every time the ship fires its weapons, it shoots a laser bolt that flies forward until it either hits something or runs out of time. If it hits something, and that other object is able to take damage, then the shot needs to convey information to that object.

We can do this by creating an object that has a collider, and travels forward at a certain rate (much like the spaceship). There are a number of different possibilities for how the shot could appear—you could create a 3D model of a missile, create a particle effect, or create a sprite. The specifics are up to you, and don't affect how the shot actually behaves in the game.

In this chapter, we'll use a *trail renderer* to display the shot. A trail renderer creates a trail behind it as it moves, which eventually disappears. This makes it especially good for representing moving objects, such as swinging sorts and flying projectiles.

The trail renderer for the shot will be a simple one: it will leave behind a thin red line, which gets thinner and thinner over time. Because the shots will always be moving forward, this will create a good-looking "blaster bolt"-like effect.

The nongraphical component of the shot will be implemented with a *kinematic rigidbody*. Ordinarily, rigidbodies respond to forces that are applied to them: gravity will pull them down, and when another rigidbody knocks into them, Newton's first law of motion means that their velocity will change. However, we don't want shots to be bumped out of the way. To tell Unity that a rigidbody ignores any forces applied to it, while still making that rigidbody collide with other bodies, you make it *kinematic*.

It's a reasonable question to ask why we're using rigidbodies at all for the shot. After all, the spaceship doesn't use one, so why do shots?

The reason is because of a limitation of Unity's physics engine. Collisions only occur when at least one of the colliding objects has a rigidbody; as a result, to ensure that the shots are always told about when they touch another object, we attach a rigidbody to it and then make it kinematic.

Let's start by creating the shot object and setting up its collision properties. We'll then add the `Shot` code to it, which is responsible for making the shot fly forward at a constant speed.

1. *Create the shot.* Create a new empty game object, and name it "Shot".

   Add a rigidbody component to the object. Then make sure Use Gravity is turned off, and Is Kinematic is turned on.

   Add a sphere collider to the object. Set its radius to 0.5, and ensure that its center is (0,0,0). The Is Trigger setting should be turned on.

2. *Add the `Shot` script.* Add a new C# script called `Shot` to the object. Open *Shot.cs*, and add the following code:

```
// Moves forward at a certain speed, and dies after a certain
// time.
public class Shot : MonoBehaviour {

    // The speed at which the shot will move forward
    public float speed = 100.0f;

    // Remove this object after this many seconds
```

```
    public float life = 5.0f;

    void Start() {
      // Destroy after 'life' seconds
      Destroy(gameObject, life);
    }

    void Update () {
      // Move forward at constant speed
      transform.Translate(
        Vector3.forward * speed  * Time.deltaTime);
    }
  }
```

The Shot code is extremely simple, and focuses on two tasks: making sure that the shot disappers after a while, and moving the shot forward at all times.

The Destroy method is usually used with only a single parameter, which is the object that you want to remove from the game. However, you can also pass in an optional second parameter, which is the number of seconds from now that you want the object to be destroyed. In the Start method, Destroy is called and the life variable is passed in, which tells Unity to destroy the object after life seconds.

The Update function simply uses the transform's Translate method to move the object forward at a constant speed. By multiplying the Vector3.forward property by speed and then by Time.deltaTime, the object will move forward at a constant speed every frame.

Next up, we'll add the shot's graphics. As we mentioned earlier, we'll be using a trail renderer to create the visual effects of the shot. A trail renderer uses a material to define exactly how the trail looks, which means that we'll need to create one.

Your material can be anything you like, but to keep the look and feel of this game simple, we'll go with a solid, unlit red color.

1. *Create a new material.* Name it "Shot".
2. *Update the shader.* To make the trail display as a solid color, without any lighting, set the material's shader to Unlit/Color.

3. *Set the color.* Once you've changed the material's shader, the parameters for the material will change to a single parameter, which is the color to use. Change it to a nice bright red color.

Once the material has been created, we can use it in a trail renderer.

1. *Create the Shot's graphics object.* Create a new empty object, and name it "Graphics". Make it a child of the Shot object, and set its position to (0,0,0).
2. *Create the trail renderer.* Add a new Trail Renderer component to the Graphics object.

   Once it's added, the Cast Shadows, Receive Shadows, and Use Light Probes should all be turned off.

   Next, set Time to 0.05, and Width to 0.2.
3. *Make the trail get narrower toward the end.* Double-click in the curve view (below the Width field), and a new control point will appear. Drag this new control point to the bottom-right of the curve view.
4. *Apply the Shot material,* Open the list of Materials, and drag in the Shot material that you just created.

   When you're done, the trail renderer's Inspector should look like Figure 11-1.

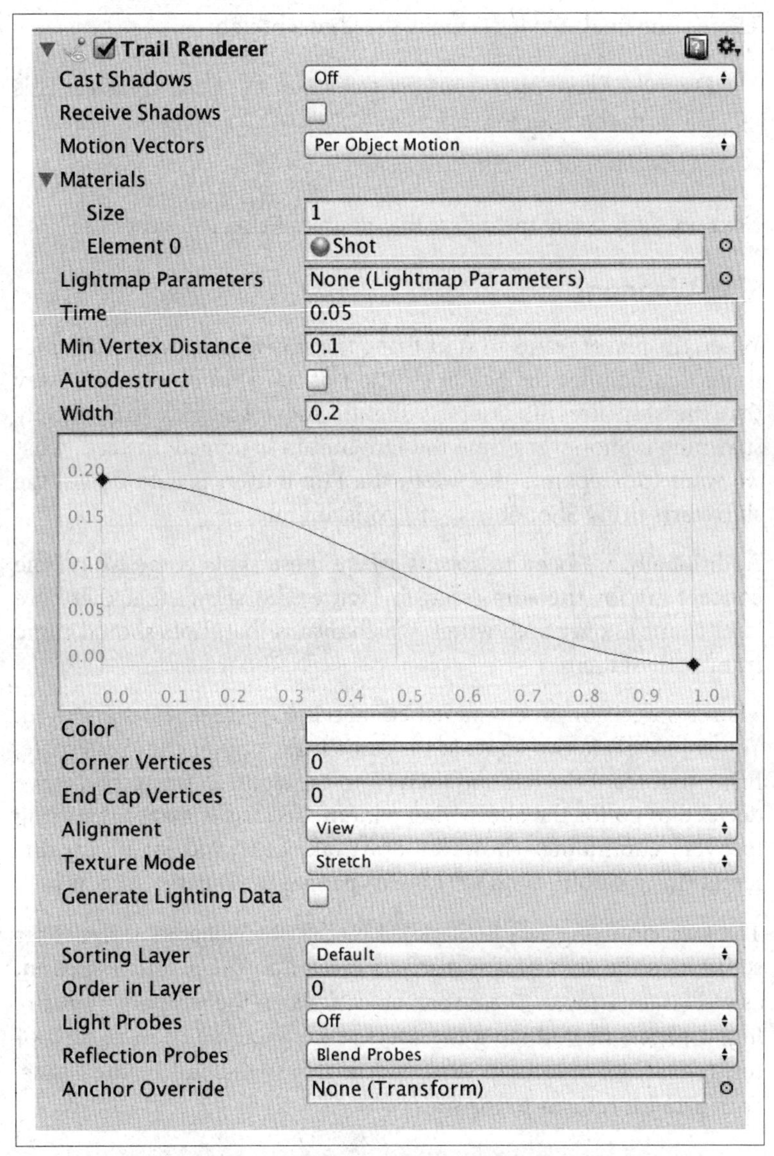

Figure 11-1. The configured trail renderer for the shot

We're not yet finished creating the Shot object: there's no way to test firing the ship's weapons (yet), but we'll be adding that very soon.

There's one final step left—make the Shot a prefab:

1. *Drag the Shot object from the scene into the Objects folder.* This will turn the Shot into a prefab.
2. *Delete the Shot from the scene.*

Next up, we'll create the object that handles firing the weapons.

Ship Weapons

When the player wants to start firing the ship's lasers, we need something that handles the actual creation of the Shot objects. The way that the ship fires its lasers is slightly more complex than simply spawning a Shot every time the Fire button is tapped; instead, what we want to happen is that when the Fire button is held down, the ship starts firing Shot objects at a constant rate.

Additionally, we need to specify *where* these shots come from. The concept art for the ship (seen in Figure 9-3) shows that there are laser cannons over both wings, which means that shots should come from both of them.

There's a decision to be made here about *how* these shots are fired. You could create two shots at the same time, or you could alternate, firing first from the left and then from the right. In this game, we've decided to go for the alternating pattern, because it makes the firing feel very continuous—however, don't take our word for it! Try different firing patterns and see how it changes the way the ship feels.

The weapon firing will be handled by the `ShipWeapons` script. This script uses the shot prefab that you created in the previous section, as well as an array of `Transform` objects; when the weapons start firing, it begins instantiating the shots at the same position as each of the `Transform` objects in turn. When it hits the end of the `Transform` array, it returns to the start.

1. *Add the `ShipWeapons` script to the ship.* Select the Ship, add a new C# script called *ShipWeapons.cs*, and add the following code to it:

```
public class ShipWeapons : MonoBehaviour {

    // The prefab to use for each shot
    public GameObject shotPrefab;
```

```
    // The list of places where a shot can emerge from
    public Transform[] firePoints;

    // The index into firePoints that the next shot will
    // fire from
    private int firePointIndex;

    // Called by InputManager.
    public void Fire() {

      // If we have no points to fire from, return
      if (firePoints.Length == 0)
        return;

      // Work out which point to fire from
      var firePointToUse = firePoints[firePointIndex];

      // Create the new shot, at the fire point's
      // position and with its rotation
      Instantiate(shotPrefab,
        firePointToUse.position,
        firePointToUse.rotation);

      // Move to the next fire point
      firePointIndex++;

      // If we've moved past the last fire point in
      // the list, move back to the start of the queue
      if (firePointIndex >= firePoints.Length)
        firePointIndex = 0;

    }

}
```

The `ShipWeapons` script keeps track of a list of positions at which shots should appear (the `firePoints` variable), as well as a prefab that represents each shot (the `shotPrefab` variable.) Additionally, it keeps track of *which* fire point the next shot should appear from (the `firePointIndex` variable); when the `Fire` button is pressed, a shot will appear from one of the fire points, and then `firePointIndex` is updated to refer to the next fire point.

2. *Create the shot fire points.* Create a new empty game object, and name it "Fire Point 1". Make it a child of the Ship object, and

then duplicate it by pressing Ctrl-D (Command-D on a Mac.) This will create another empty object called "Fire Point 1".

Set the position of Fire Point 1 to (-1.9, 0, 0). This will place it to the left of the ship.

Set the position of Fire Point 2 to (1.9, 0, 0). This will place it to the right of the ship.

When you're done, the position of Fire Points 1 and 2 should look like Figures 11-2 and 11-3.

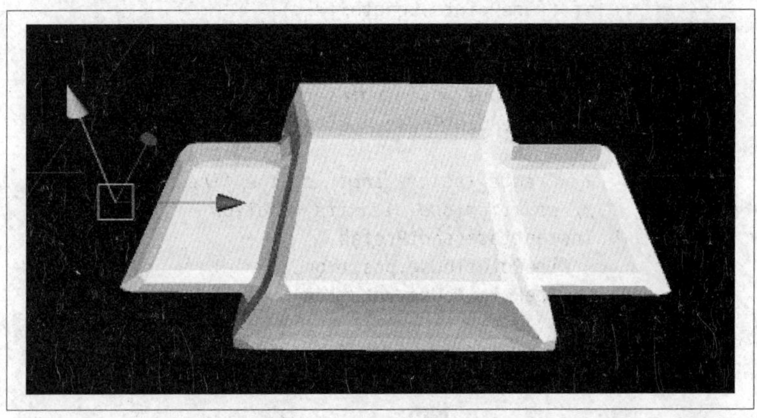

Figure 11-2. The position of Fire Point 1

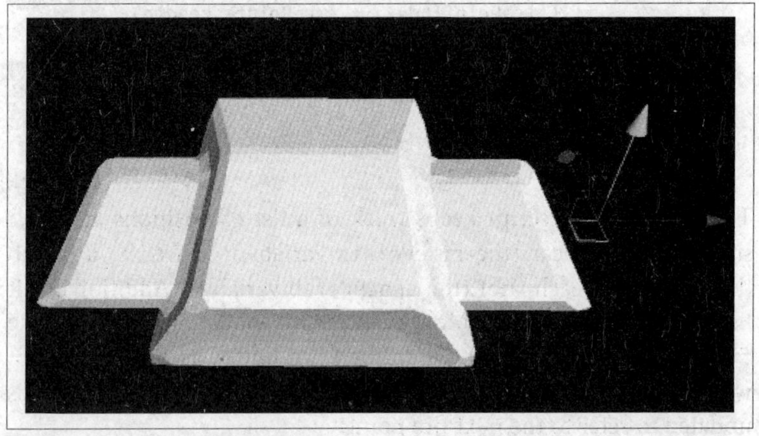

Figure 11-3. The position of Fire Point 2

3. *Configure the `ShipWeapons` script.* Drag the Shot prefab that you created in the earlier section to the `ShipWeapons`'s Shot Prefab slot.

   Next, we need to add both of the Fire Point objects to the `ShipWeapons` script. You can do this by setting the size of the Fire Points array to 2, and then dragging each object in one at a time, but there's a faster way.

   Select the Ship, and click the lock at the top right of the Inspector. This will lock the Inspector, and means that the object that the Inspector is showing won't change when you select another object.

   Next, select both Fire Point objects in the Hierarchy by clicking Fire Point 1 and then holding the Ctrl key (Command key on a Mac), and clicking Fire Point 2.

   Then, drag these two objects onto the `ShipWeapons`' Fire Points slot. Be sure to drag it onto the text "Fire Points" (and not anything below it), or it won't work.

This technique works for *any* array variable in a script. It can save you a *lot* of dragging and dropping. One thing to keep in mind is that the order of the objects may not be preserved when you drag and drop from the hierarchy into an array.

4. *Unlock the Inspector.* Now that you're done configuring the `ShipWeapons` script, unlock the Inspector by clicking the lock icon at the top right.

In this game, the spaceship only has two fire points, but the script is capable of handling more. If you want to, you can add as many more fire points as you like—just make sure that they're children of the Ship object, and that they're added to the Fire Points list in the Inspector.

Next up, we'll add the Fire button to the game's interface, which will let you actually fire the weapons.

Fire Button

We'll now add a button that makes the ship start firing its weapons when the user starts touching the button, and stop firing when they lift their finger.

There will only ever be one Fire button in the game, but there will be multiple ships. This means that we can't hook the Fire button directly to the ship; instead, we need to add support to the Input Manager to allow it to deal with multiple instances of the `ShipWeapons` script.

The way that the Input Manager will deal with it is this: because there's only ever one ship in the game at a time, there will only be one `ShipWeapons` instance in the game at a time. When a `ShipWeapons` script appears, it will contact the `InputManager` singleton, and inform it that it's the current `ShipWeapons` script. The `InputManager` will record this, and will use it as part of the firing system.

Finally, the Fire button will be connected to the Input Manager object, and will send a "firing started" message when the Fire button starts being held down, and a "firing stopped" message when the button stops being held down. The Input Manager will forward these messages to the current `ShipWeapons` script, resulting in firing.

An alternative method of doing this is to use the `FindObjectOfType` method. This method searches all objects for any component that matches a type, and returns the first one it finds. Using `FindObjectOfType`, you do away with the need to have an object *register* itself as the current object, but this comes at a cost: `FindObjects OfType` is slow, since it needs to check every component of every object in the scene. It's OK to use every now and again, but you shouldn't use it every frame.

First, we'll add support for tracking the current `ShipWeapons` instance to the `InputManager` class; we'll then add code to `ShipWeapons` that makes it register as the current instance when it appears, and de-register it when the component is removed (such as when the ship is destroyed).

We'll need to add the ShipWeapons management code to InputManager, by adding the following properties and methods to the InputManager class:

```
public class InputManager : Singleton<InputManager> {

  // The joystick used to steer the ship.
  public VirtualJoystick steering;

>   // The delay between firing shots, in seconds.
>   public float fireRate = 0.2f;
>
>   // The current ShipWeapons script to fire from.
>   private ShipWeapons currentWeapons;
>
>   // If true, we are currently firing weapons.
>   private bool isFiring = false;
>
>   // Called by ShipWeapons to update the currentWeapons
>   // variable.
>   public void SetWeapons(ShipWeapons weapons) {
>     this.currentWeapons = weapons;
>   }
>
>   // Likewise; called to reset the currentWeapons
>   // variable.
>   public void RemoveWeapons(ShipWeapons weapons) {
>
>     // If the currentWeapons object is 'weapons',
>     // set it to null.
>     if (this.currentWeapons == weapons) {
>       this.currentWeapons = null;
>     }
>   }
>
>   // Called when the user starts touching the Fire button.
>   public void StartFiring() {
>
>     // Kick off the routine that starts firing
>     // shots.
>     StartCoroutine(FireWeapons());
>   }
>
>   IEnumerator FireWeapons() {
>
>     // Mark ourself as firing shots
>     isFiring = true;
```

```
>
>     // Loop for as long as isFiring is true
>     while (isFiring) {
>
>       // If we have a weapons script, tell it
>       // to fire a shot!
>       if (this.currentWeapons != null) {
>         currentWeapons.Fire();
>       }
>
>       // Wait for fireRate seconds before
>       // firing the next shot
>       yield return new WaitForSeconds(fireRate);
>
>     }
>
>   }
>
>   // Called when the user stops touching the Fire button
>   public void StopFiring() {
>
>     // Setting this to false will end the loop in
>     // FireWeapons
>     isFiring = false;
>   }
>
  }
```

This code keeps track of the current `ShipWeapons` script that is responsible for firing the shots from the ship. The `SetWeapons` and `RemoveWeapons` methods will be called by the `ShipWeapons` script when it is created and destroyed.

When the `StartFiring` method is called, a new coroutine is started, which fires a shot by calling `Fire` on the `ShipWeapons` component and then waits for `fireRate` seconds. This then loops while `isFiring` is true; `isFiring` is set to false when the `StopFiring` method is called. The `StartFiring` and `StopFiring` methods will be called when the user starts and stops touching the Fire button, which we'll set up shortly.

We then need to add the `InputManager`-communicating code to `ShipWeapons` by adding the following methods to the `ShipWeapons` class:

```
public class ShipWeapons : MonoBehaviour {

  // The prefab to use for each shot
  public GameObject shotPrefab;
```

```
>   public void Awake() {
>     // When this object starts up, tell the input
>     // manager to use me as the current weapon
>     // object
>     InputManager.instance.SetWeapons(this);
>   }
>
>   // Called when the object is removed
>   public void OnDestroy() {
>     // Don't do this if we're not playing
>     if (Application.isPlaying == true) {
>       InputManager.instance
>         .RemoveWeapons(this);
>     }
>   }

    // The list of places where a shot can emerge from
    public Transform[] firePoints;

    // The index into firePoints that the next shot
    // will fire from
    private int firePointIndex;

    // Called by InputManager.
    public void Fire() {

      // If we have no points to fire from, return
      if (firePoints.Length == 0)
        return;

      // Work out which point to fire from
      var firePointToUse = firePoints[firePointIndex];

      // Create the new shot, at the fire point's position
      // and with its rotation
      Instantiate(shotPrefab,
        firePointToUse.position,
        firePointToUse.rotation);

      // Move to the next fire point
      firePointIndex++;

      // If we've moved past the last fire point in the list,
      // move back to the start of the queue
      if (firePointIndex >= firePoints.Length)
        firePointIndex = 0;

    }

  }
```

When a spaceship is created, the `ShipWeapons` script's `Awake` method now accesses the `InputManager` singleton and registers itself as the current weapons script. When the script is destroyed—such as when the ship collides with an asteroid, which we'll be adding later—the `OnDestroy` method makes the input manager de-register this script.

Notice how the `OnDestroy` method checks to see if `Application.isPlaying` is true before continuing? That's because, when you stop playing a game in the editor, all objects are destroyed, and as a result, all scripts that have an `OnDestroy` method have that method called on them. However, this creates a problem, since asking for the `InputManager.singleton` will result in an error, because the game is ending and that object has been destroyed.

In order to get around this problem, we check `Application.isPlaying`. This property is false after you ask Unity to stop playing, which avoids the problematic call to `InputManager.singleton` entirely.

We'll now create the Fire button that instructs the Input Manager to start and stop firing. Because we need to tell the Input Manager about the button beginning to be held down and ending, we can't use the default button behavior, which only sends a message after a "click" (finger down and then up). We'll instead need to use Event Triggers to send individual messages on both the Pointer Down and Pointer Up events.

First, though, let's create and position the button. You will need to create a new button by opening the GameObject menu, and choosing UI → Button. Name the new button "Fire Button".

Set both the anchors and the pivot of the button to Bottom Right by clicking on the Anchor button at the top-left of the Inspector, holding the Alt key (Option on a Mac), and clicking on the Bottom Right option.

Next, set the position of the button to (-50, 50, 0). This will place the button at the bottom-right of the canvas. Set both the width and height of the button to 160.

Set the Source Image of the button's Image component to the `Button` sprite. Set the Image Type to `Sliced`.

Select the Text child object of the Fire Button, and set its text to "Fire". Set its Font to CRYSTAL-Regular, and its Font Size to 28. Set its alignment to be both vertically and horizontally centered.

Finally, set the color of the Fire button to a light cyan by clicking on the Color field, and in the Hex Color field, enter 3DFFD0FF (see Figure 11-4).

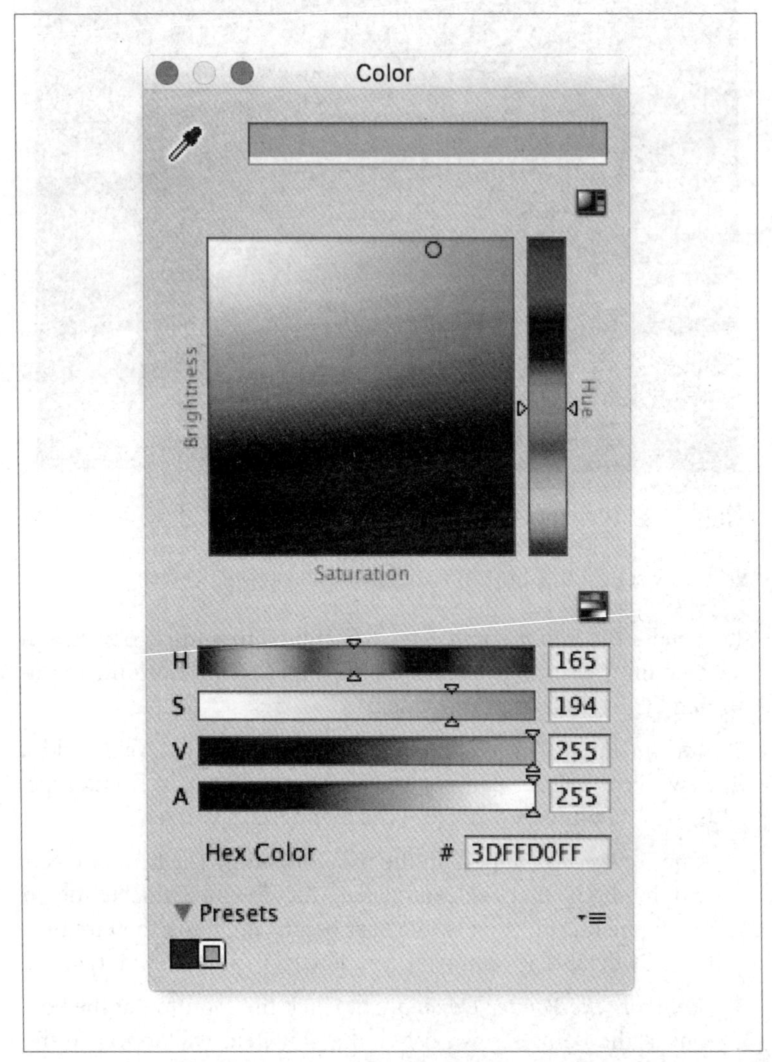

Figure 11-4. Setting the color of the fire button's label

When you're done, the button should look like Figure 11-5.

Figure 11-5. The Fire button

We'll now set up this button to behave the way we need it:

1. *Remove the Button component.* Select the Fire Button object, and click on the settings icon at the top right of the Button component. Click Remove Component.
2. *Add an Event Trigger, and add the Pointer Down event.* Add a new Event Trigger component, and then click Add Event Type. Choose "PointerDown" from the menu that appears.

   A new event will appear in the list, containing the list of objects and methods that will run when the pointer touches down inside the button (that is, the user begins touching the Fire button). By default, it's empty, so you'll need to add a new target.
3. *Configure the Pointer Down event.* Click the + button at the bottom of the PointerDown list, and a new item will appear in the list.

Drag and drop the Input Manager object from the Hierarchy panel into the slot. Next, change the method from "No Function" to "InputManager→StartFiring".

4. *Add and configure the Pointer Up event.* Next, you need to add an event for when the finger lifts up from the screen. Click the Add Event Type again, and choose "PointerUp".

   Configure this event in the same way as the PointerDown, but make the method called on the InputManager be "StopFiring".

When you're done, the Inspector should look like Figure 11-6.

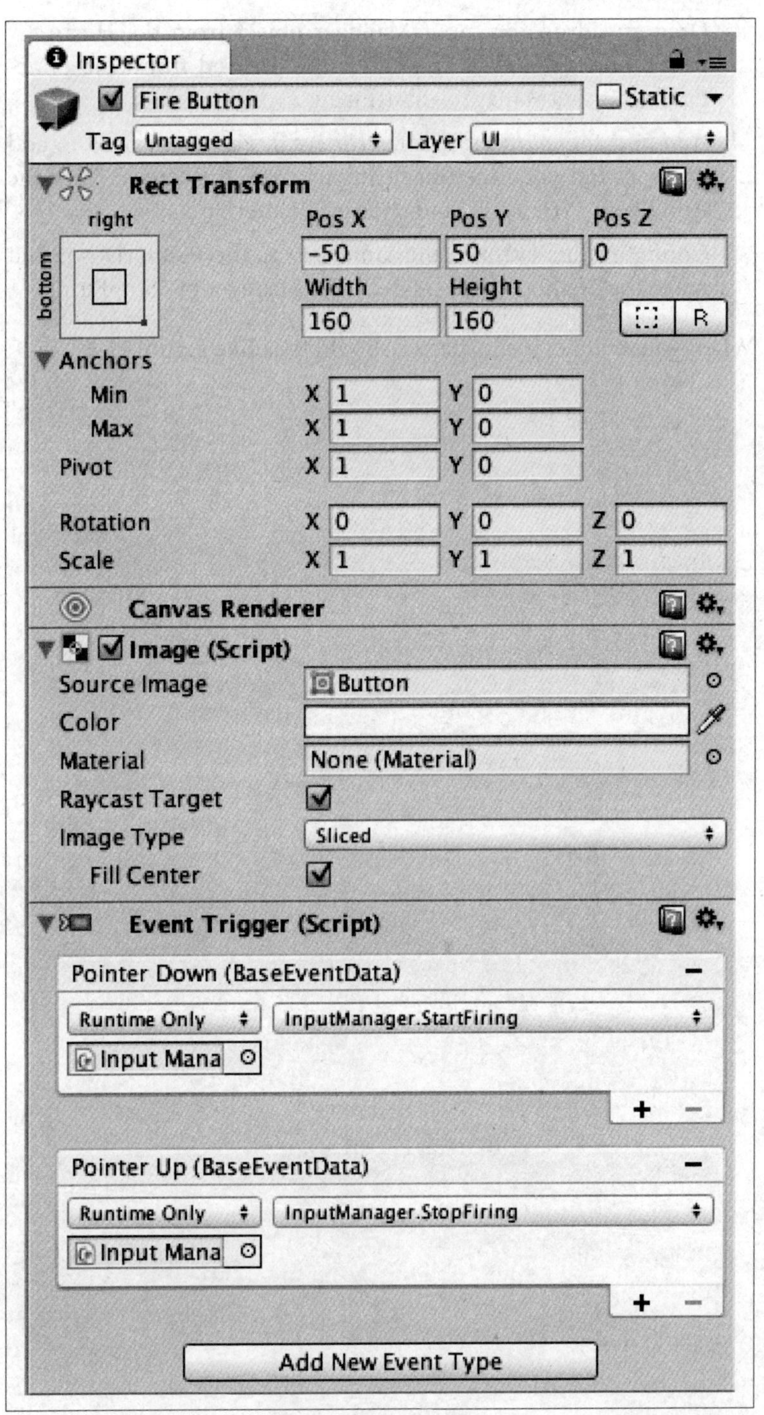

Figure 11-6. The configured Fire button

5. *Test the Fire button.* Play the game. When you hold down the Fire button, shots will appear!

Target Reticle

Currently, there's no clear way for the player to know where they're aiming at. Because both the camera and the ship can be rotating, it's actually quite tricky to aim shots correctly. To fix this, we'll use the indicator system that we created earlier to display a target reticle on the screen.

We'll create a new object that, like the Space Station, instructs the Indicator Manager to create a new indicator on screen that tracks its position. This object will be an invisible child object of the ship, and placed a distance away from the front of the ship. This will have the effect of making the indicator place itself at the point where the player's currently aiming.

Finally, this indicator should use a special icon, so that it's clear that the indicator represents the aiming point. The *Target Reticle.psd* image contains a crosshairs icon that will do the job nicely.

1. *Create the Target object.* Name this object "Target", and make it a child of the Ship.

2. *Position the Target.* Set the position of the Target object to (0,0,100). This will place the target some distance away from the ship.

3. *Add the `ShipTarget` script.* Add a new C# script to the Target object named *ShipTarget.cs*, and add the following code to it:

```
public class ShipTarget : MonoBehaviour {

  // The sprite to use for the target reticle.
  public Sprite targetImage;

  void Start () {

    // Register a new indicator that tracks this
    // object, using a yellow color and the custom
    // sprite.
    IndicatorManager.instance.AddIndicator(gameObject,
      Color.yellow, targetImage);
  }
```

```
    }
```

The `ShipTarget` code uses the `targetImage` variable to tell the Indicator Manager to use a custom sprite on the screen. This means that the Target Image slot needs to be configured.

4. *Configure the `ShipTarget` script.* Drag the "Target Reticle" sprite into the Target Image slot of the `ShipTarget` script.
5. *Play the game.* As you fly around, a target reticle will appear where the ship is aiming.

Wrapping Up

The weapons systems are all prepared. You should take the ship out for a spin, and see how it feels. You might notice that there are no targets in space; while this is a largely realistic representation of space, which famously doesn't contain much of *anything*, it's not so great for our gameplay. Turn the page, and let's fix that.

CHAPTER 12
Asteroids and Damage

Asteroids

So far, you've got a ship flying around space, you've got indicators on screen, and you've got the ability to aim and shoot your laser cannon. What you *don't* have is a legitimate target to shoot at. (The space station doesn't count.)

So, it's time to finally remedy this. We'll create the asteroids, which will do not a huge amount by themselves besides fly around. Additionally, we'll create a system that creates those asteroids, and flings them at the space station.

First, let's make the prototypical asteroid. The asteroid will be composed of two objects: the high-level, abstract object that contains the collider and all logic, and an additional "graphics" object that's responsible for providing the visible presence of the asteroid to the player.

1. *Create the object.* Make a new empty game object, and name it "Asteroid".
2. *Add the asteroid model to it.* Locate the Asteroid model in the *Models* folder. Drag it onto the Asteroid object you just created, and rename the new child object "Graphics". Reset the Position of the Graphics object's Transform component, so that it's positioned at (0,0,0).
3. *Add a rigidbody and sphere collider to the Asteroid object.* Don't add it to the Graphics object.

Once they're added, turn gravity off on the rigidbody, and make the radius of the sphere collider be 2.

4. *Add the `Asteroid` script.* Add a new C# script to the Asteroid game object, called *Asteroid.cs*, and add the following code to it:

```
public class Asteroid : MonoBehaviour {

  // The speed at which the asteroid moves.
  public float speed = 10.0f;

  void Start () {
    // Set the velocity of the rigidbody
    GetComponent<Rigidbody>().velocity
      = transform.forward * speed;

    // Create a red indicator for this asteroid
    var indicator = IndicatorManager.instance
      .AddIndicator(gameObject, Color.red);

  }

}
```

The `Asteroid` script is very simple: when the object appears, a "forward" force is applied to the object's rigidbody, which makes it start moving forward. Additionally, the indicator manager is told to add a new indicator to the screen for this asteroid.

You'll get a warning about the `indicator` variable being written to but not read. That's OK—it won't cause a bug in the game. We'll be adding a little more code later that uses the `indicator` variable, which will remove this warning.

When you're done, the Inspector for the Asteroid should look like Figure 12-1.

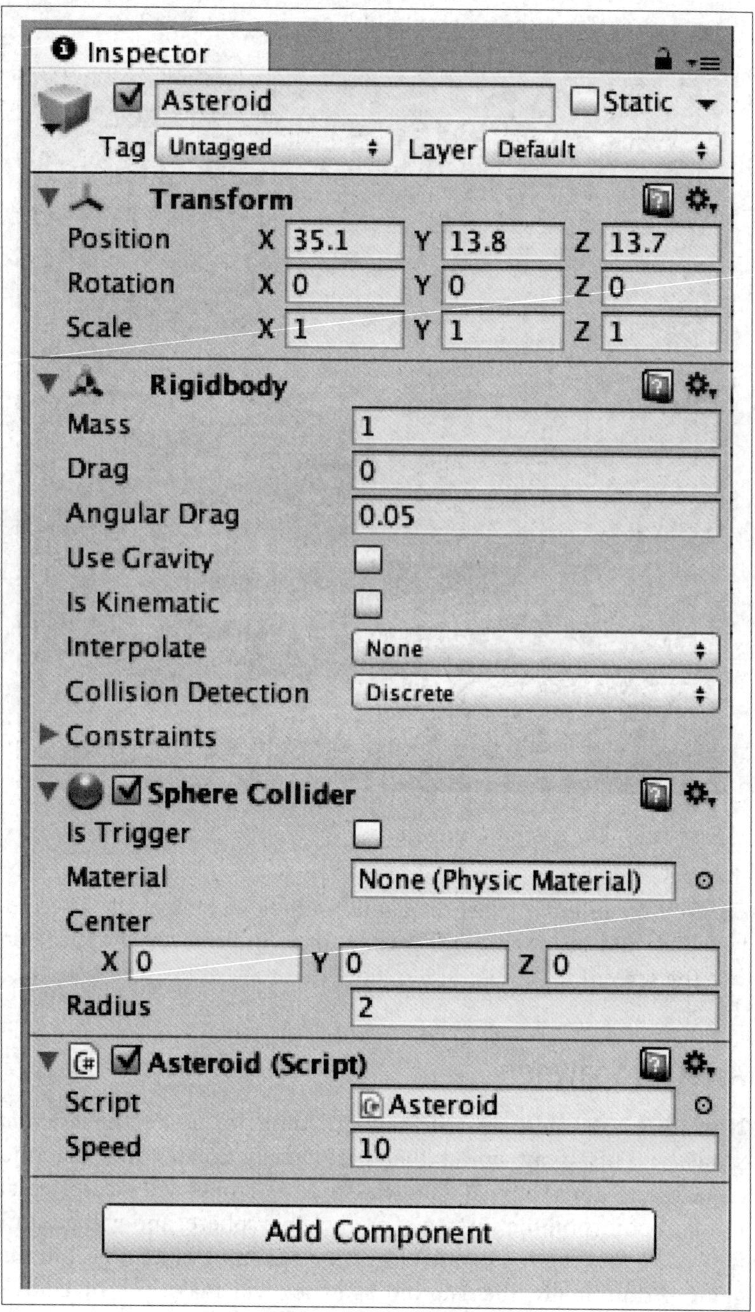

Figure 12-1. The configured asteroid

When you're done, the object should look like Figure 12-2.

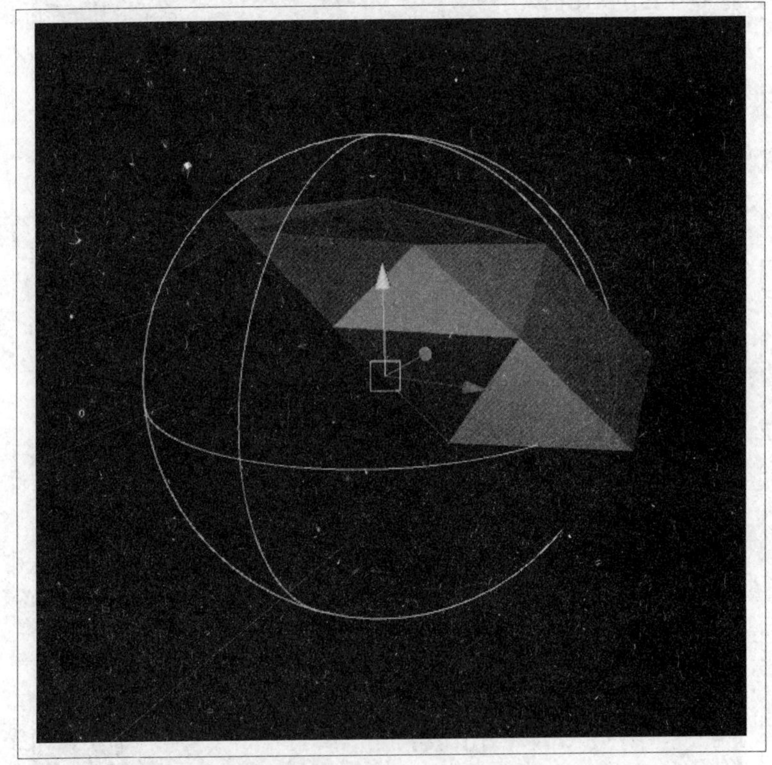

Figure 12-2. The asteroid, in-game

6. *Test the asteroid.* Start the game, and take a look at the asteroid. It should be moving forward, and an indicator will appear on the screen!

Asteroid Spawner

Now that asteroids are working, it's time to create the *asteroid spawner*. This is an object that periodically creates new asteroid objects, and aims them at a target. These asteroids will be created at random points on the surface of an invisible sphere, and will be configured so that their "forward" direction is aiming at an object in the game. Additionally, the asteroid spawner will make use of Unity's "Gizmos" feature, which allows you to show extra information in the scene view, to visualize the volume of space where asteroids will appear.

First, let's turn the prototype asteroid that you created in the last section into a prefab. We'll then create and set up the Asteroid Spawner.

1. *Make the asteroid a prefab.* Drag the Asteroid object from the Hierarchy panel into the Project panel. This will create a prefab from the object. Next, delete the Asteroid from the scene.
2. *Create the Asteroid Spawner.* Make a new empty game object, and name it "Asteroid Spawner". Set its position to (0,0,0).

   Next, add a new C# script called *AsteroidSpawner.cs*, and add the following code to it:

```
public class AsteroidSpawner : MonoBehaviour {

  // The radius of the spawn area
  public float radius = 250.0f;

  // The asteroids to spawn
  public Rigidbody asteroidPrefab;

  // Wait spawnRate ± variance seconds between each asteroid
  public float spawnRate = 5.0f;
  public float variance = 1.0f;

  // The object to aim the asteriods at
  public Transform target;

  // If false, disable spawning
  public bool spawnAsteroids = false;

  void Start () {
    // Start the coroutine that creates asteroids
    // immediately
    StartCoroutine(CreateAsteroids());
  }

  IEnumerator CreateAsteroids() {

    // Loop forever
    while (true) {

      // Work out when the next asteroid should appear
      float nextSpawnTime
        = spawnRate + Random.Range(-variance, variance);

      // Wait that much time
```

```
            yield return new WaitForSeconds(nextSpawnTime);

            // Additionally, wait until physics is about to
            // update
            yield return new WaitForFixedUpdate();

            // Create the asteroid
            CreateNewAsteroid();
        }

    }

    void CreateNewAsteroid() {

        // If we're not currently spawning asteroids, bail out
        if (spawnAsteroids == false) {
            return;
        }

        // Randomly select a point on the surface of the sphere
        var asteroidPosition = Random.onUnitSphere * radius;

        // Scale this by the object's scale
        asteroidPosition.Scale(transform.lossyScale);

        // And offset it by the asteroid spawner's location
        asteroidPosition += transform.position;

        // Create the new asteroid
        var newAsteroid = Instantiate(asteroidPrefab);

        // Place it at the spot we just calculated
        newAsteroid.transform.position = asteroidPosition;

        // Aim it at the target
        newAsteroid.transform.LookAt(target);
    }

    // Called by the editor while the spawner object
    // is selected.
    void OnDrawGizmosSelected() {

        // We want to draw yellow stuff
        Gizmos.color = Color.yellow;

        // Tell the Gizmos drawer to use our current position
        // and scale
        Gizmos.matrix = transform.localToWorldMatrix;

        // Draw a sphere representing the spawn area
        Gizmos.DrawWireSphere(Vector3.zero, radius);
```

```
        }

        public void DestroyAllAsteroids() {
            // Remove all asteroids in the game
            foreach (var asteroid in
                FindObjectsOfType<Asteroid>()) {
                Destroy (asteroid.gameObject);
            }
        }
    }
```

The `AsteroidSpawner` script uses the coroutine `CreateAsteroids` to continuously create new asteroid objects by calling `CreateNewAsteroid`, waiting a moment, and then repeating the process.

Additionally, the `OnDrawGizmosSelected` method causes a wireframe sphere to appear around it when selected. This sphere represents the locations where asteroids come from: they'll appear at the surface of the sphere and move toward the target.

3. *Flatten the Asteroid Spawner.* Set the Asteroid Spawner's Scale to (1,0.1,1). Doing this will make the asteroids mostly appear in a circle around their target, rather than in a sphere (Figure 12-3).

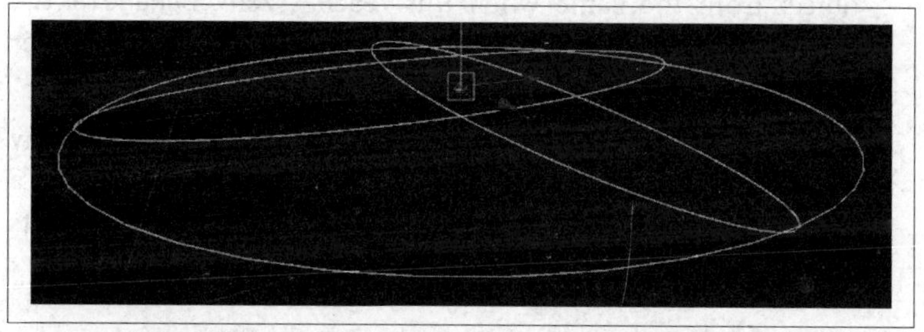

Figure 12-3. The Asteroid Spawner, in the Scene View

4. *Configure the `AsteroidSpawner`.* Drag the Asteroid prefab that you just created into the Asteroid Prefab slot, and drag the Space Station object into the Target slot. Turn Spawn Asteroids on.
5. *Test the game.* Asteroids will start appearing, and moving toward the space station!

Damage-Dealing and Taking

Your ship can now fly around the space station and all of the asteroids that are being hurled toward it, but the shots that you're firing don't actually do anything. We need to add the capability to both inflict and respond to damage.

"Damage," in this game, simply means that some objects have "hit points," which is a number representing their health. If an object's hit points are reduced to zero, the object is removed from the game.

Some objects will be able to deal damage, and some objects will be able to inflict damage. Some objects will be able to do both, such as the asteroids—they can receive damage dealt to them by the laser shots, and they can inflict damage to things they hit, like the space station.

To make this work, we'll create two separate scripts: `DamageTaking`, and `DamageOnCollide`.

- The `DamageTaking` script maintains the number of hit points that the object it's attached to has remaining, and removes the object from the game when this reaches zero. `DamageTaking` also exposes a method, `TakeDamage`, which is called by other objects to inflict damage upon it.
- The `DamageOnCollide` script runs code when it collides with any object, or enters a trigger area. If the object it collides with has a `DamageTaking` component, the `DamageOnCollide` script calls its `TakeDamage` method.

The `DamageOnCollide` script will be added to the Shot and Asteroid, while the `DamageTaking` script will be added to the Space Station and Asteroid.

Let's get started by making the asteroids take damage:

1. *Add the `DamageTaking` script to asteroids.* Select the Asteroid prefab in the Project pane, add a new C# script called *DamageTaking.cs* to it, and add the following code to the file:

```
public class DamageTaking : MonoBehaviour {

  // The number of hit points this object has
  public int hitPoints = 10;
```

```
    // If we're destroyed, create one of these at
    // our current position
    public GameObject destructionPrefab;

    // Should we end the game if this object is destroyed?
    public bool gameOverOnDestroyed = false;

    // Called by other objects (like Asteroids and Shots)
    // to take damage
    public void TakeDamage(int amount) {

      // Report that we got hit
      Debug.Log(gameObject.name + " damaged!");

      // Deduct the amount from our hit points
      hitPoints -= amount;

      // Are we dead?
      if (hitPoints <= 0) {

        // Log it
        Debug.Log(gameObject.name + " destroyed!");

        // Remove ourselves from the game
        Destroy(gameObject);

        // Do we have a destruction prefab to use?
        if (destructionPrefab != null) {

          // Create it at our current position
          // and with our rotation.
          Instantiate(destructionPrefab,
            transform.position, transform.rotation);
        }

      }

    }

  }
```

The DamageTaking script simply keeps track of the number of hit points the object has, and provides a method that other objects can call to apply damage. If the hit points ever reaches zero or below, the object is destroyed, and if a destruction prefab (such as an explosion, which we'll add in "Explosions" on page 276) was provided, it's created.

2. *Configure the Asteroid.* Change the asteroid's Hit Points variable to 1. This will make the asteroid very easy to destroy.

Next, we'll make the Shot objects deal damage to anything they hit.

3. *Add the `DamageOnCollide` script to shots.* Select the Shot prefab, add a new C# script called *DamageOnCollide.cs* to it, and add the following code to the file:

```
public class DamageOnCollide : MonoBehaviour {

  // The amount of damage we'll deal to anything we hit.
  public int damage = 1;

  // The amount of damage we'll deal to ourselves when we
  // hit something.
  public int damageToSelf = 5;

  void HitObject(GameObject theObject) {
    // Do damage to the thing we hit, if possible
    var theirDamage =
      theObject.GetComponentInParent<DamageTaking>();
    if (theirDamage) {
      theirDamage.TakeDamage(damage);
    }

    // Do damage to ourself, if possible
    var ourDamage =
      this.GetComponentInParent<DamageTaking>();
    if (ourDamage) {
      ourDamage.TakeDamage(damageToSelf);
    }
  }

  // Did an object enter this trigger area?
  void OnTriggerEnter(Collider collider) {
    HitObject(collider.gameObject);
  }

  // Did an object collide with us?
  void OnCollisionEnter(Collision collision) {
    HitObject(collision.gameObject);
  }
}
```

The `DamageOnCollide` script is also very simple; if it detects either a collision, or an object intersecting with the object's trigger collider (which is the case for the ship), the `HitObject` method is called, which determines if the hit object has a `DamageTaking` component. If it does, that component's `TakeDamage` method is called. Additionally, we do the same thing on the *current* object; the reason for this is that, if an asteroid hits the space station, we want to destroy it as well as make the space station take some damage.

4. *Test the game.* Fly around, and shoot some asteroids. When a shot hits an asteroid, the asteroid will vanish.

Next, we'll make the space station destructible.

5. *Add `DamageTaking` to the Space Station.* Select the Space Station, and add a `DamageTaking` script component.

   Turn on Game Over On Destruction. This won't do anything yet, but it will be used later to make the game end when the space station is destroyed.

   When you're done, the Inspector for the Space Station should look like Figure 12-4.

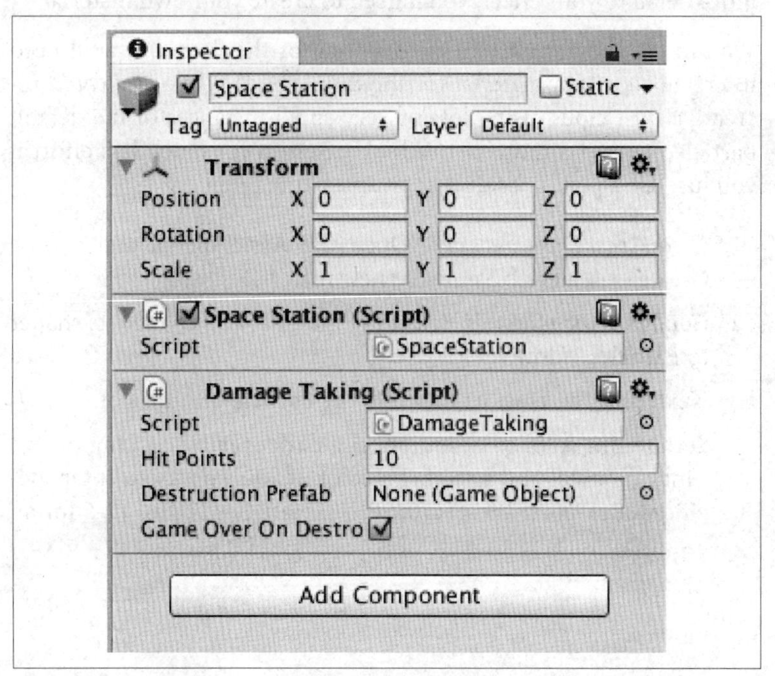

Figure 12-4. Adding the DamageTaking script to the space station

Explosions

When an asteroid is destroyed, it simply vanishes. This isn't terribly satisfying—it would be better to have the asteroid vanish in an explosion.

One of the best ways to create explosions is using a particle effect. Particle effects are great for situations where you want an element of natural-looking randomness. They're great for things like smoke, fire, wind, and (of course) explosions.

The explosion in this game will be composed of *two* particle effects. The first particle effect will create an initial bright flash. The second will leave behind some dust that eventually fades away.

When working with particle effects, it's important to have your resources already lined up. In particular, you need to decide whether your particle effect needs to use a custom material, or whether it should use the default particle material. The default material is just a blurry circle, which is useful for lots of things, but if you need to add more detail to your effect, you'll need to create your own material.

We can use the default particle material for the flash, but we'll need to create a custom material for the dust cloud. While you *could* re-create a dust cloud using lots of very small instances of the default particle, you get a much better looking effect for much less effort if you just use a picture of dust as your starting point.

1. *Create the Dust material.* Open the Asset menu, and choose Create → Material. Name the new material "Dust".
2. *Configure the material.* Select the material and change its shader to Particles/Additive.

   Next, drag the Dust texture into the Particle Texture slot.

   Set the tint color to a semiopqaue dark gray by clicking on the Tint Color slot and selecting a color. If you'd prefer to enter specific values, enter these: (70, 70, 70, 190). See Figure 12-5 for an example.

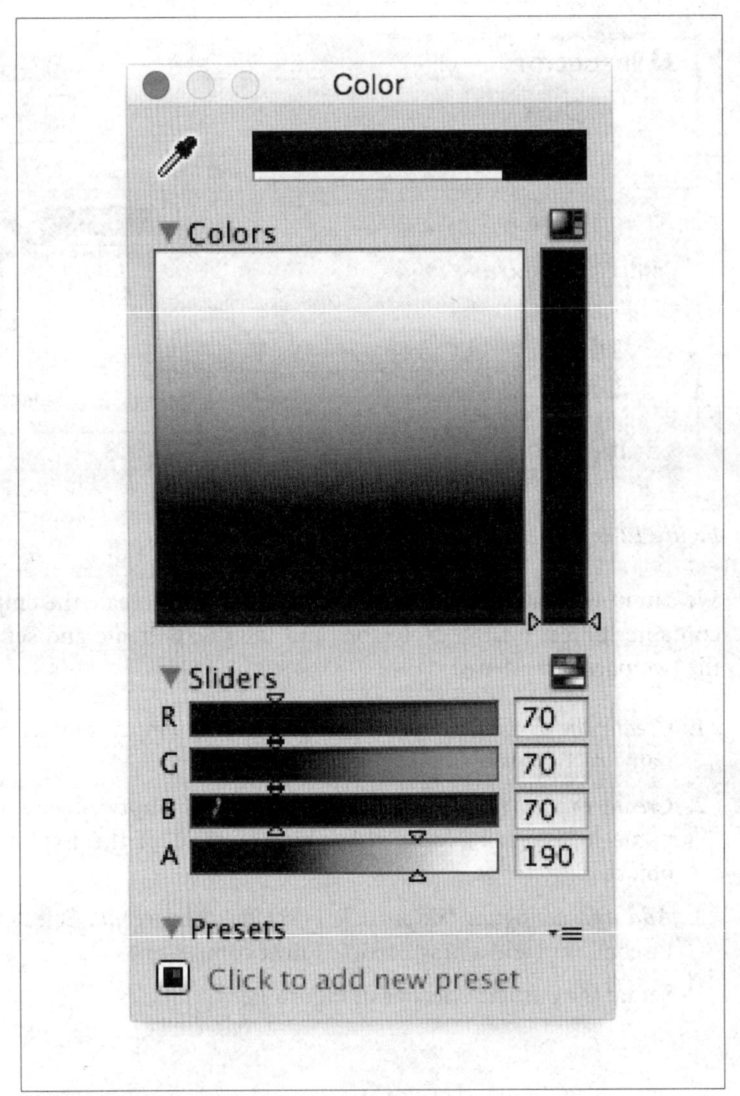

Figure 12-5. The Dust material's tint color

Finally, set the Soft Particles Factor to 0.8.

When you're done, the material's Inspector should look like Figure 12-6.

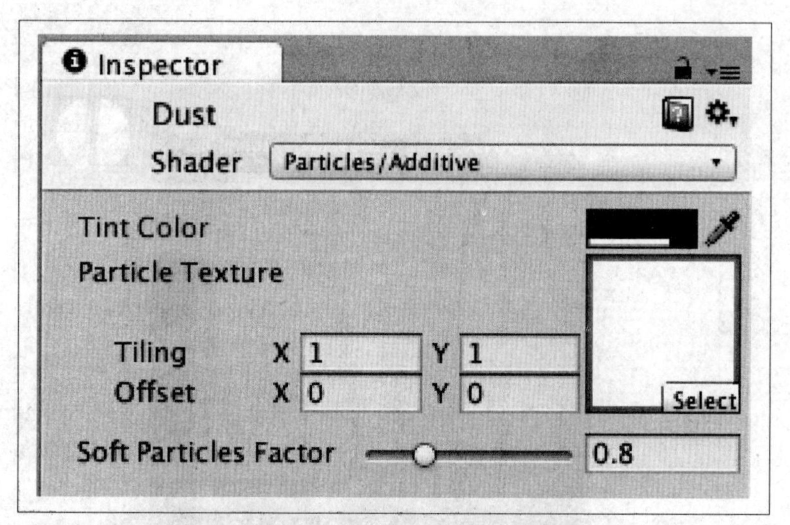

Figure 12-6. The Dust material

We can now create the particle systems. First, we'll create the empty container object for the explosion, and then we'll create and set up the two particle systems.

1. *Create the Explosion object.* Create a new empty object, and name it "Explosion".
2. *Create the Fireball object.* Create a second empty object, and name it "Fireball". Make this object a child of the Explosion object.
3. *Add and configure the particle effect for the Fireball.* Select the Fireball, and add a new Particle Effect component.

   Set up the particle effect as in Figure 12-7.

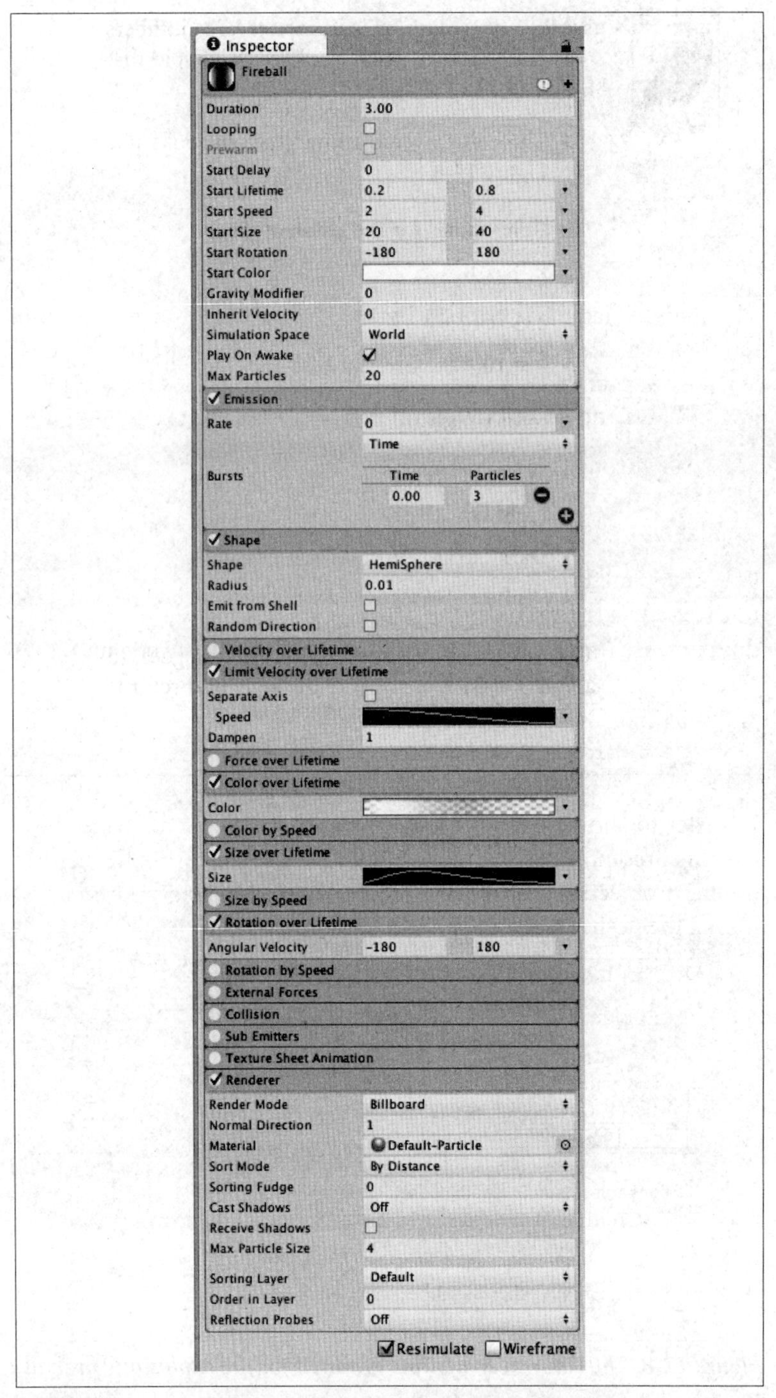

Figure 12-7. The Inspector for the Fireball's particle effect

While most of these parameters are numbers that you can just enter in, there's a couple that need some explanation:

- The color over lifetime gradient looks like Figure 12-8.

The alpha values for the gradient are:

- 0 at 0%
- 255 at 12%
- 0 at 100%

The color values are:

- White at 0%
- Light tan at 12%
- Dark tan at 57%
- White at 100%

The size over lifetime starts at zero, goes to about 3 at 35% of the way through, and returns to zero at the end (see Figure 12-9).

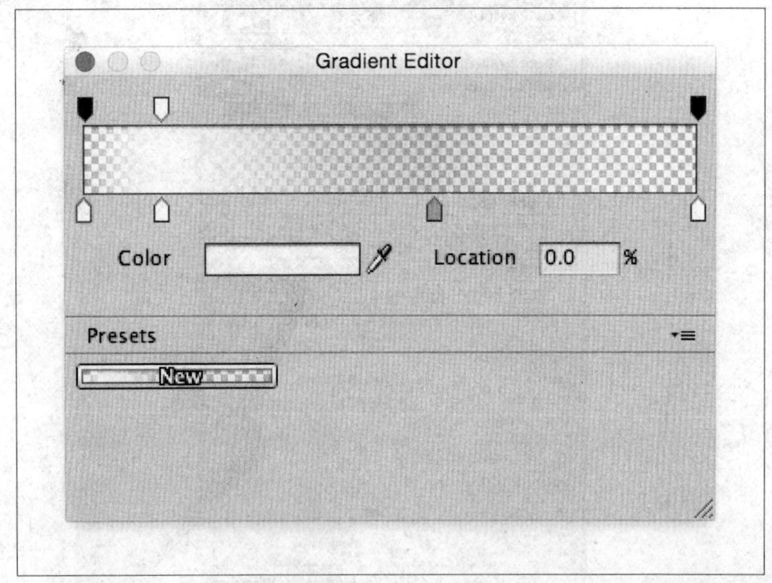

Figure 12-8. The color-over-lifetime gradient for the explosion's fireball

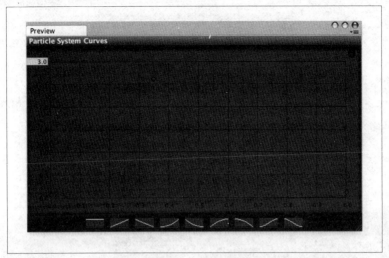

Figure 12-9. The size-over-lifetime curve for the explosion's fireball

The Fireball object creates that initial, short-lived flash for the explosion. The second particle effect, which we're about to add, will be the Dust effect.

1. *Create the Dust object.* Make an empty game object, and name it "Dust". Make it a child of the Explosion object.
2. *Add and configure the particle system.* Add a new Particle System component, and set it up as per Figure 12-10.

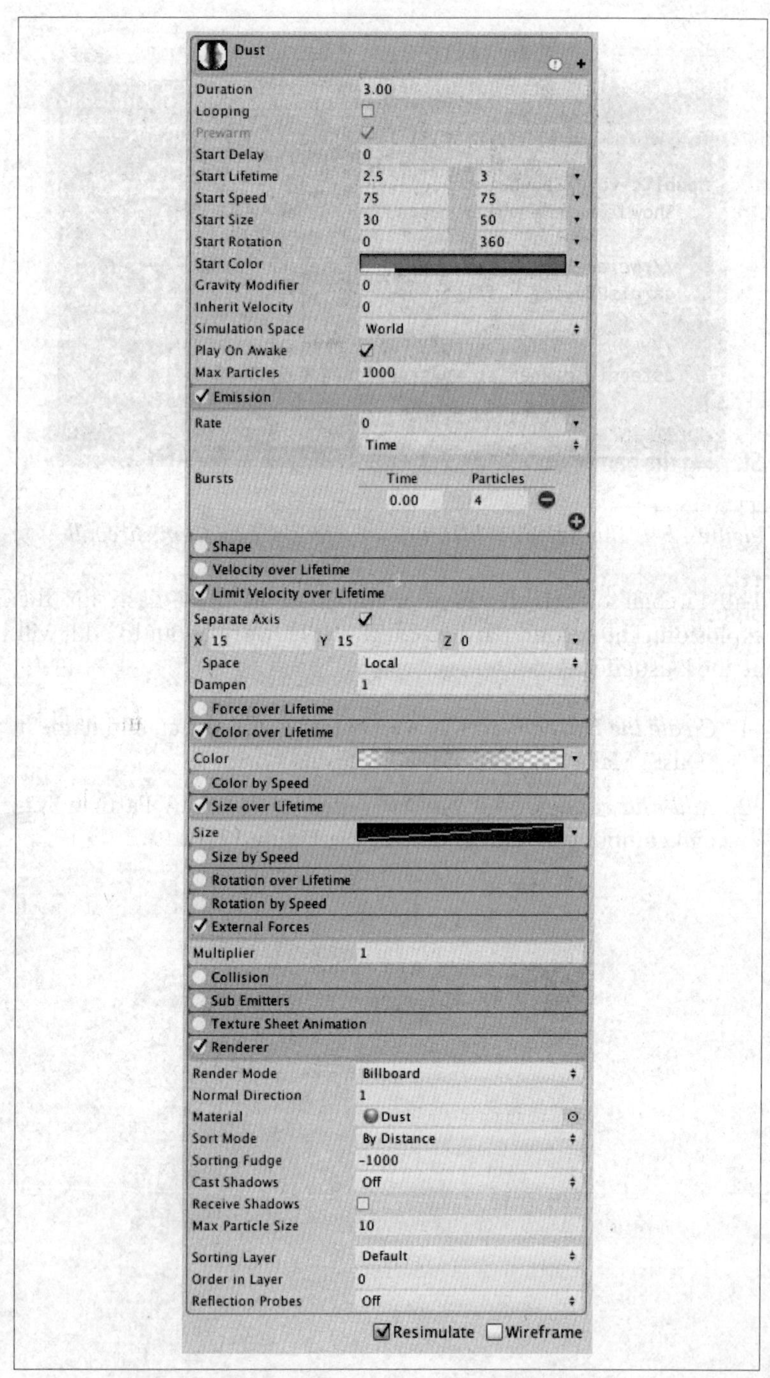

Figure 12-10. The inspector for the Dust particle

Some things that aren't immediately copyable from Figure 12-10:

- The Material used by the Renderer is the Dust material you just made. Simply drag and drop it into the Material slot.
- The start color is the RGBA value [130, 130, 120, 45]. Click on the Start Color variable, and enter these numbers.
- The size over lifetime is a straight line, going from 0% to 100%.
- The color over lifetime looks like Figure 12-11—the color is a constant tan color, and the alpha goes from 0 at 0%, to 255 at 14%, to 0 at 100%.

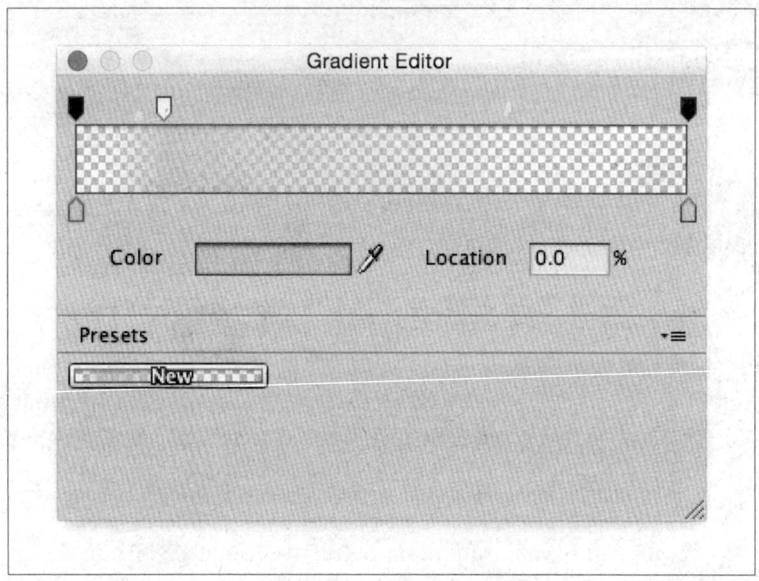

Figure 12-11. The color-over-lifetime for the explosion's dust particles

You're all done! You can now use this explosion for your asteroids:

1. *Convert the object into a prefab.* Drag the Explosion object into the Project pane, and then remove it from the scene.

2. *Make the asteroids use the explosion when they're destroyed.* Select the Asteroid prefab, and drag the Explosion into the Destruction Prefab slot.
3. *Test it out.* When you shoot down the asteroids, they'll explode!

Wrapping Up

Now that the asteroids and damage model have been created, we're close to having a full game. In the next chapter, we'll start to polish this thing, and turn it into a bigger, better experience.

CHAPTER 13

Audio, Menus, Death, and Explosions!

The core gameplay of your space shooter is done, but it's not a complete game yet. In order to be playable outside of Unity, you'll need to add menus and other controls that let the player navigate around the game as an application. Finally, we'll polish up the game by replacing the temporary art with higher-fidelity 3D models and materials.

Menus

Right now, the gameplay is entirely constrained to the editor's Play button. When you start the game, you're in the action immediately, and if your space station is destroyed, you have to stop the game and start it again.

To provide the player with a little more context for their game, we need to add menus. In particular, we need to add an especially important button: "New Game." If the space station is destroyed, we need a way to let the player start again.

Adding the menu structure to a game goes a long way toward making the game feel complete. We'll be adding four components as part of the menus:

Main Menu
: This screen presents the game's title, and shows the New Game button.

Paused screen

This screen shows the text "Paused," and contains a button to unpause.

Game Over screen

This screen shows Game Over and the New Game buttons.

In-Game UI

This screen contains the joystick, indicators, Fire button, and everything else that the player actually sees while playing the game.

Each of these UI groups will be exclusive—only one will appear at a time. The game will start with the Main Menu, and when you click on the New Game button, the menu will disappear and be replaced with the In-Game UI (in addition to the actual game action being started).

Unity's UI system lets you test your menu using your computer's mouse or touchpad. However, you should still test how the menu feels on a real touchscreen as you develop it, such as through the Unity Remote app (see "Unity Remote" on page 79.)

The first step in this process is to bring the In-Game UI components together into a single object, so that it can be managed all at once.

1. *Create the In-Game UI container.* Select the Canvas object, and make a new empty child object. Name this object "In-Game UI".
2. *Configure the container.* Make In-Game UI's anchors to stretch horizontally and vertically, and set the left, top, bottom, and right margins to zero. This makes it fill the entire canvas.

Next, we'll bring all of the existing UI elements together into the container.

3. *Group the game's UI.* Select every child of the canvas *except* the In-Game UI container, and move it into In-Game UI.

We'll now start building the other menus. Before we begin, it's helpful to turn off the In-Game UI, so that it doesn't distract from the other UI content you're about to build.

4. *Disable the In-Game UI.* Select the In-Game UI object, and disable it by clicking the checkbox at the top-left of the Inspector. When you're done, it should look like Figure 13-1.

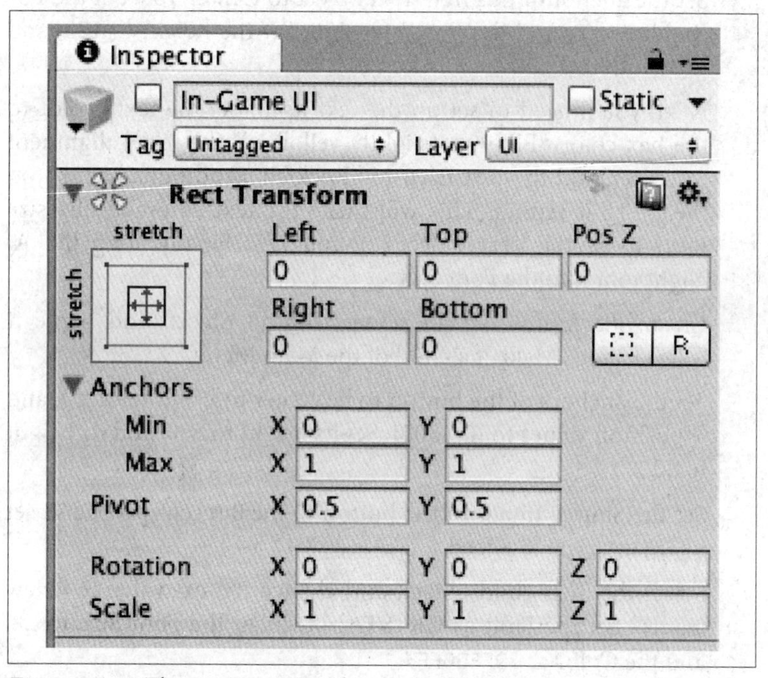

Figure 13-1. The In-Game UI, shown disabled; also note the size and position of the object, which is set to fill the entire canvas with no border

Main Menu

The content of the Main Menu is very simple—it's a text label that shows the game's title ("Rockfall"), and a button that creates a new game.

Much like the In-Game UI, the Main Menu will be composed of an empty container object, with all of the UI components that belong to the menu added as a child.

1. *Create the Main Menu container.* Create a new empty game object, and make it a child of the Canvas. Name it "Main Menu".

   Make it fill the entire canvas by setting it to stretch horizontally and vertically. Set all of the margin values to zero.

2. *Create the title label.* Create a new Text object by opening the GameObject menu, and choosing UI → Text. Make it a child of the Main Menu, and name it "Title".

   Set the anchor of this new Text object to Center Top. Set the Pos X value to 0, and the Pos Y to -100. Set the height to 120, and the width to 1024.

   Next, you'll need to set up the text itself. Set the text's color to the hex color #FFE99A (slightly yellow-y), the text's alignment to center, and the text itself to "Rockfall". Additionally, turn on the Best Fit setting. This will make the text automatically size itself to fit the Text object's boundaries. Finally, drag the At Night font into the Font Slot.

3. *Create the Button.* Create a new Button object, and name it "New Game". Make it a child of the Main Menu.

   Set the anchors of the button to be center top, and set the X and Y position values to [0, -300]. Set its height to 330, and its height to 80.

   Set the Source Image of the button to the `Button` sprite, and set its Image Type to Sliced.

   Select the Text child object, and change its text value to "New Game". Set the Font as CRYSTAL-Regular, the Font Size to 28, and the Color as `3DFFD0FF`.

When you're done, the menu should look like Figure 13-2.

Figure 13-2. The Main Menu

Before you continue, disable the Main Menu container.

Paused Screen

The Paused screen shows the text "Paused", along with a button to unpause the game. To build it, follow the same steps as you did for the Main Menu, but with the following changes:

- The container object should be called "Paused".
- The text of the Title object should be "Paused".
- The button object should be called "Unpause Button".
- The text of the button should be "Unpause".

When you're done, the Pause menu should look like Figure 13-3.

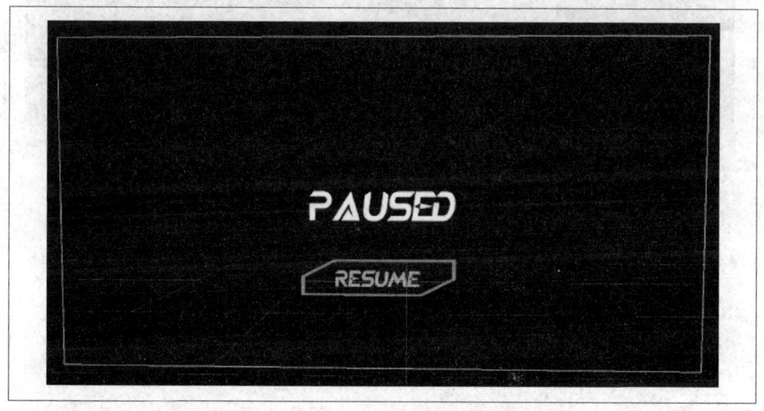

Figure 13-3. The Pause menu

Disable the Paused container before building the final menu: the Game Over screen.

Game Over Screen

The Game Over screen shows the text "Game Over", along with a button to start the game again. The Game Over screen will appear when the space station is destroyed, which will end the game.

Again, follow the same steps as for the Main Menu and Paused screens, but with the following changes:

- The container object should be called "Game Over".
- The text of the Title object should be "Game Over".
- The button object should be called "New Game Button".

- The text of the button should be "New Game".

When you're done, the Game Over screen should look like Figure 13-4.

Figure 13-4. The Game Over menu

All three of these new menus are pretty much identical, and you might be wondering why you've done the same work three times. The reason is that you'll want to customize them later, and breaking them apart now will save you work later.

There's one last UI component that we need to add to the game, and that's a way to Pause the game.

Adding a Pause Button

The Pause button will appear at the top-right of the In-Game UI, and will signal to the game that the user wants to stop the action for a moment.

To build the Pause button, you first need to create a new Button object, and make it a child of the In-Game UI container object. Name it "Pause Button".

Set the anchors of the Pause button to the top-right, and set the X and Y position values to [-50, -30]. Set the width to 80, and the height to 70.

Set the Source Image of the Image component to the `Button` sprite.

Set the text of the Text child object to "Pause". Set its font to CRYSTAL-Regular, and its size to 28. Set its color to #3DFFD0FF.

Congratulations! Your UI is all done. However, none of the buttons that you've set up work correctly. In order for it all to work, you'll need to add a Game Manager to coordinate everything.

Game Manager and Death

The Game Manager, much like the Input Manager and the Indicator Manager, is a singleton object. The Game Manager has two main jobs:

- Managing the state of the game and the menus, and
- Spawning the ship and station

When the game starts, the game will be in an unstarted state. The ship and station won't be in the scene, and the asteroid spawner won't be creating an asteroids. Additionally, the Game Manager will display the Main Menu container object, and hide every other menu.

When the user taps the New Game button, the In-Game UI will be displayed, the ship and station created, and the asteroid spawner will be told to start creating asteroids. Additionally, the Game Manager will set up some important elements of the game: the `Camera Follow` script will be told to follow the new Ship object, and the Asteroid Spawner will be told to aim its asteroids at the Space Station.

Finally, the Game Manager will handle the Game Over state. You might recall from earlier that the `DamageTaking` script has a checkbox called "Game Over On Destroyed". We'll be setting that up to instruct the Game Manager to end the game whenever the object that the script is attached to is destroyed, if the checkbox is on. Ending the game is simply a matter of turning off the asteroid spawner, and destroying the current ship (and the station, if it happens to still be around).

Before we get started building the Game Manager, we need to be able to create multiple copies of the Ship and the Station. This requires turning both of these objects into prefabs, and also defining the positions at which they'll both appear.

Turn the Ship and Space Station into prefabs. Drag and drop the Ship into the Project pane to create the prefab, and then remove it from the scene. Repeat this process for the Space Station.

Start Points

We'll now create two marker objects, which will serve as indicators for where the Ship and Space Station should be created when a new game starts. These indicators won't be visible to the player, but we'll make them visible to you inside the editor.

1. *Create the Ship position marker.* Create a new empty game object, and name it "Ship Start Point".

   Click the icon at the top-left of the Inspector and choose the red label (see Figure 13-5). The object will now appear in the scene view, despite being invisible to the player.

   Position the marker where you want the ship to appear.

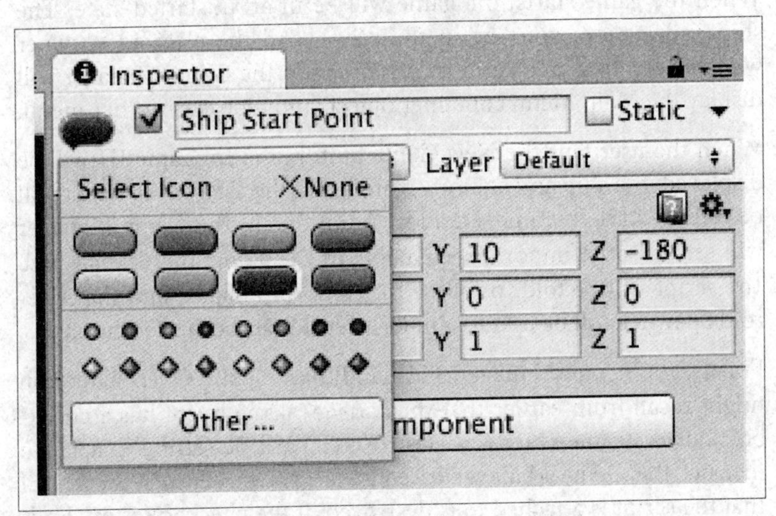

Figure 13-5. Selecting a label for the ship's start point

2. *Create the Space Station position marker.* Repeat these steps, this time creating an object called "Station Start Point". Position it where you want the space station to appear.

With this done, we're now able to create and set up the Game Manager.

Creating the Game Manager

The Game Manager largely serves as a central point for storing critical information about the game, such as references to the current Ship and Space Station, as well as changing the state of important game objects when either a button is clicked or when a `Damage Taking` script reports that the game should end.

To set up the Game Manager, create a new empty game object called Game Manager, add a new C# script to it called *GameManager.cs*, and add the following code to the file:

```
public class GameManager : Singleton<GameManager> {

  // The prefab to use for the ship, the place it starts from,
  // and the current ship object
  public Gameobject shipPrefab;
  public Transform shipStartPosition;
  public GameObject currentShip {get; private set;}

  // The prefab to use for the space station, the place
  // it starts from, and the current ship object
  public GameObject spaceStationPrefab;
  public Transform spaceStationStartPosition;
  public GameObject currentSpaceStation {get; private set;}

  // The follow script on the main camera
  public SmoothFollow cameraFollow;

  // The containers for the various bits of UI
  public GameObject inGameUI;
  public GameObject pausedUI;
  public GameObject gameOverUI;
  public GameObject mainMenuUI;

  // Is the game currently playing?
  public bool gameIsPlaying {get; private set;}

  // The game's Asteroid Spawner
  public AsteroidSpawner asteroidSpawner;

  // Keeps track of whether the game is paused or not.
  public bool paused;

  // Show the main menu when the game starts
  void Start() {
    ShowMainMenu();
  }

  // Shows a UI container, and hides all others.
```

```
void ShowUI(GameObject newUI) {

  // Create a list of all UI containers.
  GameObject[] allUI
    = {inGameUI, pausedUI, gameOverUI, mainMenuUI};

  // Hide them all.
  foreach (GameObject UIToHide in allUI) {
    UIToHide.SetActive(false);
  }

  // And then show the provided UI container.
  newUI.SetActive(true);
}

public void ShowMainMenu() {
  ShowUI(mainMenuUI);

  // We aren't playing yet when the game starts
  gameIsPlaying = false;

  // Don't spawn asteroids either
  asteroidSpawner.spawnAsteroids = false;
}

// Called by the New Game button being tapped
public void StartGame() {
  // Show the In-Game UI
  ShowUI(inGameUI);

  // We're now playing
  gameIsPlaying = true;

  // If we happen to have a ship, destroy it
  if (currentShip != null) {
    Destroy(currentShip);
  }

  // Likewise for the station
  if (currentSpaceStation != null) {
    Destroy(currentSpaceStation);
  }

  // Create a new ship, and place it
  // at the start position
  currentShip = Instantiate(shipPrefab);
  currentShip.transform.position
    = shipStartPosition.position;
  currentShip.transform.rotation
    = shipStartPosition.rotation;
```

```
    // And likewise for the station
    currentSpaceStation = Instantiate(spaceStationPrefab);

    currentSpaceStation.transform.position =
      spaceStationStartPosition.position;

    currentSpaceStation.transform.rotation =
      spaceStationStartPosition.rotation;

    // Make the follow script track the new ship
    cameraFollow.target = currentShip.transform;

    // Start spawning asteroids
    asteroidSpawner.spawnAsteroids = true;

    // And aim the spawner at the new space station
    asteroidSpawner.target = currentSpaceStation.transform;

  }

  // Called by objects that end the game
  // when they're destroyed
  public void GameOver() {
    // Show the Game Over UI
    ShowUI(gameOverUI);

    // We're no longer playing
    gameIsPlaying = false;

    // Destroy the ship and the station
    if (currentShip != null)
      Destroy (currentShip);

    if (currentSpaceStation != null)
      Destroy (currentSpaceStation);

    // Stop spawning asteroids
    asteroidSpawner.spawnAsteroids = false;

    // And remove all lingering asteroids from the game
    asteroidSpawner.DestroyAllAsteroids();
  }

  // Called when the Pause or Resume buttons are tapped
  public void SetPaused(bool paused) {

    // Switch between the in-game and paused UI
    inGameUI.SetActive(!paused);
    pausedUI.SetActive(paused);

    // If we're paused..
```

```
        if (paused) {
          // Stop time
          Time.timeScale = 0.0f;
        } else {
          // Resume time
          Time.timeScale = 1.0f;
        }
      }

    }
```

The Game Manager script is bulky, but simple. It has two main functions: managing the appearance of the menus and In-Game UI and creating and destroying the space station and ship when the game begins and ends.

Let's walk through what it does, one step at a time.

Initial setup

The `Start` method is called when the Game Manager first appears in the scene—that is, at the start of the game. The only thing that it does is cause the main menu to appear, calling `ShowMainMenu`.

```
// Show the main menu when the game starts
void Start() {
  ShowMainMenu();
}
```

In order to show any UI, we use a method called `ShowUI` that handles the presentation of the desired UI object and the dismissal of all other UI objects. It does this by hiding *all* UI objects, and then unhiding the desired UI element:

```
// Shows a UI container, and hides all others.
void ShowUI(GameObject newUI) {

  // Create a list of all UI containers.
  GameObject[] allUI
    = {inGameUI, pausedUI, gameOverUI, mainMenuUI};

  // Hide them all.
  foreach (GameObject UIToHide in allUI) {
    UIToHide.SetActive(false);
  }

  // And then show the provided UI container.
  newUI.SetActive(true);
}
```

With this implemented, ShowMainMenu can be implemented. All it does is show the main menu UI (via ShowUI), and indicates to the game that gameplay isn't currently happening, and that the asteroid spawner should not be spawning asteroids:

```
public void ShowMainMenu() {
  ShowUI(mainMenuUI);

  // We aren't playing yet when the game starts
  gameIsPlaying = false;

  // Don't spawn asteroids either
  asteroidSpawner.spawnAsteroids = false;
}
```

Starting the game

The StartGame method, which is called when the New Game button is tapped, shows the In-Game UI (which hides the other UI as a result), and sets up the scene for gameplay by removing any existing ship or station, and creating new ones. It also makes the camera start tracking the newly created ship, and tells the asteroid spawner to start throwing asteroids at the newly created station:

```
  // Called by the New Game button being tapped
  public void StartGame() {
    // Show the In-Game UI
    ShowUI(inGameUI);

    // We're now playing
    gameIsPlaying = true;

    // If we happen to have a ship, destroy it
    if (currentShip != null) {
      Destroy(currentShip);
    }

    // Likewise for the station
    if (currentSpaceStation != null) {
      Destroy(currentSpaceStation);
    }

    // Create a new ship, and place it
    // at the start position
    currentShip = Instantiate(shipPrefab);
    currentShip.transform.position
      = shipStartPosition.position;
    currentShip.transform.rotation
      = shipStartPosition.rotation;
```

```
    // And likewise for the station
    currentSpaceStation = Instantiate(spaceStationPrefab);

    currentSpaceStation.transform.position =
      spaceStationStartPosition.position;

    currentSpaceStation.transform.rotation =
      spaceStationStartPosition.rotation;

    // Make the follow script track the new ship
    cameraFollow.target = currentShip.transform;

    // Start spawning asteroids
    asteroidSpawner.spawnAsteroids = true;

    // And aim the spawner at the new space station
    asteroidSpawner.target = currentSpaceStation.transform;

  }
```

Ending the game

The `GameOver` method will be called by certain objects that, when they're destroyed, end the game. It shows the Game Over UI, stops gameplay, and destroys the current ship and station. Additionally, asteroid spawning is stopped, and all remaining asteroids are removed. Essentially, we're returning to the initial starting conditions of the game:

```
// Called by objects that end the game when they're destroyed
public void GameOver() {
  // Show the Game Over UI
  ShowUI(gameOverUI);

  // We're no longer playing
  gameIsPlaying = false;

  // Destroy the ship and the station
  if (currentShip != null)
    Destroy (currentShip);

  if (currentSpaceStation != null)
    Destroy (currentSpaceStation);

  // Stop spawning asteroids
  asteroidSpawner.spawnAsteroids = false;

  // And remove all lingering asteroids from the game
```

```
        asteroidSpawner.DestroyAllAsteroids();
    }
```

Pausing the game

The SetPaused method is called when either the Pause or Resume buttons are tapped. All it does is manage the display of the pause UI, and stop or resume the flow of time.

```
    // Called when the Pause or Resume buttons are tapped
    public void SetPaused(bool paused) {

        // Switch between the in-game and paused UI
        inGameUI.SetActive(!paused);
        pausedUI.SetActive(paused);

        // If we're paused..
        if (paused) {
            // Stop time
            Time.timeScale = 0.0f;
        } else {
            // Resume time
            Time.timeScale = 1.0f;
        }
    }
```

Setting Up the Scene

With the code written, we can now set up the Game Manager in the scene. Configuring the Game Manager is entirely a matter of connecting objects in the scene to variables in the script:

- Ship Prefab should be the ship prefab you just made.
- Ship start position should be the ship start position in the scene.
- Station prefab should be the station prefab you just made.
- Station start position should be the station start position in the scene.
- Camera Follow should be the Main Camera in the scene.
- In-Game UI, Main Menu UI, Paused UI, and Game Over UI should be their equivalent UIs in the scene.
- Asteroid Spawner should be the Asteroid Spawner object in the scene.
- Leave Warning UI for now; that's for the next section.

When you're done, the Inspector for the Game Manager should look like Figure 13-6.

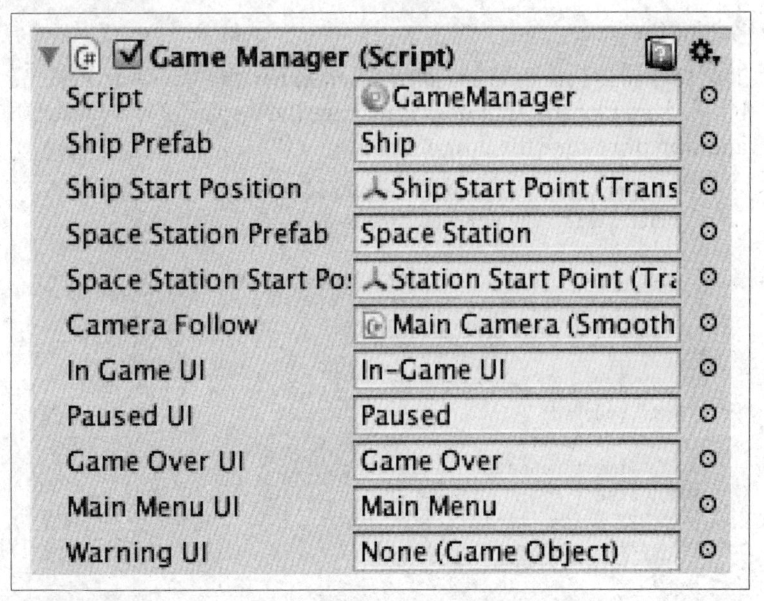

Figure 13-6. The Inspector for the Game Manager

Now that the Game Manager is set up, we need to connect the various buttons that are in the Game UI to the Game Manager.

1. *Connect the Pause button.* Select the Pause button in the In-Game UI, and click the + button at the bottom of the Clicked event. Drag the Game Manager into the slot that appears, and change the function to GameManager → SetPaused. A checkbox will appear; turn it on. This has the effect of calling the `SetPaused` method on the Game Manager, and passing in the boolean value `true`.
2. *Connect the Unpause button.* Select the Unpause button in the Paused menu. Perform the same set of steps as for the Pause button, but with one change: turn the checkbox off. This will make the button call `SetPaused` with the boolean value `false`.
3. *Connect the New Game buttons.* Select the New Game button in the Main Menu, and click the + button at the bottom of the Clicked event. Drag the Game Manager into the slot, and change the function to GameManager → StartGame.

Next, repeat these steps for the New Game button in the Game Over screen.

The buttons will now be set up! Before we're done, there are a few more minor things we need to set up to get the full gameplay experience.

First, we need to make it so that destroying the Space Station ends the game. The Space Station already has the `DamageTaking` script on it; we need to make this script call the `GameOver` function on the Game Manager.

4. *Add the call to `GameOver` in DamageTaking.cs.* Open the file and add the following code:

```
public class DamageTaking : MonoBehaviour {

    // The number of hit points this object has
    public int hitPoints = 10;

    // If we're destroyed, create one of these at
    // our current position
    public GameObject destructionPrefab;

    // Should we end the game if this object is destroyed?
    public bool gameOverOnDestroyed = false;

    // Called by other objects (like Asteroids and Shots)
    // to take damage
    public void TakeDamage(int amount) {

        // Report that we got hit
        Debug.Log(gameObject.name + " damaged!");

        // Deduct the amount from our hit points
        hitPoints -= amount;

        // Are we dead?
        if (hitPoints <= 0) {

            // Log it
            Debug.Log(gameObject.name + " destroyed!");

            // Remove ourselves from the game
            Destroy(gameObject);

            // Do we have a destruction prefab to use?
            if (destructionPrefab != null) {
```

```
            // Create it at our current position
            // and with our rotation.
            Instantiate(destructionPrefab,
                  transform.position, transform.rotation);
        }

>       // If we should end the game now, call the
>       // GameManager's GameOver method.
>       if (gameOverOnDestroyed == true) {
>           GameManager.instance.GameOver();
>       }
    }

  }

}
```

This code makes the object check to see if the `gameOverOnDestroyed` variable is set to `true`; if it is, the Game Manager's `GameOver` method is called, ending the game.

We also need to make the asteroids inflict damage when they collide. To do this, we'll add a `DamageOnCollide` script to them.

To make the asteroids apply damage, select the Asteroid prefab, and add a `DamageOnCollide` component.

Next, the asteroids should display their distance to the space station. This will help the player decide which asteroid is the most important one to go over. We'll do this by modifying the `Asteroid` script to query the Game Manager for the current space station, which is then given to the `showDistanceTo` variable of the asteroid's indicator.

To make the asteroid show the distance label, open *Asteroid.cs*, and add the following code to the `Start` function:

```
public class Asteroid : MonoBehaviour {

  // The speed at which the asteroid moves.
  public float speed = 10.0f;

  void Start () {
    // Set the velocity of the rigidbody
    GetComponent<Rigidbody>().velocity
      = transform.forward * speed;

    // Create a red indicator for this asteroid
    var indicator =
      IndicatorManager.instance.AddIndicator(
```

```
        gameObject, Color.red);

>       // Track the distance from this object to
>       // the current space station that's
>       // managed by the GameManager
>       indicator.showDistanceTo =
>         GameManager.instance.currentSpaceStation
>           .transform;
    }

}
```

This code sets up the indicator to show the distance from the asteroid to the current space station, which helps the player to prioritize the asteroids that are closest to the station.

You're done!

Play the game. You can now fly around and shoot asteroids, and asteroids will destroy the space station if too many hit it; you can also destroy the space station by shooting at it, and if the station is destroyed, game over!

Boundaries

There's one last core piece of gameplay that we need to add: we want to warn the player if they're getting too far away from the space station. If the player goes too far, we'll show a red warning border around the edges of the screen; if turn around, it's game over.

Creating the UI

First, we'll set up the UI for the warning:

1. *Add the Warning sprite.* Select the Warning texture. Change the texture's type to Sprite/UI.

   What we need to do is *slice* the sprite, so that it can be stretched over the entire screen without distorting the shape of the corners.

2. *Slice the sprite.* Click the Sprite Editor button, and the sprite will appear in a new window. In the panel at the lower-right of the window, set the border to 127 for all sides. This will make the corners not get stretched (see Figure 13-7).

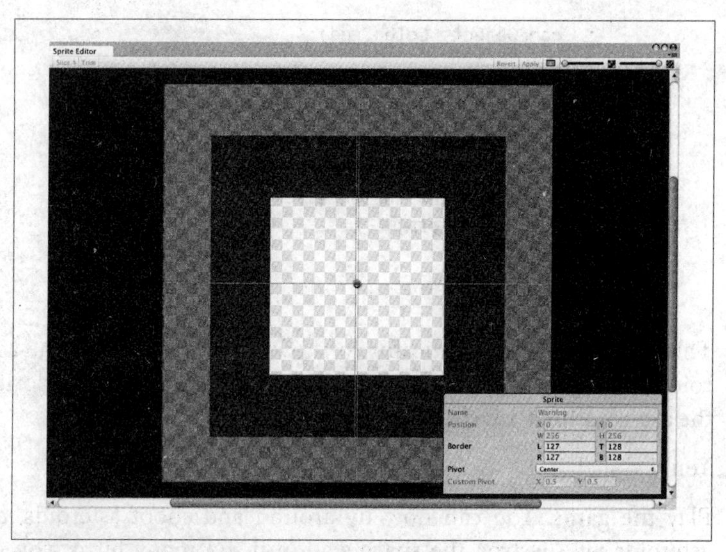

Figure 13-7. Slicing the Warning sprite

Click the Apply button.

3. *Next, we'll create the Warning UI.* This will simply be an image displayed on the UI, which will be set to stretch over the entire screen.

   To set up the warning UI, create a new empty game object, and name it "Warning UI". Make it a child of the Canvas object.

   Set the anchors to stretch horizontally and vertically, and set the margins to zero. This will make it fill the entire canvas.

   Add an Image component to it. Make the Source Image of this Image component be the Warning sprite that you just created, and set the Image Type to sliced. The image will be stretched over the entire screen.

With that done, it's time to code it up.

Coding the Boundary

The boundaries are invisible to the player, which means that they'll be invisible while editing the game. If you want to visualize the volume in which the player can fly around, you'll need to use the Gizmos feature again, just like you did for the Asteroid Spawner.

There are two concentric spheres that we care about, which we'll call the *warning* sphere and the *destroy* sphere. Both of these spheres will be centered on the same point, but they'll have different radii: the warning sphere's radius will be less than that of the destroy sphere.

- If the ship's position is within the warning sphere, then all is good, and no warning will be visible.
- If the ship is outside the warning sphere, then the warning will appear on screen, which will signal to the player that they need to turn around and head back.
- If the ship is outside the destroy sphere, the game ends.

The actual checking of the ship's position will be handled by the Game Manager, which will use the data stored inside the Boundary object (which you're about to create) to determine whether the ship is outside either of the two spheres.

Let's get started by creating the Boundary object, and adding the code that visualizes the two spheres:

1. *Create the Boundary object.* Create a new empty object, with the name "Boundary".

   Add a new C# script to the object called *Boundary.cs*, and add the following code to it:

   ```
   public class Boundary : MonoBehaviour {

     // Show the warning UI when the player is this far from the
     // center
     public float warningRadius = 400.0f;

     // End the game when the player is this far from the center
     public float destroyRadius = 450.0f;

     public void OnDrawGizmosSelected() {
       // Show a yellow sphere with the warning radius
       Gizmos.color = Color.yellow;
       Gizmos.DrawWireSphere(transform.position,
         warningRadius);

       // And show a red sphere with the destroy radius
       Gizmos.color = Color.red;
       Gizmos.DrawWireSphere(transform.position,
         destroyRadius);
     }
   }
   ```

When you return to the game editor, you'll see two wireframe spheres. The yellow sphere shows the warning radius, and the red sphere shows the destroy radius (as seen in Figure 13-8).

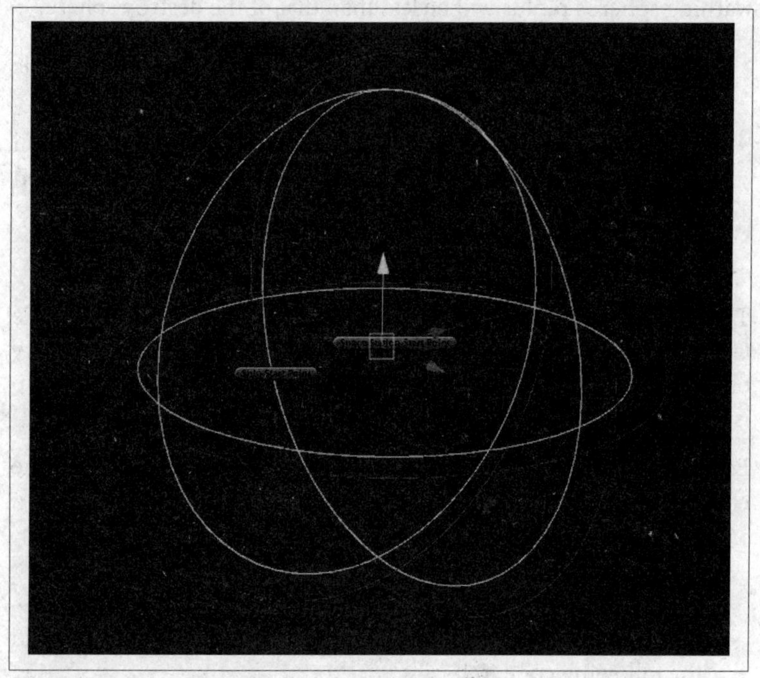

Figure 13-8. The boundaries

The Boundary script doesn't actually do any logic of its own in-game. Instead, the GameManager uses its data to determine if the player has flown beyond the boundary radii.

Now that the boundary object has been created, we just need to set up the Game Manager to use it.

2. *Add the boundary fields to the `GameManager` script, and update `GameManager` to use them.* Add the following code to *GameManager.cs*:

```
public class GameManager : Singleton<GameManager> {

  // The prefab to use for the ship, the place it starts from,
  // and the current ship object
  public GameObject shipPrefab;
```

```
public Transform shipStartPosition;
public GameObject currentShip {get; private set;}

// The prefab to use for the space station, the place it
// starts from, and the current ship object
public GameObject spaceStationPrefab;
public Transform spaceStationStartPosition;
public GameObject currentSpaceStation {get; private set;}

// The follow script on the main camera
public SmoothFollow cameraFollow;

> // The game's boundary
> public Boundary boundary;

// The containers for the various bits of UI
public GameObject inGameUI;
public GameObject pausedUI;
public GameObject gameOverUI;
public GameObject mainMenuUI;

> // The warning UI that appears when we approach
> // the boundary
> public GameObject warningUI;

// Is the game currently playing?
public bool gameIsPlaying {get; private set;}

// The game's Asteroid Spawner
public AsteroidSpawner asteroidSpawner;

// Keeps track of whether the game is paused or not.
public bool paused;

// Show the main menu when the game starts
void Start() {
  ShowMainMenu();
}

// Shows a UI container, and hides all others.
void ShowUI(GameObject newUI) {

  // Create a list of all UI containers.
  GameObject[] allUI
    = {inGameUI, pausedUI, gameOverUI, mainMenuUI};

  // Hide them all.
  foreach (GameObject UIToHide in allUI) {
    UIToHide.SetActive(false);
  }
```

```
    // And then show the provided UI container.
    newUI.SetActive(true);
  }

  public void ShowMainMenu() {
    ShowUI(mainMenuUI);

    // We aren't playing yet when the game starts
    gameIsPlaying = false;

    // Don't spawn asteroids either
    asteroidSpawner.spawnAsteroids = false;
  }

  // Called by the New Game button being tapped
  public void StartGame() {
    // Show the In-Game UI
    ShowUI(inGameUI);

    // We're now playing
    gameIsPlaying = true;

    // If we happen to have a ship, destroy it
    if (currentShip != null) {
      Destroy(currentShip);
    }

    // Likewise for the station
    if (currentSpaceStation != null) {
      Destroy(currentSpaceStation);
    }

    // Create a new ship, and place it
    // at the start position
    currentShip = Instantiate(shipPrefab);
    currentShip.transform.position
      = shipStartPosition.position;
    currentShip.transform.rotation
      = shipStartPosition.rotation;

    // And likewise for the station
    currentSpaceStation = Instantiate(spaceStationPrefab);

    currentSpaceStation.transform.position =
      spaceStationStartPosition.position;

    currentSpaceStation.transform.rotation =
      spaceStationStartPosition.rotation;

    // Make the follow script track the new ship
    cameraFollow.target = currentShip.transform;
```

```
        // Start spawning asteroids
        asteroidSpawner.spawnAsteroids = true;

        // And aim the spawner at the new space station
        asteroidSpawner.target = currentSpaceStation.transform;

    }

    // Called by objects that end the
    // game when they're destroyed
    public void GameOver() {
        // Show the Game Over UI
        ShowUI(gameOverUI);

        // We're no longer playing
        gameIsPlaying = false;

        // Destroy the ship and the station
        if (currentShip != null)
            Destroy (currentShip);

        if (currentSpaceStation != null)
            Destroy (currentSpaceStation);

>       // Stop showing the warning UI, if it was visible
>       warningUI.SetActive(false);

        // Stop spawning asteroids
        asteroidSpawner.spawnAsteroids = false;

        // And remove all lingering asteroids from the game
        asteroidSpawner.DestroyAllAsteroids();
    }

    // Called when the Pause or Resume buttons are tapped
    public void SetPaused(bool paused) {

        // Switch between the in-game and paused UI
        inGameUI.SetActive(!paused);
        pausedUI.SetActive(paused);

        // If we're paused..
        if (paused) {
            // Stop time
            Time.timeScale = 0.0f;
        } else {
            // Resume time
            Time.timeScale = 1.0f;
        }
    }
```

```
>   public void Update() {
>
>     // If we don't have a ship, bail out
>     if (currentShip == null)
>       return;
>
>     // If the ship is outside the Boundary's Destroy Radius,
>     // game over. If it's within the Destroy Radius, but
>     // outside the Warning radius, show the Warning UI. If
>     // it's within both, don't show the Warning UI.
>
>     float distance =
>       (currentShip.transform.position
>         - boundary.transform.position)
>           .magnitude;
>
>     if (distance > boundary.destroyRadius) {
>       // The ship has gone beyond the destroy radius,
>       // so it's game over
>       GameOver();
>     } else if (distance > boundary.warningRadius) {
>       // The ship has gone beyond the warning radius,
>       // so show the warning UI
>       warningUI.SetActive(true);
>     } else {
>       // It's within the warning threshold, so don't
>       // show the warning UI
>       warningUI.SetActive(false);
>     }
>
>   }

}
```

This new code makes use of the Boundary class that you just created to check to see if the player has gone beyond either the warning radius or the destroy radius. Every frame, the distance from the player to the center of the boundary spheres is checked; if they're beyond the warning radius, the warning UI is made to appear, and if they're beyond the destroy radius, the game ends. If the player is *within* the warning radius, they're fine, so the warning radius is disabled. This means that if the player flies outside the warning radius and then returns to safety, they'll see the warning UI appear and then disappear.

Next, you just need to connect up the slots. The Game Manager needs a reference to the Boundary object you created a moment ago, as well as a reference to the Warning UI.

3. *Configure the Game Manager to use the boundary objects.* Drag Warning UI into the Warning UI slot, and the Boundary object into the Boundary slot.
4. *Play the game.* When you get close to the boundary, the warning will appear, and if you don't turn around, game over!

Final Polish

Congratulations! You've now finished setting up the core gameplay of a rather sophisticated space shooter. As you followed along in the preceding sections, you set up a space environment, created spaceships, space stations, asteroids, and laser beams; set up their physics; and set up all of the various logical components that connect them together. On top of that, you've created the UI that's necessary for actually playing the game outside of the Unity editor.

The core of the game is done, but there's still room for some visual improvements. Because the visuals of the game are quite sparse, we don't have many visual reference points to give the player a sense of traveling at speed. Additionally, we'll add a little more color to the game by adding trail renderers to the ship and asteroids.

Space Dust

If you've ever played a spaceflight game before, like *Freelancer* or *Independence War*, you might have noticed how, when the player flies around, small pieces of dust, debris, and other small objects move past the player.

To improve our game, we'll add small dust motes that provide a sense of depth and perspective as the player moves past them. We'll achieve this with a particle system that moves with the player, continuously creating dust particles in a sphere surrounding the player. Importantly, these dust particles will not move relative to the player. This means that, as the player flies, dust particles will appear that the player then flies past. This creates a much better impression of speed in the game.

To create the dust particles, follow these steps:

1. *Drag the Ship prefab into the scene.* We'll be making some changes to the prefab.

2. *Create the Dust child object.* Create a new empty game object, and name it "Dust". Make it a child of the Ship game object you just dragged out.
3. *Add a Particle System* component to it. Copy the settings in Figure 13-9 to it.

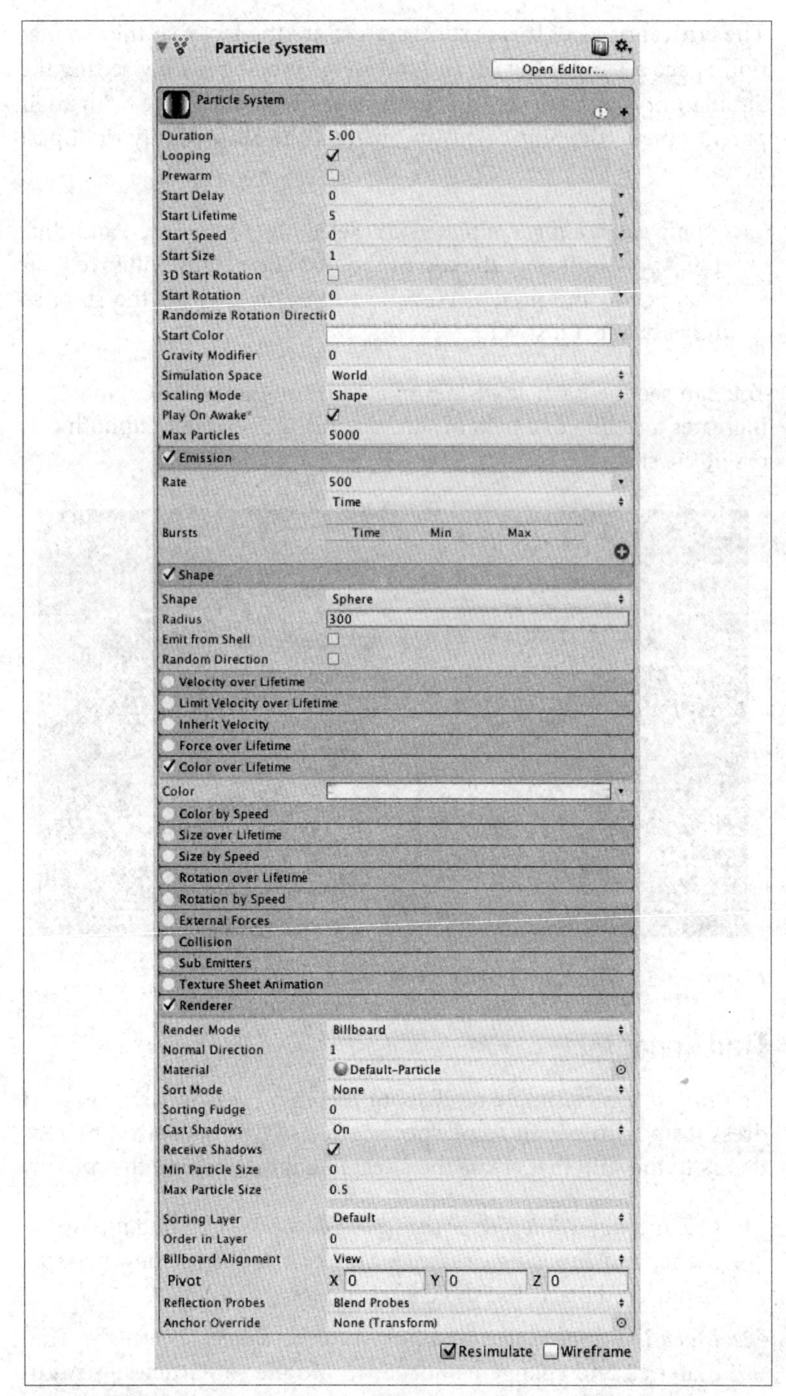

Figure 13-9. The settings for the dust particles

The critical parts of this particle system are the fact that the Simulation Space setting is `World`, and the Shape is a `Sphere`. By setting the Simulation Space to `World`, the particles will not move with their parent object (the Ship). This means that the Ship will fly right past them.

4. *Apply your changes to the prefab.* Select the Ship object, and click the Apply button at the top of the Inspector. This will save your changes to the prefab. We're not quite done with the ship, so don't delete it just yet.

You can see the particle system in action in Figure 13-10. Note how it creates a feeling of a field of stars against the relatively smooth colors of the skybox.

Figure 13-10. The dust particle system

Trail Renderers

The ship is a very simple model, but there's no reason why you can't dress it up a little with some special effects. We'll add two line renderers to the ship that create the effect of engines behind them.

1. *Create a new Material for the trail.* Do this by opening the Assets menu, and choosing Create → Material. Name the new material "Trail", and place it in the *Objects* folder.
2. *Make the Trail material use an Additive shader.* Select the Trail material, and change its Shader to Mobile → Particles → Addi-

tive. This is a simple shader that simply adds its color to the background. Leave the Particle Texture empty—it won't be needed.

3. *Add a new child object to the Ship.* Name it "Trail 1". Position it at (-0.38,0,-0.77).
4. *Add a Trail Renderer component.* Make it use the settings in Figure 13-11. Note that the Material it's using is the new Trail material you just created.

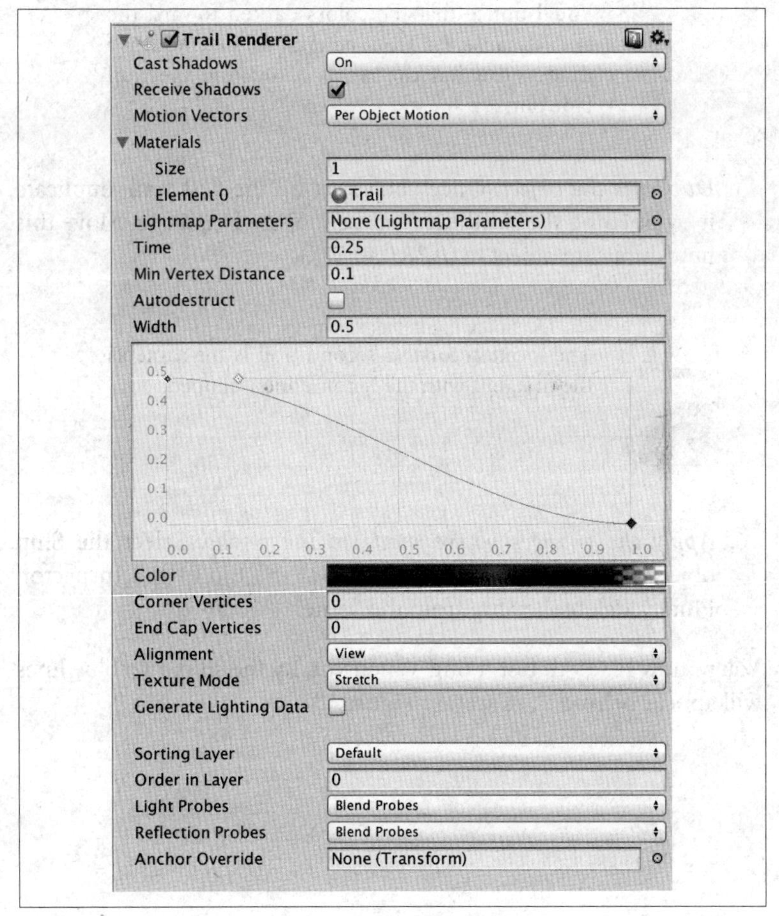

Figure 13-11. The settings for the ship's trail renderer

The colors used in the trail renderer's gradient are:

- #000B78FF
- #061EF9FF
- #0080FCFF
- #000000FF
- #00000000

You'll notice that the colors darken toward the end. Because the Trail material uses an Additive shader, this has the result of making the trail fade out.

5. *Duplicate the object.* Once you've set up the first trail, duplicate it by opening the Edit menu and choosing Duplicate. Move this new duplicate object to (0.38,0,-0.77).

The location for this second trail is the same as the first, but with the X component flipped.

6. *Apply the changes you've made to the prefab.* Select the Ship object, and click the Apply button at the top of the Inspector. Finally, delete the Ship from the scene.

You're now ready to test it out! When you fly the ship, two blue lines will appear behind it, as seen in Figure 13-12.

Figure 13-12. The engine trails behind the ship

We'll now apply a similar effect to the asteroids. The asteroids in the game are quite dark, and while they have indicators to help the player keep track of where they are, they could do with a little more color. To improve things, we'll add a trail renderer to them.

1. *Add an Asteroid to the scene.* Drag out the Asteroid prefab into the scene, so that you can make changes.
2. *Add a Trail Renderer component to the Graphics child object.* Use the settings you see in Figure 13-13.

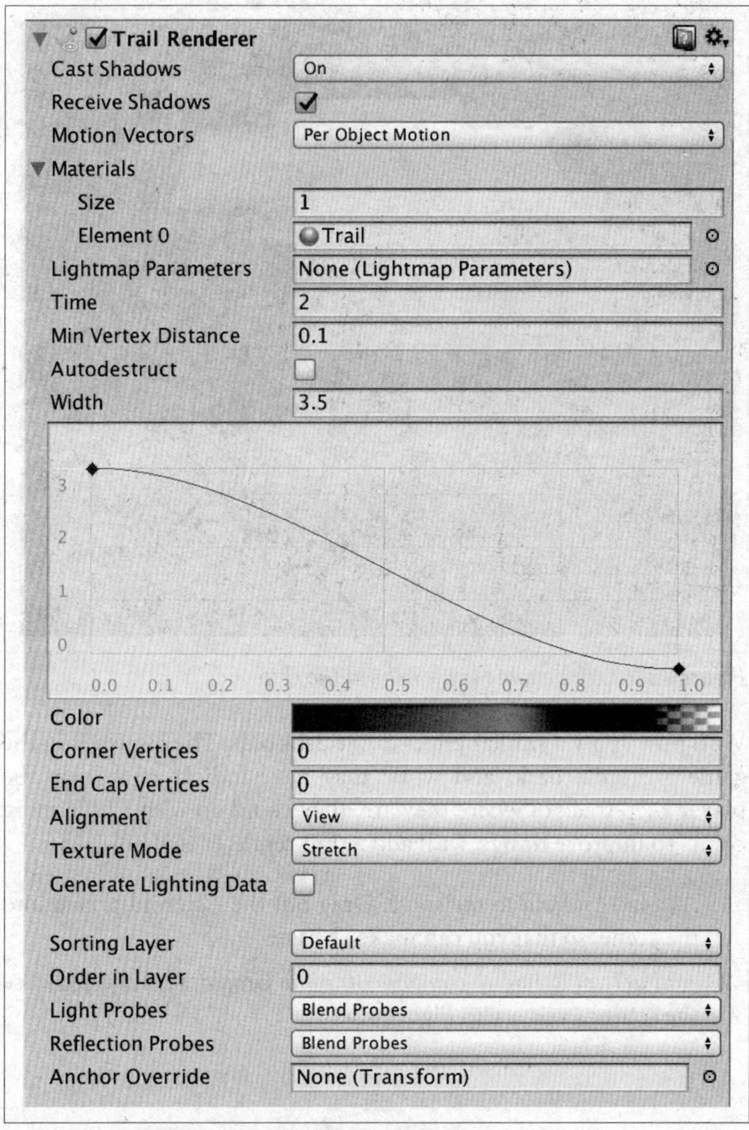

Figure 13-13. The settings for the asteroid's trail

3. *Apply the changes to the Asteroid prefab, and remove it from the scene.*

The asteroids will now have a bright trail behind them. You can see the full game in action in Figure 13-14.

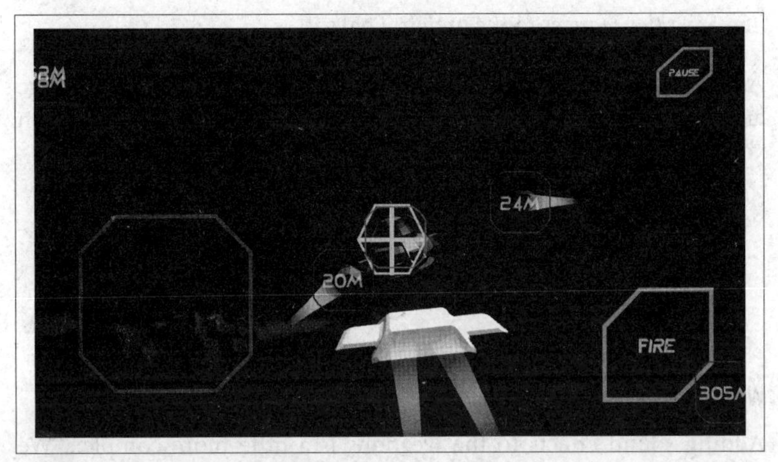

Figure 13-14. The game in action

Audio

There's one last part that we need to add: audio! Even though there's no sound in real space, video games are seriously improved with the addition of sound. There are three sounds that we need to add: the roar of the ship's engines, the zap of laser blasts, and the boom of asteroids exploding. We'll add each one, one at a time.

We've included some public-domain sound effects in the book's files, which you'll find in the *Audio* folder.

Spaceship

First, we'll add a looping sound effect to the spaceship.

1. *Add the Ship to the scene.* We're about to make some changes to it.
2. *Add an Audio Source component to the Ship.* Audio Sources are how you make sound happen.
3. *Turn on the Loop setting.* We want the engine noises to play continuously as the player flies the ship.
4. *Add the rocket sound.* Drag the Engine audio clip into the AudioClip slot.

5. *Save the changes to the prefab.* That's it!

Adding looping sounds is incredibly easy, and gives you a huge amount of overall improvement to the game for very little effort on your part.

Don't delete the Ship from the scene yet—we'll add some more to it in a moment.

Weapon effects

Adding sound effects to the weapons is a little more complex. We want to play a sound effect every time the weapon fires, which means we'll need to make the code aware of sound effects.

First, we'll need to add audio sources to the two weapon points:

1. *Add Audio Sources to the weapon fire points.* Select both of the weapon fire points. With both of them selected, add an Audio Source.
2. *Add the Laser effect to the audio sources.* Once you've done that, turn off the Play On Awake setting—we only want to play sound when we fire a shot.
3. *Add code to play the sound effect when shots are fired.* Add the following code to *ShipWeapons.cs*:

```
public class ShipWeapons : MonoBehaviour {

  // The prefab to use for each shot
  public GameObject shotPrefab;

  public void Awake() {
    // When this object starts up, tell the input manager
    // to use me as the current weapon object
    InputManager.instance.SetWeapons(this);
  }

  // Called when the object is removed
  public void OnDestroy() {
    // Don't do this if we're not playing
    if (Application.isPlaying == true) {
      InputManager.instance.RemoveWeapons(this);
    }
```

```
    }

    // The list of places where a shot can emerge from
    public Transform[] firePoints;

    // The index into firePoints that the next
    // shot will fire from
    private int firePointIndex;

    // Called by InputManager.
    public void Fire() {

      // If we have no points to fire from, return
      if (firePoints.Length == 0)
        return;

      // Work out which point to fire from
      var firePointToUse = firePoints[firePointIndex];

      // Create the new shot, at the fire point's position
      // and with its rotation
      Instantiate(shotPrefab,
        firePointToUse.position,
        firePointToUse.rotation);

>     // If the fire point has an audio source
>     // component, play its sound effect
>     var audio
>       = firePointToUse.GetComponent<AudioSource>();
>     if (audio) {
>       audio.Play();
>     }

      // Move to the next fire point
      firePointIndex++;

      // If we've moved past the last fire point in the list,
      // move back to the start of the queue
      if (firePointIndex >= firePoints.Length)
        firePointIndex = 0;

    }
  }
```

This code checks to see if the fire point that the shot is being fired from has an AudioSource component. If it does have one, it's made to play the shot sound.

4. *Save your changes to the Ship prefab, and remove it from the scene.*

You're done. Now, you'll hear a sound effect every time a shot is fired!

Explosions

There's one last sound effect to add: an explosion sound effect, for when explosions appear. This one's easy: we just need to add an audio source to the explosion object, and set it to play on awake. When an explosion appears, it will automatically play the Explosion sound.

1. *Add an Explosion to the scene.*
2. *Add an Audio Source component to the explosion.* Drag in the Explosion audio clip, and turn on Play On Awake.
3. *Save your changes to the prefab and remove it from the scene.*

You now get explosions whenever one appears!

Wrapping Up

You're now all done. Congratulations! *Rockfall* is now complete, and in your hands. It's up to you to decide what to do next with it!

Some ideas:

Add new weapons
: Maybe a rocket that turns to face its target?

Add enemies that attack the player
: The asteroids are pretty simple, and fly straight at the space station, while nothing actually goes after the player.

Add damage effects to the space station
: When an asteroid hits, add a particle system that emits smoke and flames at the point of impact. It's not realistic, but that hasn't stopped us adding any of the other features to the game.

PART IV

Advanced Features

In this part, we'll take a closer look at some specific features of Unity, ranging from a more detailed examination of the UI system to some deeper Unity plumbing through extending the editor. We'll also examine the lighting and shading system, and conclude with a tour of the extended Unity ecosystem, and discuss how to get your games onto your devices and out into the world.

CHAPTER 14

Lighting and Shaders

In this chapter we'll look at lighting and materials, which are the primary thing—besides the textures that you use—that determine how your game looks. In particular, we'll take a closer look at the Standard shader, which is a shader designed to make it simple to create good-looking materials. We'll also talk about how to write your own shaders, which give you a huge amount of control over how objects appear in game. Finally, we'll discuss how to use global illumination and lightmapping, which can create a great-looking environment by realistically modeling how light bounces in a scene.

Materials and Shaders

In Unity, the appearance of objects is defined by the *material* attached to it. A material is composed of two things: a *shader* and data that's used by that shader.

A shader is a very small program that runs on the graphics card. Every single thing that you see on the screen is the result of a shader calculating the correct color value to show for each pixel.

In Unity, there are two main different types of shaders: *surface* shaders and *vertex-fragment* shaders.

Surface shaders are responsible for calculating the color of the surface of an object. As we already discussed in "Materials and Shaders" on page 325, there are multiple components that define the color of a surface, including its albedo, smoothness, and so on. A surface shader's job is to compute the values of each of these properties for every pixel of the object; this surface information is then returned to Unity, which combines the surface information with information about each of the lights in the scene, and determines the final lit color for that pixel.

By contrast, vertex-fragment shaders are much simpler. This kind of shader is responsible for calculating the final colour of the pixel; if your shader needs to include lighting information, you'll need to do it yourself. Vertex-fragment shaders give you low-level control, which means that they're great for effects. Because they're generally simpler, they're also typically a lot faster than surface shaders.

Surface shaders actually get compiled down to vertex-fragment shaders by Unity, which does much of the heavier lifting for you by implementing the lighting calculations needed to achieve realistic lighting. Anything you can do in a surface shader can also be done in a vertex-fragment shader, but it will take more effort.

Unless you have a very specific use case, surface shaders are generally your best choice. In this chapter, we'll look at both.

Unity also provides a third type of shader, called a *fixed-function* shader. Fixed function shaders work by combining predefined operations together, rather than letting you write your own custom shaders. Fixed function shaders were the main way things got done before the wider adoption of custom shaders; they're less complex than custom shaders, but tend to not look as great, and their use is now discouraged. We won't be talking about fixed-function shaders in this chapter, but if you really want to learn about them, Unity's documentation includes a tutorial on writing fixed-function shaders (*http://docs.unity3d.com/Manual/ShaderTut1.html*).

Let's get started by creating our own custom surface shader that's quite similar to the standard shader, but adds the ability to show rim lighting—that is, a highlight around the edges of the object. You can see an example of this effect in Figure 14-1.

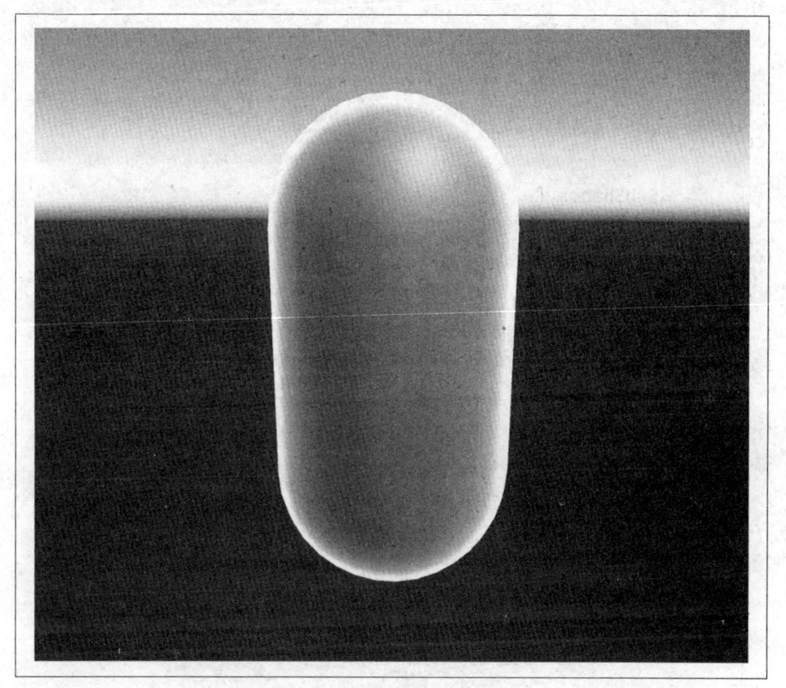

Figure 14-1. Rim lighting, using a custom shader

To begin creating the effect, follow these steps:

1. *Create a new project.* Name it whatever you like and select 3D mode.
2. *Create a new shader* by choosing Create → Shader → Surface Shader. Name the new shader "SimpleSurfaceShader".
3. *Double-click it.*
4. *Replace it with the following code:*

```
Shader "Custom/SimpleSurfaceShader" {

    Properties {
        // The color to tint the object with
        _Color ("Color", Color) = (0.5,0.5,0.5,1)

        // The texture to wrap the object in;
        // defaults to a plain white texture
        _MainTex ("Albedo (RGB)", 2D) = "white" {}

        // How smooth the surface should be
        _Smoothness ("Smoothness", Range(0,1)) = 0.5
```

```
        // How metallic the surface should be
        _Metallic ("Metallic", Range(0,1)) = 0.0

    }

    SubShader {
        Tags { "RenderType"="Opaque" }
        LOD 200

        CGPROGRAM
            // Physically based Standard lighting model, and
            // enable shadows on all light types
            #pragma surface surf Standard fullforwardshadows

            // Use shader model 3.0 target, to get nicer-
            // looking lighting
            #pragma target 3.0

            // The following variables are "uniform" - the
            // same value is used for every pixel

            // The texture to use for the albedo
            sampler2D _MainTex;

            // The color to tint the albedo with
            fixed4 _Color;

            // The smoothness and metallicness properties
            half _Smoothness;
            half _Metallic;

            // 'Input' contains variables whose values are
            // different for every pixel
            struct Input {
                // Texture coordinates at this pixel
                float2 uv_MainTex;

            };

            // This single function computes the properties
            // of this surface
            void surf (Input IN,
              inout SurfaceOutputStandard o) {

                // Using the data stored in IN and the
                // variables above, compute the values and
                // store them in 'o'

                // Albedo comes from a texture tinted by
                // color
```

```
                fixed4 c =
                  tex2D (_MainTex, IN.uv_MainTex) * _Color;
                o.Albedo = c.rgb;

                // Metallic and smoothness come from slider
                // variables
                o.Metallic = _Metallic;
                o.Smoothness = _Smoothness;

                // Alpha value for this comes from the
                // texture we're using for albedo
                o.Alpha = c.a;

            }
        ENDCG
    }

    // If the computer running this shader isn't capable of
    // running at shader model 3.0, fall back to the built-in
    // "Diffuse" shader, which doesn't look anywhere as good
    // but is guaranteed to work
    FallBack "Diffuse"
}
```

4. *Create a new material,* called "SimpleSurface".
5. *Select the new material,* and open the Shader menu at the top of its Inspector. Choose Custom → SimpleSurfaceShader.

The properties for your surface shader will now appear in the Inspector (Figure 14-2).

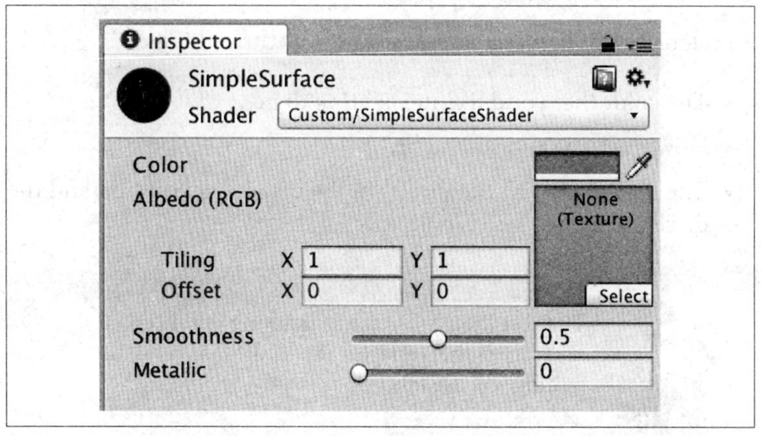

Figure 14-2. The Inspector for the custom shader

6. *Create a new capsule* by choosing GameObject → 3D Object → Capsule.
7. *Drag the SimpleShader material* onto the capsule, and it will start using the new material (Figure 14-3).

Figure 14-3. A capsule, using the custom shader

At the moment, the object looks very similar to the standard shader. Let's now add rim lighting!

To calculate rim lighting, you need to know three things:

- The color that you'd like the lighting to be.
- How thick the rim should be.
- The angle between the direction the camera is pointing and the direction the surface is pointing.

The direction a surface is pointing is called the *normal* of the surface. The code that we'll be writing will use this term.

The first two items are *uniform*—that is, their value applies to every pixel of the object. The third item is *varying*, which means that its value depends on where you're looking; the angle between the camera's direction and the surface's normal depends on whether you're looking at the middle of the cylinder or the edges.

Varying values are calculated at runtime by the graphics card for your surface shader to use, while uniform values are exposed as properties of the material that you can modify through the Inspector. To add support for rim lighting, therefore, we first need to add the two uniform variables to the shader.

1. *Modify the `Properties` section of the shader* to include the following code:

```
Properties {
    // The color to tint the object with
    _Color ("Color", Color) = (0.5,0.5,0.5,1)

    // The texture to wrap the object in;
    // defaults to a plain white texture
    _MainTex ("Albedo (RGB)", 2D) = "white" {}

    // How smooth the surface should be
    _Smoothness ("Smoothness", Range(0,1)) = 0.5

    // How metallic the surface should be
    _Metallic ("Metallic", Range(0,1)) = 0.0

>   // The color the rim lighting should be
>   _RimColor ("Rim Color", Color) = (1.0,1.0,1.0,0.0)
>
>   // How thick the rim lighting should be
>   _RimPower ("Rim Power", Range(0.5,8.0)) = 2.0

}
```

This code makes the shader show two new fields in the Inspector. We now need to make these properties available to the shader's code, so that the `surf` function can make use of them.

2. *Add the following code to the shader:*

```
// The smoothness and metallicness properties
half _Smoothness;
half _Metallic;

> // The color for the rim lighting
```

```
> float4 _RimColor;
>
> // How thick the rim lighting should be - closer to
> // zero means thicker rim
> float _RimPower;
```

Next, we need to make the shader able to get the direction that the camera is looking. All of the varying values that the shader uses are included in the `Input` structure, which means that the look direction needs to be added there.

The `Input` structure can have several fields added to it, and Unity will automatically fill them with the relevant information. If you add a `float3` variable called `viewDir`, Unity will put the direction that the camera is looking into it.

`viewDir` isn't the only variable name that Unity will automatically use for varying information. For a full list, see Unity's Surface Shader documentation (*http://docs.unity3d.com/Manual/SL-SurfaceShaders.html*).

3. *Add the following code to the `Input` structure:*

```
struct Input {
    // Texture coordinates at this pixel
    float2 uv_MainTex;

>   // The direction the camera is looking at this vertex
>   float3 viewDir;
};
```

The material's Inspector will now be showing the additional fields (Figure 14-4).

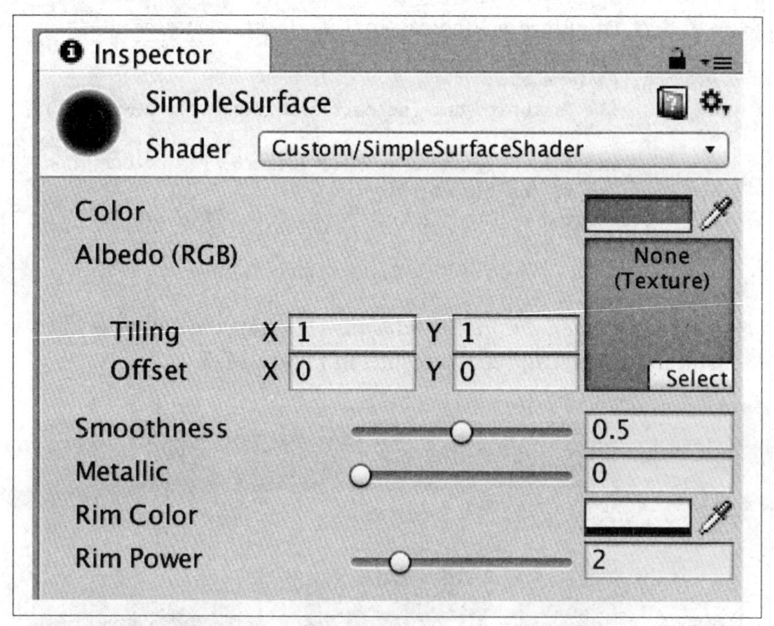

Figure 14-4. The Inspector, showing the newly added fields

We now have all of the information we need to calculate the rim lighting; the last step is to actually perform the calculation, and add it to the surface's information.

4. *Add the following code to the* `surf` *function:*

```
// This single function computes the properties of this
// surface
void surf (Input IN, inout SurfaceOutputStandard o) {

    // Using the data stored in IN and the variables above,
    // compute the values and store them in 'o'

    // Albedo comes from a texture tinted by color
    fixed4 c = tex2D (_MainTex, IN.uv_MainTex) * _Color;
    o.Albedo = c.rgb;

    // Metallic and smoothness come from slider variables
    o.Metallic = _Metallic;
    o.Smoothness = _Smoothness;

    // Alpha value for this comes from the texture we're
    // using for albedo
    o.Alpha = c.a;
```

```
>       // Calculate how bright the rim light should be at this
>       // pixel
>       half rim =
>         1.0 - saturate(dot (normalize(IN.viewDir), o.Normal));
>
>       // Use this brightness to calculate the rim colour, and
>       // use it for the emission
>       o.Emission = _RimColor.rgb * pow (rim, _RimPower);

    }
```

5. *Save the shader, and return to Unity.* The capsule will now have a rim light! You can see the results in Figure 14-5.

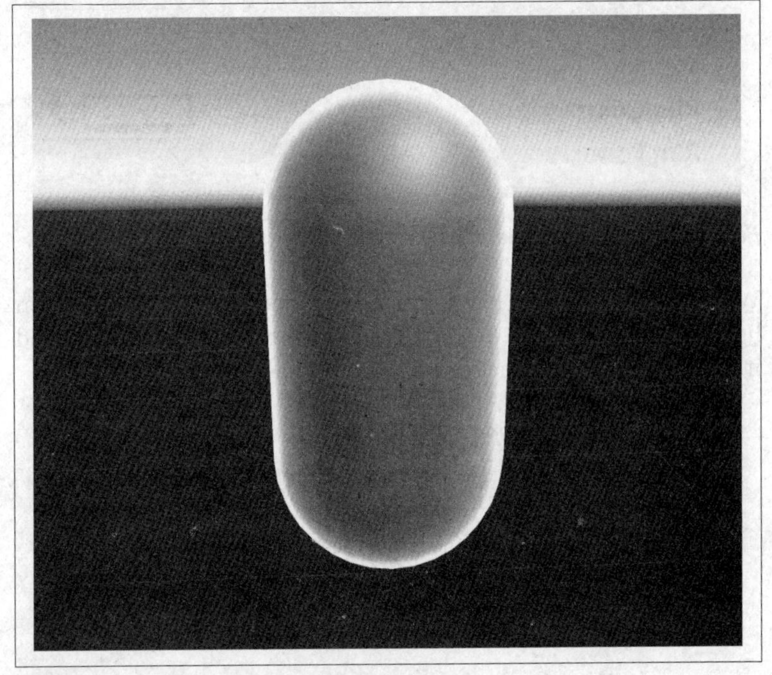

Figure 14-5. The capsule, with rim lighting

You can also tweak the rim lighting through the material's properties. Try changing the Rim Color setting to change the brightness and tint of the rim lighting, and the Rim Power setting to adjust how thick the rim appears.

Surface shaders are great for building upon the existing shading system, and they're your best choice if you want your surface to respond to the lights you've added to your scene. However, there are some situations where you either don't care about lighting, or need very specific control over the appearance of the surface. In these situations, you can create your own completely custom fragment-vertex shaders.

Fragment-Vertex (Unlit) Shaders

A fragment-vertex shader is so named because it's really two shaders in one: a *fragment shader* and a *vertex shader*. These are two separate functions that control the appearance of how a surface is rendered.

A *vertex shader* is a function that transforms each *vertex*—that is, each point in the object's space—from world-space to view-space, in preparation for rendering. World-space means the world that you see in the Unity editor: objects are positioned in space, and you can move them around. However, when Unity needs to render the scene using a camera, that camera must first convert the positions of every object in the scene into *view-space*: a space in which the entire world, and every object in it, is repositioned such that the camera is in the middle of the world. Additionally, in view-space, the entire world is reshaped to make objects further away from the camera smaller. The vertex shader is also responsible for calculating the value of the varying variables that should be passed to the fragment shader.

You almost never need to write your own vertex shader, but there are situations where it can be useful. For example, if you want to distort an object's shape, you can write your own vertex shader that modifies the position of each vertex.

The *fragment shader* is the other half of the pair. Fragment shaders are responsible for calculating the final color of each fragment—that is, pixel—of the object. The fragment shader receives the value of the varying variables calculated by the vertex shader; this value is *interpolated*, or blended, based on the proximity of the fragment being rendered to its nearest vertices.

Because a fragment shader has full control over the final color of the object, it's up to the shader itself to calculate the effect of nearby lights. If your shader doesn't perform the calculation itself, the surface won't appear lit.

It's for this reason that surface shaders are the recommended way to make your surfaces lit; lighting calculation can get complex, and it's often a lot easier to not have to think about it.

In fact, surface shaders are actually fragment-vertex shaders. Unity converts surface shaders to the lower-level fragment-vertex code for you, and adds in the lighting calculations.

The downside is that surface shaders are designed for the general case, and can be less efficient than a handcoded shader.

To demonstrate how fragment-vertex shaders work, we'll create a simple one that renders objects in a single flat color. We'll then modify it to render a gradient depending on where the object is on the screen.

1. *Create a new shader* by opening the Assets menu, and choosing Create → Shader → Unlit Shader. Name the new shader "SimpleUnlitShader".
2. *Double-click it to open it.*
3. *Replace the contents of the file with the following code:*

```
Shader "Custom/SimpleUnlitShader"
{
    Properties
    {
        _Color ("Color", Color) = (1.0,1.0,1.0,1)

    }
    SubShader
    {
        Tags { "RenderType"="Opaque" }
        LOD 100

        Pass
        {
            CGPROGRAM

            // Define which functions should be
            // used in this shader.

            // The 'vert' function will be used as
            // the vertex shader.
            #pragma vertex vert
```

```
// The 'frag' function will be used as
// the fragment shader.
#pragma fragment frag

// Include a number of useful utilities from
// Unity.
#include "UnityCG.cginc"

float4 _Color;

// This structure is given to the
// vertex shader for each vertex
struct appdata
{
    // The position of the vertex in world space.
    float4 vertex : POSITION;

};

// This structure is given to the
// fragment shader for each fragment
struct v2f
{
    // The position of the fragment in
    // screen space
    float4 vertex : SV_POSITION;
};

// Given a vertex, transform it
v2f vert (appdata v)
{
    v2f o;

    // Convert the vertex from world space to
    // view space by multiplying it with a matrix
    // provided by Unity. (This comes from
    // UnityCG.cginc)
    o.vertex = UnityObjectToClipPos(v.vertex);

    // Return it, passing it to the fragment
    // shader
    return o;
}

// Given interpolated information about
// nearby vertices, return the final color
fixed4 frag (v2f i) : SV_Target
{
   fixed4 col;
```

```
                // Render the provided color
                col = _Color;

                return col;
            }
            ENDCG
        }
    }
}
```

4. *Create a new material* by opening the Assets menu, and choosing Create → Material. Name the material "SimpleShader".
5. *Select the new material,* and change the shader to Custom → SimpleUnlitShader.
6. *Create a sphere in the scene* by opening the GameObject menu, and choosing 3D Object → Sphere. Drag the SimpleShader material onto it.

The sphere will now be a single flat color. You can see the result in Figure 14-6.

Figure 14-6. The sphere, rendered using a flat color

Flat-shading an object with a color is a common enough task that a very similar shader to the one you just wrote comes bundled with Unity. You can find it in the Shader menu, under Unlit → Color.

Next, we'll build upon this shader to make it dynamically animate. This will involve zero scripting; instead, all of the animation will be done inside the graphics shader itself.

1. *Add the following code to the frag function:*

```
fixed4 frag (v2f i) : SV_Target
{
    fixed4 col;

    // Render the provided color
    col = _Color;

>   // Fade over time - start black, fade up to _Color
>   col *= abs(_SinTime[3]);

    return col;
}
```

2. *Return to Unity,* and notice that the object has changed to black. This is expected.
3. *Hit the Play button,* and watch the object fade in and out. You can see an example of this in action in Figure 14-7.

Figure 14-7. The object fading in and out

As we've seen, vertex-fragment shaders give us signficant control over the appearance of our objects, and this section has given you a taste of them. Discussing this in all of its detail can fill an entire book; if you'd like a more in-depth discussion of how to use shaders, check out Unity's detailed documentation (*http://docs.unity3d.com/Manual/SL-Reference.html*).

Global Illumination

When an object is lit, the shader that's responsible for rendering that object needs to perform several complex calculations to determine the amount of light that the object is receiving, and use that to calculate the color of the object that the camera can see. This is usually fine, but certain things are very difficult to compute at runtime.

For example, if a sphere is resting on a white surface in direct sunlight, the sphere should be lit from below, because light is bouncing up off the ground. However, the shader can only know about the direction of the sun itself, and as a result it doesn't show this lighting. It's certainly possible to calculate it, but this quickly becomes a very challenging problem to solve every frame.

A better solution is using *global illumination* and *lightmapping*. Global illumination is the name given to a number of related techniques that compute the amount of light received by every surface in a scene, taking into account how light bounces off objects.

Global illumination results in very realistic lighting, but it is very processor intensive; as a result, the lighting calculations can also be done ahead of time, in the Unity editor. The results can then also be stored in a *lightmap*, which records the final amount of light received by every part of every surface in the scene.

Because the global illumination calculation takes place ahead of time, it can only think about objects that are guaranteed to never move (and thus change the way that light works in the scene.) Any moving objects in your game can't directly use global illumination; instead, a different solution is needed, which we'll talk about shortly.

Using lightmapping can significantly improve the performance of the realistic lighting in a scene, because the lighting calculations have been performed ahead of time and are stored in textures. However, if you use lightmaps, these textures must be loaded into memory in order to be used by the renderer. This can be a problem if your scene is already complex, or already uses a lot of textures. A way to mitigate this problem is to reduce the resolution of the lightmap, but this reduces the visual quality of the lighting.

With this in mind, let's dive into using global illumination by creating a scene. We'll first set up a few materials of different colors, to help us to see how light will bounce around the scene; we'll then create some objects, and make them use the global illumination system.

1. *Create a new scene in Unity.*
2. *Create a new material* called "Green". Keep the shader as Standard, and change the Albedo color to green. You can see the Inspector settings for this material in Figure 14-8.

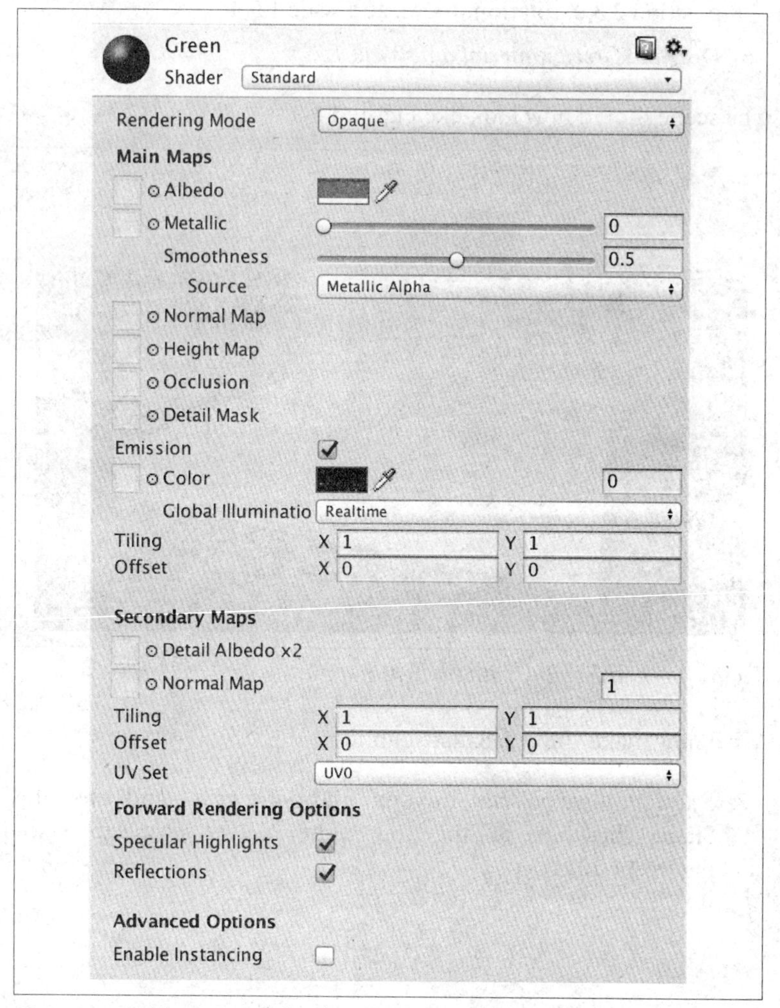

Figure 14-8. The settings for the Green material

Next, let's create the objects in the world.

3. *Create a cube* by opening the GameObject menu, and choosing 3D Object → Cube. Name the cube "floor", and set its position to 0,0,0, and set its scale to 10,1,10.
4. *Create a second cube,* and name it "Wall 1". Set its position to -1,3,0, and its rotation to 0,45,0. Additionally, set its scale to 1,5,4.
5. *Create a third cube,* and name it "Wall 2". It should have the position 2,3,0, rotation 0,45,0, and scale 1,5,4.
6. *Drag the Green material onto Wall 1.*

The scene should now look like Figure 14-9.

Figure 14-9. The scene, with no lightmapping

We'll now make Unity calculate lighting.

7. *Select all three objects*—the floor and both walls—and select the Static checkbox at the top right of the Inspector (see Figure 14-10).

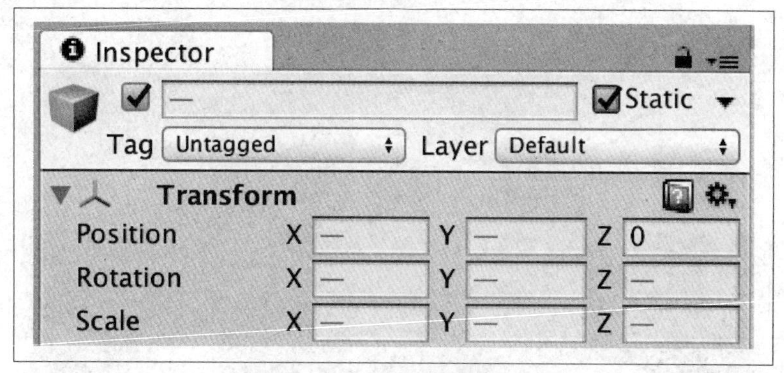

Figure 14-10. Setting the objects to be static

With static objects in the scene, Unity will immediately begin calculating lighting information. After a few moments, the lighting will shift subtly. The most significant effect that you'll notice is the fact that the green wall will bounce some of its light onto the back of the white wall. Compare the difference between Figures 14-11 and 14-12.

Figure 14-11. The scene with global illumination inactive

Figure 14-12. The scene with global illumination active—note the green reflection on the back of the wall

This will make the lighting use real-time global illumination. This looks good, but it causes a signficant performance hit, because only part of the lighting calculation is being done ahead of time. To improve performance at the cost of higher memory usage, you can *bake* the lighting into a lightmap.

8. *Select the directional light* and set the Baking setting to `Baked`.

After a moment, the lighting will be calculated and stored in a lightmap.

While global illumination is good for statical objects, it won't affect your nonstatic objects. To improve this, you can use light probes.

Light Probes

A light probe is an invisible object that picks up the lighting coming from all directions and records it. Nearby nonstatic objects can then use this lighting information to illuminate themselves.

Light probes don't work in isolation. Instead, you create them in groups; at runtime, objects that need lighting information combine their nearest probes based on how close they are to the object. This allows an object to reflect more light as it gets closer to a surface that's bouncing light, for example.

Let's add a nonstatic object to the scene, and then add some light probes, to see how they affect the lighting.

1. *Add a new capsule to the scene* by choosing GameObject → 3D Object → Capsule. Place the capsule somewhere near the green wall.

You'll notice that the capsule doesn't pick up any of the reflected green light from the wall. In fact, it's receiving *too much* light in the direction of the wall, because the light from the sky is illuminating it in that direction. This light should be blocked by the wall. You can see this in Figure 14-13.

2. *Add some light probes* by opening the GameObject menu and choosing Light → Light Probe Group.

You'll see a collection of spheres appear; each sphere represents a single light probe, and shows the lighting that is being received at that point in space.

3. *Reposition the probes* so that none of them are embedded in the scene—that is, they're all floating in space, and not stuck inside the floor or walls.

You can adjust the position of individual probes in the group by selecting the light probe group and clicking Edit Light Probes; you can then select individual probes and move them.

With this done, your capsule will now pick up the reflected green light from the wall—compare the difference between Figures 14-13 and 14-14.

Figure 14-13. The scene with no light probes

Figure 14-14. The scene with light probes--note the green light reflecting onto the capsule

The more probes you have, the longer your lighting will take to calculate. Additionally, if you have a sudden change in lighting (that is, between areas) you should cluster the probes more densely near the transition, to prevent your objects looking out of place.

Thinking About Performance

Before we wrap up the work in this chapter, now's a good opportunity to talk about the performance tools built into Unity. The lighting setup that you use in your game can significantly affect the performance on the user's device; additionally, marking objects as static has performance implications in addition to making global illumination and lightmapping possible.

However, not all of your game's performance depends upon the graphics of the game: the amount of time that your scripts take up on the CPU can have as much of an impact.

Helpfully, Unity comes with several tools and features that you can take advantage of to boost your productivity.

The Profiler

The profiler is a tool that records data about your game as it's being played. It gathers information from several different locations every frame, such as:

- The script methods that are called every frame, and the amount of time that's taken in calling them
- The number of "draw calls"—that is, instructions to the graphics chip that cause it to do drawing work—needed to draw the frame
- The amount of memory consumed by the game, in both the scripts and in graphics memory
- The amount of CPU time taken in playing audio
- The number of active physics bodies, and the amount of physical collisions that need to be processed in the frame
- The amount of data being sent and received over the network

The Profiler is split into two halves. The top half is itself divided up into several rows—one for each of these data recorders. You can see an image of the profiler in Figure 14-15. As the game is played, each recorder fills with information. The bottom half shows detailed information about the specific frame that you're inspecting, from the currently selected recorder.

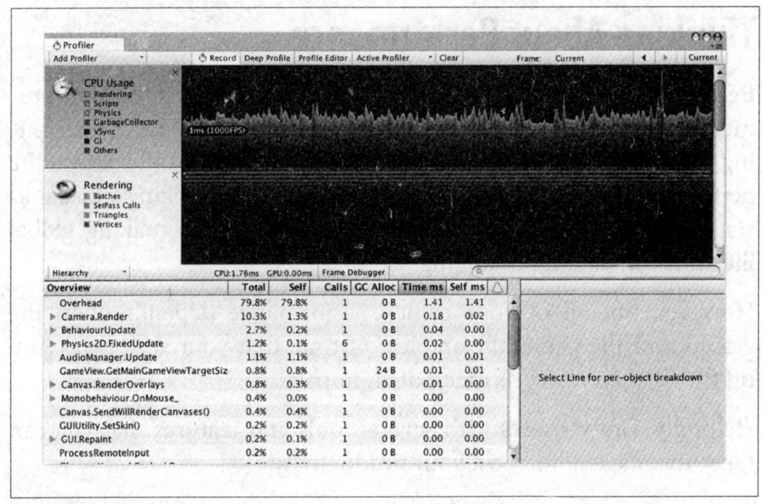

Figure 14-15. The Profiler

The specific results that you'll see here in this book won't necessarily be identical to what you see in your game. It depends on the hardware you're using, of both the computer you're using Unity on and the mobile device you're testing your game on, as well as the specific version of Unity you're using. Unity Technologies is always changing the engine behind the scenes, so you'll likely see different results.

That said, the steps you follow to gather data about your game's performance *will* be the same, and you can apply the techniques to just about any game.

1. *To open the Profiler, open the Window menu, and choose Profiler.* Alternatively, you can press Command-7 on a Mac, or Ctrl-7 on a PC. When you do, the Profiler will appear.

To begin using the Profiler, you just need to have it open while the game is running.

2. *Start the game* by pressing the Play button, or by pressing Ctrl-P (Command-P on a Mac).

The Profiler will begin to fill up with information. It's a lot easier to analyze the game when it's not in the middle of running, so you'll need to take the game out of Play mode before you continue.

3. *After a few moments, stop or pause the game.* The data in the Profiler won't go away.

Now that the Profiler has stopped filling with data, you can now take a closer look at individual frames.

4. *Click and drag with the left mouse button over the top row in the Profiler.* A vertical line will appear as you do so, and the data being shown in the bottom half of the Profiler will update to show the selected frame.

Different recorders show different information. In the case of the CPU, it defaults to showing the Hierarchy, which is the list of all of the methods that were called in that frame, sorted by the amount of time each method took to call (Figure 14-16). You can also click the triangles at the left of each row to open them and see information about the methods that that row called.

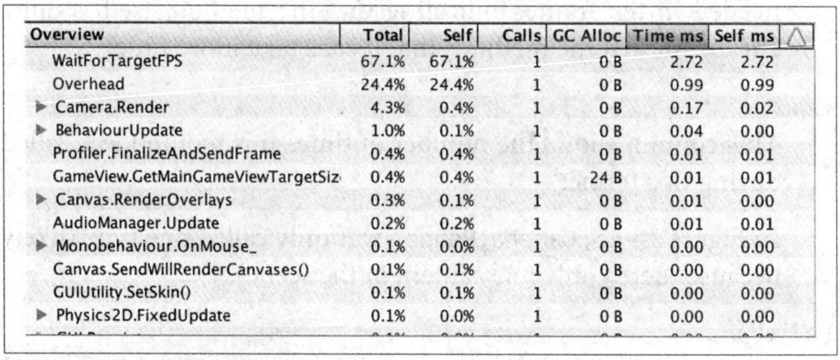

Overview	Total	Self	Calls	GC Alloc	Time ms	Self ms
WaitForTargetFPS	67.1%	67.1%	1	0 B	2.72	2.72
Overhead	24.4%	24.4%	1	0 B	0.99	0.99
▶ Camera.Render	4.3%	0.4%	1	0 B	0.17	0.02
▶ BehaviourUpdate	1.0%	0.1%	1	0 B	0.04	0.00
Profiler.FinalizeAndSendFrame	0.4%	0.4%	1	0 B	0.01	0.01
GameView.GetMainGameViewTargetSiz	0.4%	0.4%	1	24 B	0.01	0.01
▶ Canvas.RenderOverlays	0.3%	0.1%	1	0 B	0.01	0.00
AudioManager.Update	0.2%	0.2%	1	0 B	0.01	0.01
▶ Monobehaviour.OnMouse_	0.1%	0.0%	1	0 B	0.00	0.00
Canvas.SendWillRenderCanvases()	0.1%	0.1%	1	0 B	0.00	0.00
GUIUtility.SetSkin()	0.1%	0.1%	1	0 B	0.00	0.00
▶ Physics2D.FixedUpdate	0.1%	0.0%	1	0 B	0.00	0.00

Figure 14-16. The Hierarchy view of the CPU profiler

We're going to spend a bit of time focusing on the CPU profiler, because understanding what it's telling you can help you identify and fix a large number of possible performance issues in your game.

The columns of the Hierarchy show different information for each row:

Total
: This column shows the percentage amount of time that calls to this method, and the methods that were called as a result, took when rendering this frame.

  In Figure 14-16, for example, calls to the `Camera.Render` method (a method internal to the Unity engine) took 4.3% of the time needed to render the entire frame.

Self
: This column shows the percentage amount of time that calls to this method, and *only* this method, took when rendering this frame. This helps to identify whether a method is responsible for taking a lot of time, or that the methods that it calls are responsible. If the value of Self is close to the value of Total, it indicates that the method itself is responsible for the time taken, and not the methods that it calls.

  In Figure 14-16, `Camera.Render` only takes 0.4% of the time needed in the frame, indicating that the method itself is quite cheap, but that the methods that it calls take more time.

Calls
: This column shows the number of times this method was called during this frame.

  In Figure 14-16, `Camera.Render` was only called one time (likely because there's only one camera in the scene).

GC Alloc
: This column shows the amount of memory that this method had to allocate during this frame. If memory is frequently allocated, it increases the chance that the memory garbage collector has to run later, which causes lag.

In Figure 14-16, the call to `GameView.GetMainGameViewTargetSize` allocated 24 bytes. While this might seem like a small number, don't forget that the game is rendering as many frames as it can; over time, if a small amount of memory is allocated every frame, it can build up, necessitating the garbage collector to step in and clean up, which harms your game's performance.

Time ms
: This column shows the amount of time, in milliseconds, that the calls to this method (and all of the calls that this method made) took to execute. In Figure 14-16, the call to `Camera.Render` took 0.17 milliseconds.

Self ms
: This column shows the amount of time, in milliseconds, that the calls to this method (and *only* this method) took to execute. In Figure 14-16, the call to `Camera.Render` spent only 0.02 milliseconds in that method; the other 0.15 milliseconds were spent in methods that were called as a result.

Warnings
: This column shows any issues that the Profiler identified. The Profiler is capable of performing some analysis of the data it records, and can give you limited amounts of advice.

Getting Data from Your Device

When you follow the steps in the previous section, the data you'll be gathering comes from the Unity Editor. However, playing your game in the Editor doesn't have the same performance characteristics as playing the game on your device. A PC or Mac generally has a much faster CPU, more RAM, and a better GPU than a mobile device. As a result, the results that you get from the Profiler will be different, and optimizing for what you see when you run the game in the Editor may not improve the performance for the end user.

To address this, you can use the Profiler to gather data from the game when it's running on the device. To do so, follow these steps:

1. *Build and install the game on your phone by following the steps in "Deployment" on page 424.* Importantly, make sure that Development Build and Autoconnect Profiler are both turned on.

2. *Make sure that your device and computer are both on the same WiFi network,* and that your device is connected to your computer with a USB cable.
3. *Launch the game on your device.*
4. *Open the Profiler, and open the Active Profiler menu.* Choose your device from the list that appears.

The Profiler will start collecting data directly from your device.

General Tips

There are several things you can do to improve your game's performance:

- In the `Rendering` profiler, try to keep the `Verts` count below 200,000 per frame.
- When selecting shaders for use in the game, choose ones from the Mobile or Unlit categories. These shaders are simpler, and take less time to run per frame than others.
- Keep the number of different materials that you're using in the scene as low as you can. Additionally, try to make as many objects as you can use the *same* material. This makes it easier for Unity to draw those objects at the same time, which is a performance gain.
- If an object will never move, scale, or rotate in the scene, turn on the Static checkbox at the top right of the Inspector. This will enable a number of internal optimizations in the engine.
- Reduce the number of lights in your scene. The more lights you have, the more work the engine has to do.
- Using baked lighting instead of real-time lighting is more efficient. Keep in mind, however, that baked lights can't move, and the baked light information will take up memory.
- Use compressed textures instead of uncompressed textures as much as you can. Compressed textures take up less memory, and take less time for the engine to access (because there's less data to read).

You can find a collection of other useful performance tips (*http://docs.unity3d.com/Manual/OptimizingGraphicsPerformance.html*) in the Unity manual.

Wrapping Up

Lighting can make your scene look signficantly better. Even if your game isn't intended to look entirely realistic, putting some effort into how your scenes are lit can make the entire game just feel better.

It's also important to keep an eye on the performance of your game. Using the Profiler, you can take a closer look at what the game is actually doing, and adjust your game using this information.

CHAPTER 15

Creating GUIs in Unity

Games are software, and all software needs a user interface. Even if it's as simple as a button that starts a new game, or a label that shows the player's current score, your game still needs a way to show the more mundane, "nongame" stuff for the user to interact with.

The good news is that Unity has a really great UI system. Introduced in Unity 4.6, the UI system is extremely flexible and powerful, and is designed for the situations that games typically encounter. For example, the UI system supports PC, console, and mobile platforms; allows a single UI to scale to multiple sizes of screens; is capable of responding to input from the keyboard, mouse, touchscreen, and game controllers; and supports displaying the UI in both screen-space and in world-space.

In short, it's a pretty incredible toolkit. While we've been building GUIs in the games discussed in Parts II and III, we'd like to look at some finer points of the GUI system, so that you're ready to take full advantage of the features that it offers.

How GUIs Work in Unity

Fundamentally, a GUI in Unity is not terribly different from the other visible objects in your scene. A GUI is a mesh that's constructed at runtime by Unity, with textures applied to it; additionally, the GUI contains scripts that respond to mouse movement, keyboard events, and touches to update and modify that mesh. The mesh is displayed via the camera.

The GUI system in Unity has several different pieces that work together. At its core, a GUI is composed of several objects with `Rect Transforms` that draw their content and respond to *events*, all contained within a Canvas.

Canvas

The *Canvas* is the object that's responsible for drawing all of the UI elements on screen. As a result, it's also the total space in which the canvas is drawn.

All UI elements are child objects of the Canvas—if the button isn't a child of the canvas, it won't appear.

The Canvas lets you decide *how* the UI is drawn. Additionally, by attaching a Canvas Scaler component, you can control how UI elements are scaled. We'll talk more about Canvas Scalers in "Scaling the Canvas" on page 367.

The Canvas can be used in one of three modes—*Screen Space - Overlay*, *Screen Space - Camera*, and *World Space*:

- When the Canvas is in the Screen Space - Overlay mode, the entire Canvas is drawn on top of the game. That is, all Cameras in the scene render their view of the game onto the screen, and then the Canvas is drawn on top of it all. This is the default mode for the Canvas.
- In Screen Space - Camera mode, the contents of the Canvas are rendered into a plane, which is positioned in 3D space some distance in front of a specified Camera. When the Camera moves, the Canvas is repositioned to keep it at the same point relative to the Camera. When being used in this mode, the Canvas is effectively a 3D object, which means that objects between the Canvas and the Camera will occlude the Canvas.
- In the World Space mode, the Canvas is a 3D object in the scene, with its own position and rotation that's independent of any Camera in the scene. This means that you can, for example, create a Canvas that contains a keypad for a door, and position it next to the door.

If you ever played the games *DOOM* (2016) or *Deus Ex: Human Revolution*, you've interacted with a world-space GUI. In these games, the player interacts with in-game computer screens by walking up to them and "clicking" on the on-screen buttons that are being displayed.

RectTransform

Unity is a 3D engine, which means that all objects have a Transform component that determines their position, rotation, and scale in 3D space. However, a GUI in Unity is 2D. This means that all UI elements are 2D rectangles that have a position, a width, and a height.

In order to control this, UI objects have a `RectTransform` object. The `RectTransform` represents a rectangle in which UI content can appear. Importantly, if a `RectTransform` is the child of *another* `RectTransform`, then the child can be positioned and sized relative to that parent.

For example, the Canvas object has a `RectTransform` that defines, at a minimum, the size of the GUI; additionally, all of the GUI elements that make up your game's GUI will have their own `RectTransform`. Because these GUI elements are child objects of the Canvas, the GUI elements' `RectTransform` will be positioned *relative* to the Canvas.

The Canvas's `RectTransform` can also define the position of the GUI, but this depends on whether the Canvas is a *screen-space*, *camera-space*, or *world-space* one or not. If the Canvas is anything but a *world-space* one, the position of the Canvas will be determined automatically.

You can take this further when you nest multiple child objects. If you create an object with a `RectTransform`, and add child objects (each with their own `RectTransform`), then those child objects will be positioned relative to the parent.

RectTransforms aren't limited to UI elements. You can add a RectTransform to any object; if you do, the RectTransform will replace the Transform component at the top of the Inspector.

The Rect Tool

The Rect tool provides you with a simple way to move and resize objects that have a RectTransform component. To activate the Rect tool, you press the T key, or choose the Rect tool from the toolbar at the top-left of the Unity window (Figure 15-1).

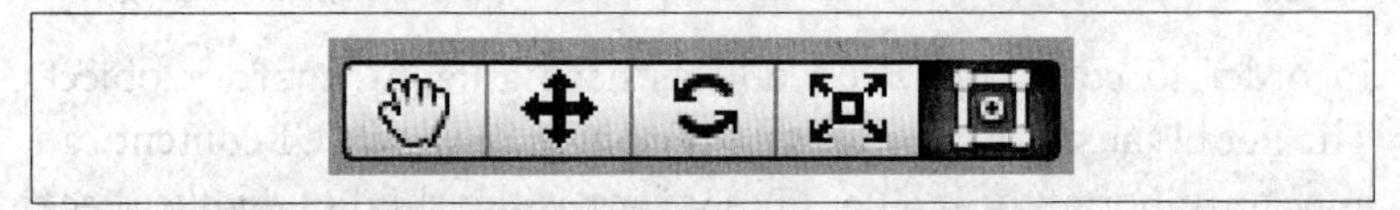

Figure 15-1. Selecting the Rect tool in the toolbar

When the Rect Tool is enabled, a rectangular set of handles will surround the selected object (Figure 15-2). When you drag these handles, the object will be resized and repositioned.

Additionally, if you move your mouse cursor near a handle, outside the rectangle, the cursor will change to show that it's in rotation mode. When you click and drag, the object will rotate around its pivot point. The pivot point is the circle in the middle of the object; if the selected object has a RectTransform, you can click and drag the pivot point to move it.

Figure 15-2. The Rect tool handles, with the pivot point in the center

The Rect tool isn't limited to UI elements! It can also be used with 3D objects; when you have one selected, the rectangle and handles will be placed depending on how you're looking at the object in the Scene view. See Figure 15-3 for an example of how it looks.

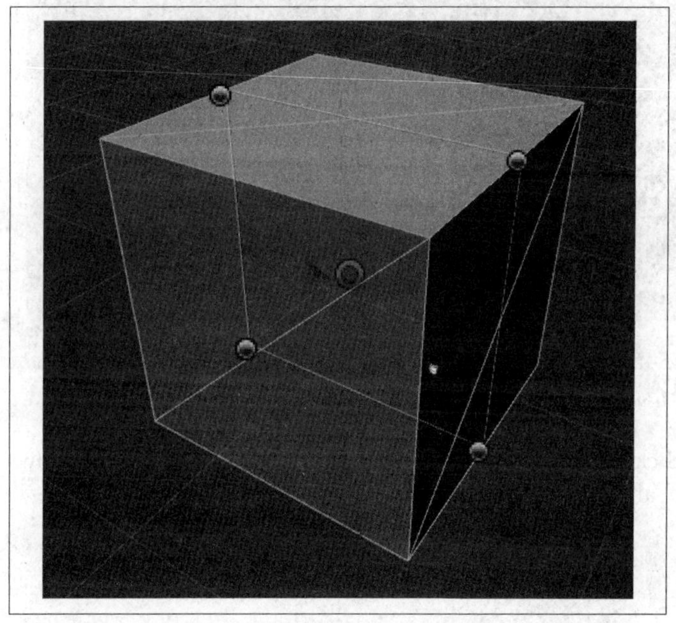

Figure 15-3. The Rect tool handles, surrounding a 3D cube

Anchors

When a `RectTransform` is a child of another `RectTransform`, it's positioned relative to its *anchors*. This allows you to define a relationship between the size of a parent rect, and the position and size of its children. For example, you can make a rectangle be positioned at the bottom of its parent, filling its entire width; when the parent changes size, the child rect's position and size will be updated.

In the Inspector for the `RectTransform`, you'll see a box that allows you to select preset values for the anchors (Figure 15-4).

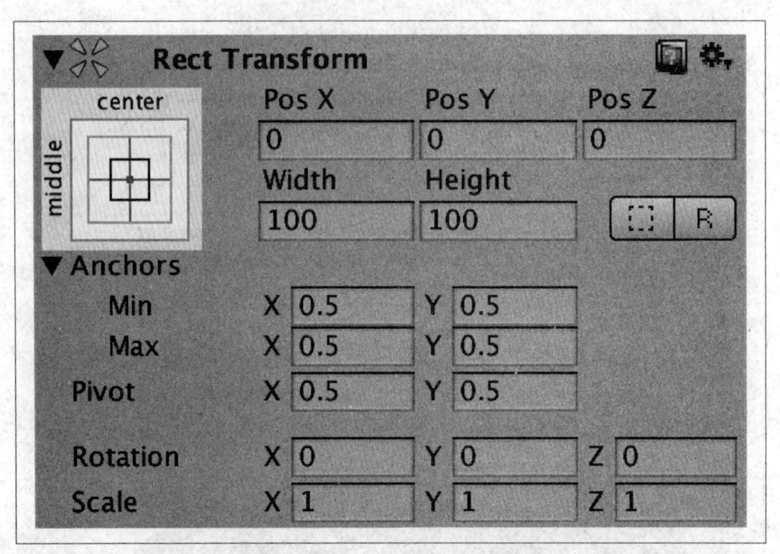

Figure 15-4. The box showing the currently selected preset for the Rect Transform's anchors

If you click this box, a small pop-up window will appear that allows you to change the preset (Figure 15-5).

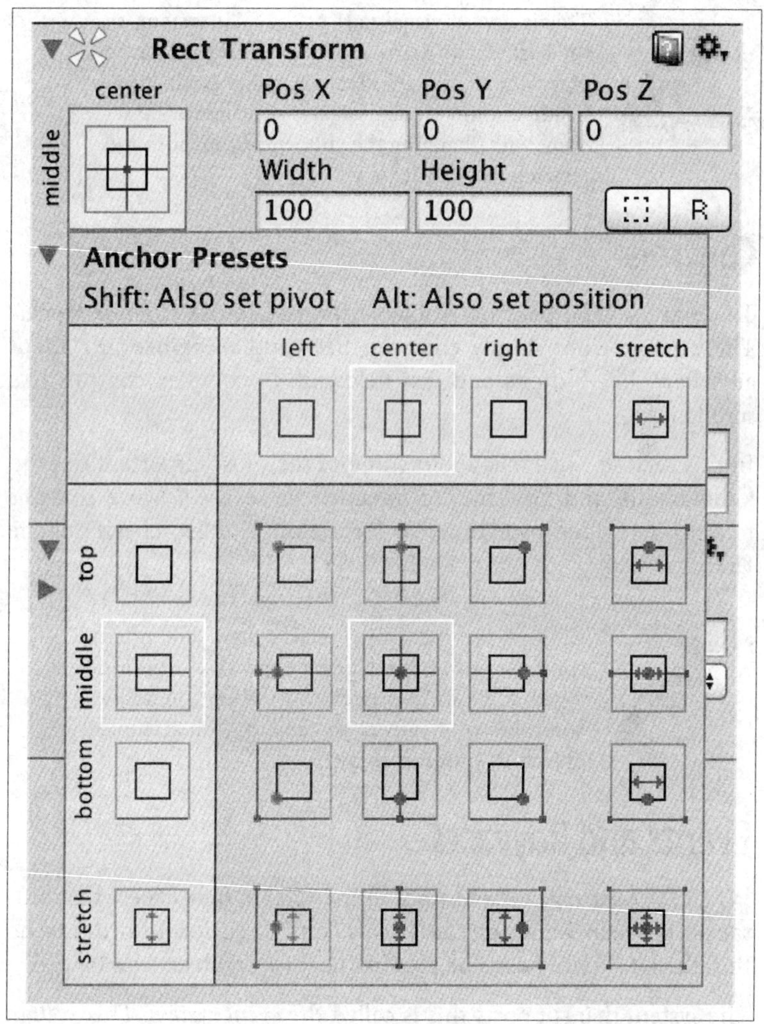

Figure 15-5. The anchor preset selection panel

Clicking on any of these presets changes the anchors of the `Rect Transform`. It won't change the position or size of the rectangle, but it will change how the rectangle changes size when its *parent* changes size.

This is a very visual part of the GUI system, so the best way to learn about how this works is to play with it yourself. Place an Image game object inside another Image object, and then experiement with changing the child view's anchors and resizing the parent view.

Controls

There are several controls available for you to use in your scenes. These range from simple controls that you'd recognize just about anywhere, like buttons and text fields, up to complex controls like scroll views.

In this section, we'll talk about some of the most important ones to know about, and how they're intended to be used. More controls tend to get added over time, so for a complete list, check out the Unity manual.

Controls in the Unity GUI system are often composed of multiple game objects, all working together. Don't be surprised if, when you add a control to your canvas, you end up with multiple objects in your hierarchy.

Events and Raycasts

When the user taps on a button on the screen, they expect that button to perform whatever task it was configured for. In order to do this, the UI system has to be able to know *which* object was tapped.

The system that supports this is called the *event system*. This system is quite sophisticated: in addition to providing input to the GUI, it can also be used as a general solution for identifying when any object in your game is clicked, tapped, or dragged.

The event system is represented by the Event System object that appears when you create your Canvas.

The event system works on the principle of *raycasts*. In a raycast, a *ray*—an invisible line—is sent from the point where the user tapped the screen. This ray continues until it hits something, at which point the event system knows what's "under" the user's finger.

Because the raycast system works in 3D space, just like the rest of the engine does, the event system is able to work with both 2D GUIs and 3D GUIs. When an event happens, such as a finger tap or mouse click, every *raycaster* in the scene fires its ray. There are three different types of raycast colliders, each of which looks for different things for the ray to hit—*graphic raycasters, 2D physics raycasters*, and *3D physics raycasters*:

- Graphic raycasters check to see if their ray has collided with any Image component in the canvas.
- 2D physics raycasters check to see if their ray has collided with any 2D collider in the scene.
- 3D physics raycasters check to see if their ray has collided with any 3D collider in the scene.

When the user taps on a button GUI, the graphic raycaster component attached to the Canvas fires a ray out from the finger's location on screen, and checks to see if it hit any Image. Because buttons have an Image component, the raycaster reports to the event system that the button was tapped.

While they're not used in the GUI system, you can use the 2D and 3D physics raycasters to detect clicks, taps, and drags on 2D and 3D objects in your scene. For example, you can use a 3D physics raycaster to detect when the user clicks on a cube.

Responding to Events

When you're building custom UI, it's often extremely useful to be able to add custom behavior to your UI elements. Generally, this involves being notified about input events like clicks and drags.

To make a script that can respond to these, you make the class conform to certain interfaces, and then implement the required methods for those interfaces. For example, the `IPointerClickHandler` interface requires its implementors to have a method with the signature `public void OnPointerClick (PointerEventData eventData)`. This method is run when the event system detects that the current pointer (either the mouse cursor or a finger touching the screen) performed a "click"—that is, the mouse button was pressed and released, or a finger was pressed and lifted, within the bounds of the image.

To demonstrate this, here's a quick little tutorial on how to respond to pointer clicks on a GUI object:

1. *In an empty scene, create a new Canvas* by opening the GameObject menu and choosing UI → Canvas. A Canvas will be added to the scene.
2. *Create a new Image* by opening the GameObject menu and choosing UI → Image. An Image object will be added as a child of the Canvas.
3. *Add a new C# script to the Image object* named *EventResponder.cs*, and add the following code to the file:

```
// Necessary for access to 'IPointerClickHandler' and
// 'PointerEventData'
using UnityEngine.EventSystems;

public class EventResponder : MonoBehaviour,
  IPointerClickHandler {

    public void OnPointerClick (PointerEventData eventData)
    {
        Debug.Log("Clicked!");
    }

}
```

4. *Run the game.* When you click on the image, the word "Clicked!" will appear in the Console.

Using the Layout System

When you create a new UI element, you generally add it directly into the scene and manually set its position and size. However, this quickly becomes untenable in two important situations:

- When you don't know the size of the canvas, because the game will be shown on different sized screens; and
- When you're going to be adding and removing content from the UI at runtime.

In these situations, you can take advantage of the layout system built into the Unity GUI system.

To illuminate how it works, we'll quickly put together a vertical list of buttons:

1. *Select a Canvas object* by clicking on it in the Hierarchy. (If you don't have one, create one by opening the GameObject menu and choosing UI → Canvas.)
2. *Create a new empty child object* by opening the GameObject menu and choosing Create Empty Child, or by pressing Ctrl-Alt-N (Command-Option-N on a Mac).
3. *Name the new object "List"*.
4. *Create a new Button* by opening the GameObject menu and choosing UI → Button. Make this new Button be a child of the List object.
5. *Add a Vertical Layout Group component to the List object* by selecting it, clicking the Add Component button, and choosing Layout → Vertical Layout Group. (You can also type the first few letters of "vertical layout group" to quickly select this object.)

You'll notice that the moment the Vertical Layout Group is added to the List object, the Button is resized to fill the entire space of the List's rectangle. You can see a before and after view of this in Figures 15-6 and 15-7.

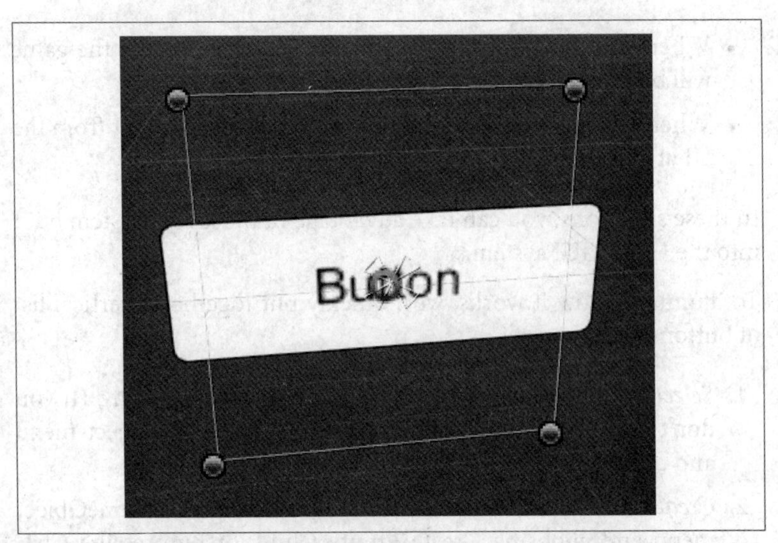

Figure 15-6. The button, before adding a Vertical Layout Group to the List

Figure 15-7. The button, after adding a Vertical Layout Group to the List

Next, watch what happens when there are *multiple* buttons in the layout group.

6. *Select the Button, and duplicate it* by pressing Ctrl-D (Command-D on a Mac).

When you do this, both the original button and the duplicate will immediately be repositioned and resized so that they both fit in the List object (Figure 15-8).

Figure 15-8. The two buttons, laid out in a vertical arrangement

In addition to the Vertical Layout Group, the GUI system also includes the Horizontal Layout Group, which works exactly the same as its Vertical counterpart, only sideways; additionally, the Grid Layout Group lays out content in a regular grid, allowing you to display multiple lines of content that wrap around as needed.

Scaling the Canvas

In addition to the fact that the various different types of screens that your game will be shown on will differ in size, the screens will likely also differ in *display density*. Display density refers to the size of the individual pixels; on more modern mobile devices, the screen is typically of a higher density.

A high-profile example of this is the Retina display that's used in all iPhones since the iPhone 4, and in all iPads since the third-generation iPad. These devices have screens that are the same physical size as the previous model, but have double the display density: on the iPhone 3GS, the screen is 320 pixels wide, but on the iPhone 4, the screen is 640 pixels wide. The content shown on the screen is designed to remain the same physical size, while the increase in display density means that content is much smoother and better-looking.

Because Unity deals in individual pixels, rendering a GUI on a high-density display will result in your GUI content being shown at half the size.

To address this problem, the Unity GUI system includes a component called the Canvas Scaler. The Canvas Scaler's role is to automatically adjust the scale of all GUI elements to ensure that they're an appropriate size for the display that the game is currently being played on.

When you create a Canvas object through the GameObject menu, a Canvas Scaler component is automatically added. The Canvas Scaler can work in one of three modes—*Constant Pixel Size*, *Scale With Screen Size*, and *Constant Physical Size*:

Constant Pixel Size
: The default mode. In this mode, the Canvas won't scale based on screen size or density.

Scale With Screen Size
: This mode makes the Canvas scale its contents based on its size compared to a "reference resolution," which you specify in the Inspector. For example, if you set the reference resolution to 640 by 480, and then play the game on a device that happens to be 1280 by 960, then every UI element will be scaled by a factor of 2.

Constant Physical Size
: This mode makes the Canvas scale its contents based on the DPI (dots per inch) reported by the device the game is on, if this is available.

In our experience, we've found that the Scale With Screen Size mode is the most useful in the majority of situations.

Transitioning Between Screens

Most game GUIs can be divided into two types: menu and in-game. The menu GUI is what the player interacts with in order to prepare the game for play—that is, choosing to start a new game or continue a previous one, configuring settings, or browsing for a multiplayer game to join. The in-game GUI is overlaid on top of the player's view of the game world.

In-game GUIs tend to not change their structure very much, and usually contain readouts on important information: how many arrows are in the player's quiver, how many hit-points they have, and the distance to the next objective. Menus, however, tend to change significantly; the main menu will usually be very different in appearance from the settings screen, because they have different structural requirements.

Because your GUI is just an object that's rendered by the camera, Unity doesn't really have the concept of a "screen" of content. There's just the current collection of objects that are present in the canvas. If you want to be able to move from one screen to another, you need to do one of two things: change the canvas that the camera is currently looking at, or move the camera to look at something different.

Changing the canvas works well for when you want to change a subset of the GUI elements. For example, if you want to keep most of the decorative GUI elements visible, but swap out a portion of the GUI, then it can make more sense to not adjust the camera and to make changes to the canvas. However, if you're doing a complete replacement of GUI elements, adjusting the position of the camera can be more effective.

One thing that's important to keep in mind is that moving the camera separately from the canvas requires that the canvas mode be set to World Space; in both Screen Space - Overlay and Screen Space - Camera, the UI always appears directly in front of the camera.

Wrapping Up

As we've seen, the Unity GUI system is extensive and powerful. You can use it in a variety of different ways, and in a variety of different contexts; additionally, its flexible design allows you to build exactly the GUI that you need.

It's important to remember that the UI for your game is one of the most important components. The UI is how your user works with your game, and on mobile devices, it's a fundamental part of your game's controls. Be prepared to spend a lot of time polishing and refining it.

CHAPTER 16

Editor Extensions

Building games in Unity means working with a lot of game objects, and dealing with all of the components that those game objects are composed of. The Inspector in Unity already takes care of a lot: by automatically exposing all of the variables in your scripts as easy-to-use text fields, checkboxes, and slots for dropping in assets and scene objects, the process of assembling a scene is made a lot faster.

However, sometimes the Inspector isn't enough of a solution. Unity was designed to make it as easy as possible to build things like 2D and 3D environments, but the developers of Unity can't possibly predict all of the things that will go in your game.

Custom editors allow you to take control of the editor itself. This can range from very small add-on windows that let you automate common tasks in the editor, all the way up to completely overriding Unity's Inspector.

When you're creating a game more complex than the games we've built, we've found that it can be incredibly time-saving to write tools for yourself to automate repetitive tasks. That's not to say that your main task as a game developer should be writing software to help make your game—your main task is to make your game! However, if you find yourself doing something repetitious or difficult to do with the existing Unity features, consider writing an editor extension to take care of it for you.

This chapter goes behind the scenes of Unity somewhat. In fact, we'll be using classes and code that the Unity editor itself uses. As a result, the code in this can get a little more complex and trickier than the earlier code we've been writing.

There are several ways that you can extend Unity. In this chapter, we'll look at four of them, each of which is slightly more complex and powerful than the last:

- Custom wizards give you a simple way to ask for input and perform some action in the scene, such as creating a complex object.
- Custom editor windows allow you to create your own windows and tabs, which can contain whatever controls you need to have.
- Custom property drawers let you create a custom user interface for your own types of data in the Inspector.
- Custom editors let you completely override the Inspector for an object.

To get started working through the examples in this chapter, it will help to be working with a new project:

1. *Create a new project* called "Editor Extensions". Make it a 3D project, and save it wherever you like (Figure 16-1).

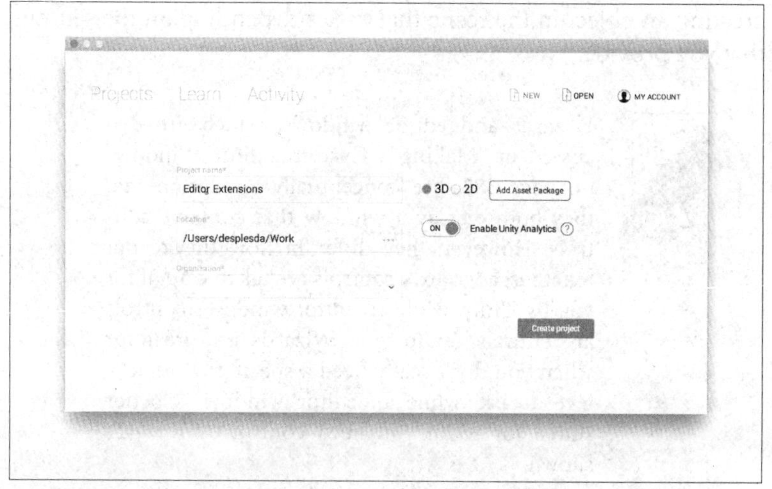

Figure 16-1. Creating a new project

2. *After Unity has loaded, create a new folder within the Assets folder.* Name this new folder *Editor*. You'll be putting your editor extension scripts in here.

Note that it's very important that the folder be named *Editor*, with the correct spelling and capitalization. Unity will be specifically looking for a folder with this name.

This folder can actually be anywhere—it doesn't need to be a direct child of the *Assets* folder, it just needs to be called *Editor*. This is useful, because it means that in a larger project, you can have multiple *Editor* folders throughout your project, which can make it easier to deal with having lots of scripts.

With that done, you're ready to start making your own custom editor scripts!

Making a Custom Wizard

We'll start by creating a custom wizard. Wizards are a simple way to show a window that lets you get input from the user and then use that to do something in the scene. A very common example is creating an object in the scene that varies depending on the settings that you provide.

Wizards and editor windows, which are discussed in "Making a Custom Editor Window" on page 382, are conceptually similar in that they both display a window that contains controls. However, they differ in how they're put together; a wizard's controls are taken care of for you by Unity, while an editor window's controls are entirely up to you. Wizards are great for when you don't really need a specific UI to achieve a task, while an editor window is better suited for when you need control over what's shown.

The best way to understand how wizards can help you in your day-to-day use of Unity is to make one. We'll create a wizard that creates game objects that show a tetrahedron—a triangular pyramid, seen in Figure 16-2—in the scene.

Figure 16-2. A tetrahedron created by the wizard

Creating objects like this involves manually creating a `Mesh` object. Usually, these objects are imported from a file, such as the *.blend* file used in Chapter 9; however, you can also create one in code.

With a `Mesh`, you can create an object that renders that mesh. To do this, you first create a new `GameObject`, and then attach two components: a `MeshRenderer` and a `MeshFilter`. Once that's done, the object is ready to use in the scene.

These steps are very easy to automate, which means that they're perfect for a wizard:

1. *Create a new C# script* called *Tetrahedron.cs* in the *Editor* folder, and add the following code to it:

```
using UnityEditor;

public class Tetrahedron : ScriptableWizard {

}
```

The `ScriptableWizard` class defines the base behavior of the wizard. We'll implement some methods that override this behavior, and get us to the good stuff.

To start, we'll implement a method that displays the wizard. This involves two tasks: first, a menu item needs to be added to Unity's menu, which the user will use to call the method; second, inside this method, we need to instruct Unity to display the wizard.

2. *Add the following code to the* `Tetrahedron` *class:*

```
using UnityEngine;
using System.Collections;
using System.Collections.Generic;

    // This method can be called anything - the important thing
    // is that it's static and has the MenuItem attribute
    [MenuItem("GameObject/3D Object/Tetrahedron")]
    static void ShowWizard() {
      // First parameter is title, second is the label on
      // the Create button
      ScriptableWizard.DisplayWizard<Tetrahedron>(
        "Create Tetrahedron", "Create");
    }
```

The `MenuItem` attribute, when attached to a `static` method, makes Unity add an entry to the application menu. In this case, it creates a new entry in the GameObject → 3D Object menu called "Tetrahedron"; when this menu item is selected, the `ShowWizard` method is called.

The method doesn't actually need to be called `ShowWizard`. You can call it whatever you like—Unity is only looking for the `MenuItem` attribute.

3. *Return to Unity,* and open the GameObject menu. Choose 3D Object → Tetrahedron, and an empty wizard window will appear (Figure 16-3).

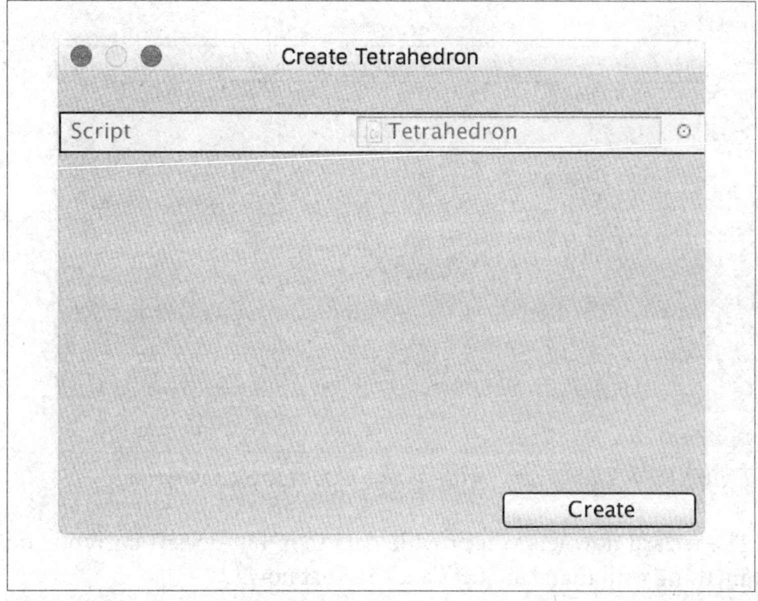

Figure 16-3. An empty wizard window

Next, we'll add a variable to the wizard's class. Doing this will make Unity show the appropriate control for this variable in the wizard's window, just like it does in the Inspector. This will be a `Vector3`, which represents the height, width, and depth of the object.

4. *Add the following variable to the `Tetrahedron`, to represent the size of the tetrahedron:*

```
// This variable will appear just like it would in the
// Inspector
public Vector3 size = new Vector3(1,1,1);
```

5. *Return to Unity.* Close and reopen the Wizard, and you'll see a slot for the Size variable (Figure 16-4).

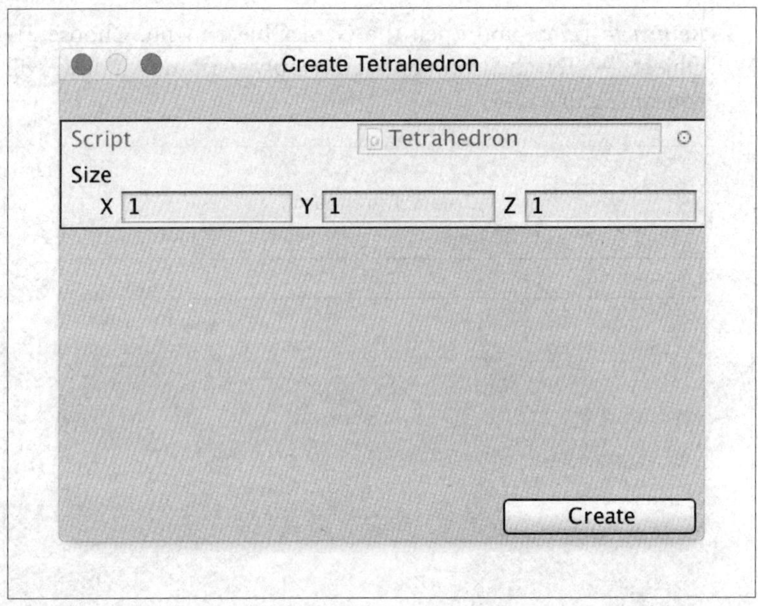

Figure 16-4. The wizard, with a control for the Size variable

The wizard now lets you provide data to it, but doesn't currently do anything with that data. Let's address that now!

When you call the `DisplayWizard` method, you provide two strings. The first is the title of the menu, and the second is the text that should appear in the wizard's Create button. When this button is tapped, your wizard's class will receive a call to the `OnWizardCreate` method, which indicates that the user has finished providing information to the wizard; after the `OnWizardCreate` method returns, Unity will close the window.

We'll now implement the `OnWizardCreate` method, which will do the bulk of the actual work of the wizard. It will create the `Mesh` that uses the Size variable, and construct a game object that renders that mesh.

6. *Add the following method to the `Tetrahedron` class:*

```
// Called when the user clicks the Create button
void OnWizardCreate() {

  // Create a mesh
  var mesh = new Mesh();
```

```
        // Create the four points
        Vector3 p0 = new Vector3(0,0,0);
        Vector3 p1 = new Vector3(1,0,0);
        Vector3 p2 = new Vector3(0.5f,
                                 0,
                                 Mathf.Sqrt(0.75f));
        Vector3 p3 = new Vector3(0.5f,
                                 Mathf.Sqrt(0.75f),
                                 Mathf.Sqrt(0.75f)/3);

        // Scale them based on size
        p0.Scale(size);
        p1.Scale(size);
        p2.Scale(size);
        p3.Scale(size);

        // Provide the list of vertices
        mesh.vertices = new Vector3[] {p0,p1,p2,p3};

        // Provide the list of triangles that connect each of
        // the vertices
        mesh.triangles = new int[] {
          0,1,2,
          0,2,3,
          2,1,3,
          0,3,1
        };

        // Update some additional data on the mesh, using this
        // data
        mesh.RecalculateNormals();
        mesh.RecalculateBounds();

        // Create a game object that uses this mesh
        var gameObject = new GameObject("Tetrahedron");
        var meshFilter = gameObject.AddComponent<MeshFilter>();
        meshFilter.mesh = mesh;

        var meshRenderer
          = gameObject.AddComponent<MeshRenderer>();
        meshRenderer.material
          = new Material(Shader.Find("Standard"));

    }
```

This method works by first creating a new `Mesh` object, and then figuring out the locations of the four points that make up the tetrahedron. These are then scaled based on the `size` vector, which means that they're repositioned such that they're in the right location to make up a tetrahedron of `size`'s width, height, and depth.

These points are provided to the `Mesh` via its `vertices` property; once that's done, a list of triangles is provided by providing a list of numbers. Each number represents one of the points provided to the `vertices`.

For example, in the triangle list, `0` refers to the first point, `1` refers to the second, and so on. The triangle list uses groups of three numbers to define a triangle; so, for example, the numbers `0, 1, 2` mean that the mesh will contain a triangle made up of the first, second, and third point in the `vertices` list. A tetrahedron is made of four triangles: the base, and the three sides. As a result, the `triangles` list is made up of four groups of three numbers.

Finally, the mesh is told to recalculate some internal information, based on the `vertices` and `triangles` data it now contains. It's then ready for use in the scene: a new `GameObject` is created, a `MeshFilter` is attached and is given the `Mesh` we just built, and a `MeshRenderer` is attached to actually show the `Mesh`. Finally, the `MeshRenderer` is given a new `Material`, which is created using the Standard shader—just like all the other built-in objects you create via the GameObject menu do.

7. *Return to Unity, and close and reopen the wizard window.* When you click the Create button, a new tetrahedron will be added to the scene. If you change the Size variable, the tetrahedron's dimensions will be different.

There's one last feature to add to the wizard. Currently, the wizard doesn't check to see if the Size variable has been set to something reasonable; for example, the wizard should refuse to create a tetrahedron that's *negative two* units in height.

Strictly speaking, this would actually can be totally fine, since Unity can handle it. However, it's useful to know how to do this kind of input validation.

For this example, we'll make the wizard refuse to create tetrahedrons when any of the Size variable's components—X, Y, or Z—have a value of zero or less.

We'll do this by implementing the `OnWizardUpdate` method, which is called every time the user makes a change to any of the variables in the wizard. This gives you a chance to check the values, and enable or disable the Create button. Importantly, you can add explanation text that tells the user *why* the wizard is refusing the input.

8. *Add the following method to the `Tetrahedron` class:*

```
// Called whenever the user changes anything in the wizard
void OnWizardUpdate() {

  // Check to make sure that the values provided are
  // valid
  if (this.size.x <= 0 ||
    this.size.y <= 0 ||
    this.size.z <= 0) {

    // When isValid is true, the Create button can
    // be clicked
    this.isValid = false;

    // Explain why this is the case
    this.errorString
      = "Size cannot be less than zero";

  } else {

    // The user can click create, so enable it and
    // clear any error message
    this.errorString = null;
    this.isValid = true;
  }
}
```

When the `isValid` property is set to `false`, the Create button will be disabled, meaning that the user cannot click it. Additionally, if you set the `errorString` property to anything besides `null`, an error message will appear in the window. You can use this to explain to the user what the problem is (Figure 16-5).

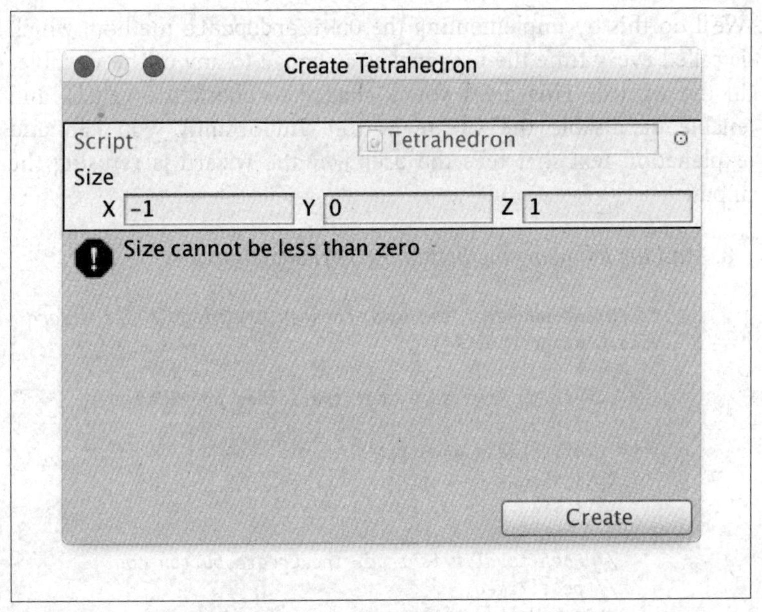

Figure 16-5. The wizard, showing an error

Wizards can let you save a lot of time in doing repetitive work, or work that's difficult to accomplish in the Unity editor alone. They're quick to code, because the Unity editor takes care of the majority of the user interface for you. However, sometimes you'll need more than what the wizard system can provide; next, we'll look at entirely custom editor windows.

Making a Custom Editor Window

A *window* is what Unity calls a region that can be either a separate, floating window, or a tab that's docked to a part of the main Unity editor's interface.

Just about every single part of Unity that you can see is an editor window.

When you create an editor window, you have complete control over its contents. This is quite different to how wizards and the Inspector work, where Unity will automatically draw the user interface for you; in an editor window, nothing will appear unless you specifically tell it to. This gives you significant power, since you can add entirely new features to Unity that are specific to your needs.

In this section, we'll create a new editor window that simply counts the number of textures in the project. Before we can get to this functionality, though, we need to learn how to draw *anything* in an editor window.

First, let's create a new empty editor window.

1. *Create a new script* called *TextureCounter.cs*, and put it in the *Editor* folder.
2. *Open it*, and replace the contents of the file with the following code:

```
using UnityEngine;
using System.Collections;
using UnityEditor;

public class TextureCounter : EditorWindow {

  [MenuItem("Window/Texture Counter")]
  public static void Init() {
    var window = EditorWindow
      .GetWindow<TextureCounter>("Texture Counter");
    // Stops this window from being unloaded when a
    // new scene is loaded
    DontDestroyOnLoad(window);
  }

  private void OnGUI() {
    // Editor GUI goes here
    EditorGUILayout.LabelField("Current selected size is "
      + sizes[selectedSizeIndex]);
  }

}
```

This code adds a new menu item to the Window menu, which creates and displays a new window that uses the `TextureCounter` class. It also marks this window as something that should not be unloaded by Unity if the current scene changes.

3. *Save this file,* and go back to Unity.

4. *Open the Window menu,* and you'll see a "Texture Counter" menu item. Click it, and an empty window will appear!

Now that we've got the empty window going, it's time to start adding controls to it. To do that, we need to know about how to work with the Editor GUI system.

The Editor GUI API

The GUI system used by the editor is quite different from the GUI system that you use to build your game.

In the game's GUI system (we'll call it *Unity GUI*), you create game objects that represent things like text labels and buttons, and you position them in your scene.

In the GUI system used to create editor GUI (we'll call it the *immediate mode GUI*, for reasons that we'll get to in a moment), you call special functions that cause a label or button to appear at a certain point; these functions are called by Unity repeatedly, every time Unity needs to redraw the screen.

The term *immediate mode* refers to the fact that the act of calling these special GUI functions causes a button to be shown on screen immediately; the screen will then later be cleared, removing the button from the screen along with everything else, and then the GUI function will be called again on the next frame. This process repeats forever.

For efficiency, Unity doesn't continuously call these editor GUI functions. Instead, it only does so when potentially needed: when the user clicks the mouse or types a key, when the window containing the GUI content resizes, or other screen-related events.

The other main difference between the immediate mode and Unity GUI systems is the way that layout works. In the Unity GUI, objects are positioned relative to their parent objects, as well as to their anchor; in the immediate mode GUI, you either provide a specific rectangle that describes the position and size of the thing you want to draw, or you make use of a managed layout system called `GUILayout`, which we'll describe in a moment.

The best way to explain this difference is through an example. For the next several pages, we'll go through the fundamentals of how to use the GUI system, as well as the different controls available to you.

Rects and layout

The simplest possible control to add in the window is a simple text label. We'll add some code that does the task, and then we'll explain it.

1. *Add the following code to the `OnGUI` method:*

```
GUI.Label(                          ❶
        new Rect(50,50,100,20),     ❷
        "This is a label!"          ❸
);
```

2. *Return to Unity, and open the editor window.* You'll now see the text "This is a label" appear in the window (Figure 16-6).

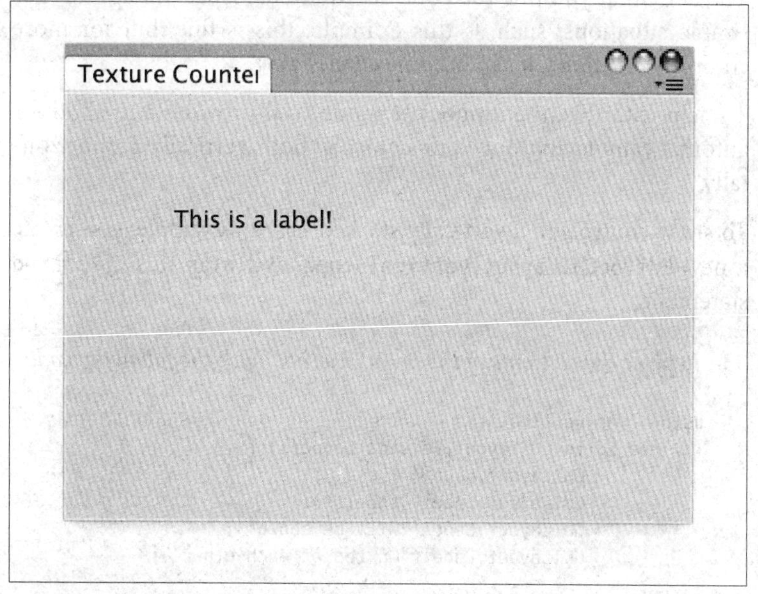

Figure 16-6. A manually placed label in the editor window

Let's walk through this code:

❶ Calls the `Label` method of the `GUI` class, which will end up displaying the text in the window.

❷ Creates a new `Rect`, which defines the x position, y position, width, and height of the label. In this case, it's positioned 50 pixels from the top and the left, is 100 pixels wide, and is 20 pixels high.

❸ Provides the actual text that should appear in the label: "This is a label!"

This code is run every time Unity needs to update what's being shown in the window. When the `GUI.Label` method is called, the label is added to the window.

Every call to a `GUI` function must take place inside the `OnGUI` method. You'l get problems if you call `GUI.Label` from anywhere else.

The `Rect` that you provide controls where the label will appear. For simple situations, such as this example, this is fine, but for more complex situations, it can become challenging.

To help with this, the immediate mode GUI provides a method for automatically laying out your controls, both vertically or horizontally.

To show controls in a vertically stacked list, for example, you create a new `EditorGUILayout.VerticalScope`, and wrap this in a `using` statement.

3. *Replace the contents of the `OnGUI` method with the following code:*

```
using (var verticalArea
  = new EditorGUILayout.VerticalScope()) {
          GUILayout.Label("These");
          GUILayout.Label("Labels");
          GUILayout.Label("Will be shown");
          GUILayout.Label("On top of each other");
}
```

There are two main differences about the labels in this example.

First, you'll notice that the `Label` method being called comes from the `GUILayout` class, not the `GUI` class. This version of the label is able to make use of the fact that they're being called within the context of the `VerticalScope`, and position themselves correctly.

Second, you don't need to provide a `Rect` to define their position and size. They'll make use of the `VerticalScope` to determine that.

Using the layout system like this is much faster, and leads to a much better experience for you as a programmer. As a result, we'll be using the layout system for almost all of the rest of this chapter.

The one exception to this is property drawers, where the GUI layout system doesn't work. In this section, we'll fall back to manually laying out controls and specifying their rectangles.

How controls work

As mentioned earlier, a control in the immediate GUI system is a function call. This is easy to understand for simple controls like labels, but for controls where the user can provide input, such as buttons and text fields, it might get a little complex.

Given that a control is the result of calling a function, how is it possible to get back any information from the user? The answer is actually quite clever: the functions for displaying a control also *return* information back to their caller.

Again, the best way to explain this is through an example.

Buttons

Let's start by creating a button using the immediate GUI system:

1. *Replace the `OnGUI` method with the following code:*

```
private void OnGUI() {
    using (var verticalArea
      = new EditorGUILayout.VerticalScope()) {
        var buttonClicked = GUILayout.Button("Click me!");
        if (buttonClicked) {
          Debug.Log("The custom window's " +
            "button was clicked!");
        }
    }
}
```

When the `GUILayout.Button` method is called, two things happen. A button appears on screen; additionally, if a mouse click just finished in this area, the method returns `true`.

This system works because `OnGUI` is called repeatedly. When the window first appears, the call to `Button` causes the button to appear on screen. When the user moves the mouse over the button and holds down the mouse button, `OnGUI` is again called, and the GUI system draws the button in a "down" state. When the user lifts the mouse button, `OnGUI` is called once more; because a click has completed, this third call to `Button` returns true.

In practical terms, you can think of this style of programming like this: `GUILayout.Button` simultaneously draws a button on the screen, and returns `true` if the user clicked it.

2. *Return to Unity,* and note that a button now appears. When you click that button, "The custom window's button was clicked!" appears in the Console tab.

Yes, this is a bit weird. However, ¯_(ツ)_/¯.

Text fields

A button is the simplest possible type of control that the user can provide information through—the user is either clicking the button, or not. However, there are more complex types of controls that the GUI system supports. For example, a text field has two tasks: showing some text to the user, *and* letting the user edit that text.

The method that you call to display a text field is `EditorGUILayout.TextField`. When you call this method, you provide a string that should be displayed in the text field; the method then returns what the user has *entered* in the text field, which may be different.

For this to work, the variable that you store the text in must not be a local variable. That is, the following code will *not* work correctly:

```
private void OnGUI() {
  using (var verticalArea
    = new EditorGUILayout.VerticalScope()) {
          string textValue = "";

          textValue
            = EditorGUILayout.TextField(textValue);
```

```
    }
  }
```

TextField is inside the EditorGUILayout class, not the GUILayout class. GUILayout *does* include a TextField method, but it doesn't have quite the same functionality.

If you test this code in Unity, it will let you type in it, but when you leave the text field, it will reset to the empty string.

To do this correctly, the variable that you store the text in must be a variable that belongs to the class:

```
private string stringValue;
private void OnGUI() {
  using (var verticalArea
    = new EditorGUILayout.VerticalScope()) {

    this.stringValue
      = EditorGUILayout.TextField(this.stringValue);
  }
}
```

This works because stringValue's contents are preserved between the different calls to OnGUI.

The TextField control displays a single line of text. If you want to display multiple lines of text, use a TextArea:

```
this.stringValue = EditorGUILayout.TextArea(
  this.stringValue,
  GUILayout.Height(80)
);
```

You'll notice that because these two controls are working with the same variable, they'll show the same text—additionally, when you make changes to one, you'll automatically make changes to the other. It's pretty cool.

In this previous example, the height of the text area was overridden by providing a GUILayout option. This can be added to *any* control; if you need a tall button, you can just add a call to GUILayout.Height(80) to any button, and it will be 80 pixels high.

Delayed text fields. An additional type of text field is the *delayed text field*. These work like regular text fields, except that the value that they return doesn't change away from the original value you put in, until they lose focus—that is, the user moves to a different text field or clicks something else.

This is useful for situations where you need to do some validation on the data the user has entered, but it doesn't make sense to do so until the user indicates that they're done typing.

You create a delayed text field using the `DelayedTextField` method, like so:

```
this.stringValue
  = EditorGUILayout.DelayedTextField(this.stringValue);
```

Special text fields. In addition to handling regular text, text fields can also be used for numbers. In particular, there are four very useful variants on the `TextField` control: integer fields, float fields, `Vector2D` fields, and `Vector3D` fields.

For example, given these backing fields in your class:

```
private int intValue;

private float floatValue;

private Vector2 vector2DValue;

private Vector3 vector3DValue;
```

You can create fields that provide data to them:

```
this.intValue
  = EditorGUILayout.IntField("Int", this.intValue);

this.floatValue
  = EditorGUILayout.FloatField("Float", this.floatValue);

this.vector2DValue
  = EditorGUILayout.Vector2Field("Vector 2D",
                                 this.vector2DValue);
this.vector3DValue
  = EditorGUILayout.Vector3Field("Vector 3D",
                                 this.vector3DValue);
```

Note the strings used for the first parameter: if you provide this, then a label will appear before the text field.

Sliders

In addition to using text fields for numeric input, you can also provide a graphical slider. For example, you can use an `IntSlider` like so:

```
var minIntValue = 0;
var maxIntValue = 10;
this.intValue
  = EditorGUILayout.IntSlider(this.intValue,
                              minIntValue,
                              maxIntValue);
```

Sliders are especially useful when combined with an `IntField` or `FloatField` control that uses the same variable, since you can use the slider to quickly set a value, but if you need to set a very specific value, you can just type it in.

You can also use min-max sliders, which let you present a way to define a minimum and maximum value. For example, given the two class variables used to store the minimum and maximum range:

```
private float minFloatValue;
private float maxFloatValue;
```

You can draw a min-max slider using the `MinMaxSlider` method:

```
var minLimit = 0;
var maxLimit = 10;
EditorGUILayout.MinMaxSlider(ref minFloatValue,
                             ref maxFloatValue,
                             minLimit,
                             maxLimit);
```

Note that this method doesn't return a value; instead, it modifies the `minFloatValue` and `maxFloatValue` variables that you pass in. Additionally, the `minLimit` and `maxLimit` values limit the minimum and maximum values that both `minFloatValue` and `maxFloatValue` can be set to.

Space

The `Space` control is entirely invisible, and simply adds space to your UI. It's useful for visually breaking up your controls into different groups:

```
EditorGUILayout.Space();
```

Lists

So far, all of the controls that we've discussed that allow for user input are quite open-ended: the user can enter whatever text or number they like. However, you'll sometimes encounter situations where you want the user to choose from a list of predefined options.

To support this, you can use a Popup. A Popup works using an array of string options, and an integer that represents the current selection from that array; when the user changes the current selection, the current selection number changes.

For example, if you add this variable to your class:

```
private int selectedSizeIndex = 0;
```

And then add this code to your `OnGUI` method:

```
var sizes = new string[] {"small","medium","large"};

selectedSizeIndex
  = EditorGUILayout.Popup(selectedSizeIndex, sizes);
```

However, it can be annoying to have to remember the association between the numbers stored in `selectedSizeIndex` and the values that they represent. A better way to do this is with *enumerations*, which are also known as *enums*.

Enums are better because they're checked by the compiler—in the previous example, you'd need to remember that "0" means "small," but it would be nicer to simply say `Small`. Enums let you do this!

Let's define an enum that defines a few different types of damage; we'll also add a variable that stores the currently selected damage type.

1. *Add the following code to your `TextureCounter` class:*

```
private enum DamageType {
  Fire,
  Frost,
```

```
        Electric,
        Shadow
    }

    private DamageType damageType;
```

Using this enum and the `damageType` variable, we can now create a Popup that shows values from this list.

2. *Add the following code to the `OnGUI` method:*

```
damageType
    = (DamageType)EditorGUILayout.EnumPopup(damageType);
```

Doing this will show a Popup containing all of the possible values representable by the `DamageType` enum, set to the currently selected value of the `damageType` variable.

You need to cast to the correct enum type, because the `EnumPopup` method doesn't know what type of enum it's using.

Scroll views

If you've been adding all of the different controls presented so far in this chapter, you might notice that the controls are starting to spill beyond the bounds of the editor window. To solve this, you can use *scroll views* to let the user scroll around.

A scroll view needs to keep track of its scroll position. As a result, you need to create a variable to store the scroll position, just like you do with other controls.

1. *Add the following variable to the `TextureCounter` class:*

```
private Vector2 scrollPosition;
```

You create a scrolling view in a very similar way as creating a vertical list: you create a new `EditorGUILayout.ScrollViewScope`, inside a `using` statement.

2. *Add the following code to your `OnGUI` method:*

```
using (var scrollView =
    new EditorGUILayout.ScrollViewScope(this.scrollPosition)) {
```

```
    this.scrollPosition = scrollView.scrollPosition;

    GUILayout.Label("These");
    GUILayout.Label("Labels");
    GUILayout.Label("Will be shown");
    GUILayout.Label("On top of each other");
  }
```

3. *Return to Unity,* and the labels will be contained within a scrolling area. You may need to resize the window to see the effect.

The Asset Database

To wrap up our discussion of editor windows, we'll return to the goal of the `TextureCounter` window: we'll make it count the number of textures in the project, and show it in a label.

To do this, we'll use the `AssetDatabase` class. This class acts as your gateway to all of the assets currently in the project, and you can use it to get information about and make changes to all files under Unity's control.

We don't have space to discuss all of the different things that the `AssetDatabase` class can do; instead, we strongly recommend that you check out the Unity manual's page on `AssetDatabase` (*http://docs.unity3d.com/Manual/AssetDatabase.html*).

1. *Replace the `OnGUI` method in `TextureCounter` with the following code:*

```
private void OnGUI() {
  using (var vertical = new EditorGUILayout.VerticalScope()) {
    // Get the list of all textures
    var paths = AssetDatabase.FindAssets("t:texture");

    // Get the count
    var count = paths.Length;

    // Show a label
    EditorGUILayout.LabelField("Texture Count",
      count.ToString());

  }
```

```
}
```

2. *Return to Unity, and add some images to your project.* It doesn't matter what images they are—drag any files in. If you're stuck for ideas, go to Flickr (*https://flickr.com/*) and search for "cats."

The editor window will now display the number of textures that you added.

Making a Custom Property Drawer

In addition to creating entirely custom editor windows, you can also extend the behavior of the Inspector window.

The role of the Inspector is to provide a user interface for configuring each of the components attached to the currently selected game object. For each component, the Inspector shows a control that represents each of its variables.

The Inspector already knows how to present the appropriate controls for common types, like strings, integers, and floats. However, if you define a custom type, the Inspector won't necessarily know how to present it correctly. This is usually fine, but it can get cluttered.

This is where *property drawers* come in. You can provide code to Unity that determines how different types of data should appear to the user.

The GUI layout system doesn't work inside custom property drawers. Instead, you'll need to manually lay out your controls. Don't worry—it's not as scary as it sounds, and we'll be doing this in the example code.

To demonstrate this, we'll create a custom class that represents a range of values, which can then be used in any script. We'll then define a custom property drawer for this custom class. To do so, follow these steps:

1. *Create a new C# script* called *Range.cs*, and put it in the *Assets* folder.
2. *Add the following code to Range.cs:*

```
[System.Serializable]
public class Range  {

  public float minLimit = 0;
  public float maxLimit = 10;

  public float min;
  public float max;

}
```

The `System.Serializable` attribute marks this class as able to be saved to disk. This also indicates to Unity that its values should appear in the Inspector.

3. *Create a second C# class* called *RangeTest*, and put it in the *Assets* folder as well. This will be a simple script component that uses a `Range`. Add the following code to *RangeTest.cs*:

   ```
   public class RangeTest : MonoBehaviour {

     public Range range;

   }
   ```

4. *Create an empty game object* and drag the `RangeTest` script onto it.

When the game object is selected, the Inspector will show the raw values (Figure 16-7).

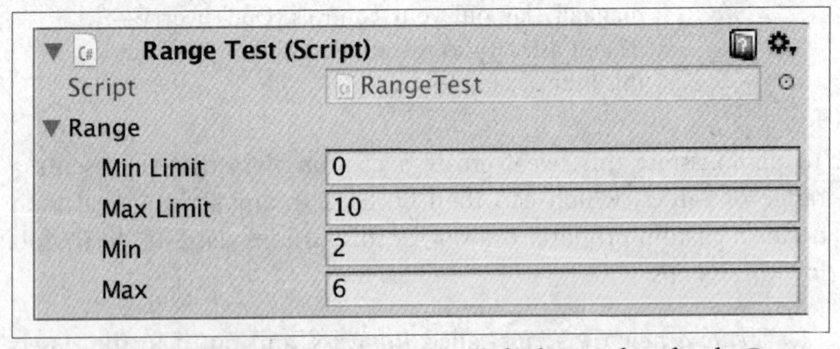

Figure 16-7. The Inspector, showing the default interface for the Range class

To override this, we'll implement a new class that replaces the default interface that Unity provides.

5. *Create a new script* called *RangeEditor.cs*, and place it in the *Editor* folder.
6. *Replace the contents of RangeEditor.cs with the following code:*

```
using UnityEngine;
using System.Collections;

using UnityEditor;

[CustomPropertyDrawer(typeof(Range))]
public class RangeEditor : PropertyDrawer {

  // This property drawer will be two lines high - one for the
  // slider, and one for the text fields that let you change
  // the values directly
  const int LINE_COUNT = 2;

  public override float GetPropertyHeight (
    SerializedProperty property, GUIContent label)
  {
    // Return the number of pixels of height that
    // this property takes up
    return base.GetPropertyHeight (property, label)
      * LINE_COUNT;
  }

  public override void OnGUI (Rect position,
    SerializedProperty property, GUIContent label)
  {

    // Get the objects that represent the fields inside this
    // Range property
    var minProperty = property.FindPropertyRelative("min");
    var maxProperty = property.FindPropertyRelative("max");

    var minLimitProperty
      = property.FindPropertyRelative("minLimit");
    var maxLimitProperty
      = property.FindPropertyRelative("maxLimit");

    // Any controls inside the PropertyScope will work
    // correctly with prefabs - values that have been
    // changed from the prefab will be bold, and you can
    // right-click on a value and choose to reset it back
    // to the prefab
    using (var propertyScope
```

```
        = new EditorGUI.PropertyScope(
            position, label, property)) {

                // Show the label; this method returns a rect
                // that stuff next to it can contain
                Rect sliderRect
                    = EditorGUI.PrefixLabel(position, label);

                // Construct rectangles for each of the controls:

                // Calculate how big a single line is
                var lineHeight = position.height / LINE_COUNT;

                // The slider needs to be one line high
                sliderRect.height = lineHeight;

                // The area for the two fields is the same shape
                // as the slider, but shifted down one line
                var valuesRect = sliderRect;
                valuesRect.y += sliderRect.height;

                // Work out rects for the two text fields
                var minValueRect = valuesRect;
                minValueRect.width /= 2.0f;

                var maxValueRect = valuesRect;
                maxValueRect.width /= 2.0f;
                maxValueRect.x += minValueRect.width;

                // Get the float values out
                var minValue = minProperty.floatValue;
                var maxValue = maxProperty.floatValue;

                // Start a change check - we do this to
                // correctly support multi-object editing
                EditorGUI.BeginChangeCheck();

                // Show the slider
                EditorGUI.MinMaxSlider(
                    sliderRect,
                    ref minValue,
                    ref maxValue,
                    minLimitProperty.floatValue,
                    maxLimitProperty.floatValue
                );

                // Show the fields
                minValue
                    = EditorGUI.FloatField(minValueRect, minValue);
                maxValue
                    = EditorGUI.FloatField(maxValueRect, maxValue);
```

```
            // Was a value changed?
            var valueWasChanged = EditorGUI.EndChangeCheck();

            if (valueWasChanged) {
                // Store the modified values
                minProperty.floatValue = minValue;
                maxProperty.floatValue = maxValue;
            }
        }

    }
}
```

This is a large piece of code, so we'll step through it in chunks.

Creating the Class

First, we need to define the class, and indicate to Unity that it should be used for drawing the interface for any `Range` property that the Inspector encounters. We do this using the `CustomPropertyDrawer` attribute, which takes as a parameter the `Range` class type.

Additionally, the `RangeEditor`'s superclass is set to `PropertyDrawer`.

```
[CustomPropertyDrawer(typeof(Range))]
public class RangeEditor : PropertyDrawer {
```

Setting the Height of the Property

A property takes up a certain amount of vertical space in the Inspector. By default, this amount is around 20 pixels; however, range properties will need more space, because we want to draw the range slider, as well as two text fields underneath it.

The `GetPropertyHeight` method is responsible for returning the height of the property, in pixels. You can override this method to change this height.

Rather than hardcode a certain value in, which might change between different versions of Unity, we define the number of lines that we want to have as a constant called `LINE_COUNT`; we then call the `base` implementation to get the size of one line, and then multiply it by `LINE_COUNT`.

```
// This property drawer will be two lines high - one for the
// slider, and one for the text fields that let you change
// the values directly
```

```
const int LINE_COUNT = 2;

public override float GetPropertyHeight (
  SerializedProperty property, GUIContent label)
{
  // Return the number of pixels of height that this
  // property takes up
  return base.GetPropertyHeight (
    property, label) * LINE_COUNT;
}
```

Overriding OnGUI

It's now time to start implementing the main method in this class: `OnGUI`. For property drawers, this method takes three parameters:

- The `position` parameter is a `Rect` that defines the position and size of available area that the `OnGUI` method has to draw its controls.
- The `property` parameter is a `SerializedProperty` object, which is the way you interact with the specific `Range` property of the component that this particular instance of the class provides.
- The `label` parameter is a `GUIContent` object that represents some graphical content—usually some text—that should appear as the label for this property.

```
public override void OnGUI (Rect position,
  SerializedProperty property, GUIContent label)
{
```

Getting the Properties

A property drawer's job is to present and modify a single property inside a component. You don't directly modify the component itself; instead, the `property` parameter mediates your access. Doing this means that Unity can provide additional functionality, such as automatic support for undo.

In the case of the `Range` object, it's a property that contains *other* properties. The `min`, `max`, `minLimit`, and `maxLimit` variables are all themselves properties, so we need to access them:

```
// Get the objects that represent the fields inside this
// Range property
```

```
var minProperty = property.FindPropertyRelative("min");
var maxProperty = property.FindPropertyRelative("max");

var minLimitProperty
  = property.FindPropertyRelative("minLimit");
var maxLimitProperty
  = property.FindPropertyRelative("maxLimit");
```

Creating a property scope

In addition to getting the objects that represent these properties, we need to indicate to the GUI system that the controls that we're drawing relate to a specific property.

Doing this means that Unity is able to customize the appearance of the controls when needed; some important examples of this include when the object that the property belongs to is a modified instance of a prefab, in which case the property should appear as bold; additionally, when you right-click on a modified property, Unity will open a menu that lets you revert its value back to the prefab.

To support all of this, we wrap all of the controls inside a `PropertyScope`:

```
using (var propertyScope
  = new EditorGUI.PropertyScope(position, label, property)) {
```

Drawing the Label

We now draw the label, using the `PrefixLabel` control. This control draws the `label` text inside the `position` rectangle; it then returns a new `Rect`, which represents the *remaining area* that controls can be drawn in, next to the label.

Doing this means that the layout of the property will follow the style established by the rest of Unity: properties have their label at the top-left corner, and their fields to the right; the area beneath the label is left empty:

```
Rect sliderRect = EditorGUI.PrefixLabel(position, label);
```

Calculating the Rectangles

Now that we know how much space is available to draw the controls in, we need to start calculating the rectangles for each of the three controls: the slider and the two text fields.

We do this by first calculating the height of a single line, in pixels, by dividing the *total* space by `LINE_COUNT`. We then set `sliderRect`'s `height` to this new `lineHeight`, while leaving its `width` alone. This means that the slider will take up the entire top line.

We can then calculate the rectangles for the two text fields. These will be shown side by side on the line underneath the slider. To calcualte this, we figure out the rectangle that represents the entire second line, and then divide it in half:

```
var lineHeight = position.height / LINE_COUNT;

// The slider needs to be one line high
sliderRect.height = lineHeight;

// The area for the two fields is the same shape as
// the slider, but shifted down one line
var valuesRect = sliderRect;
valuesRect.y += sliderRect.height;

// Work out rects for the two text fields
var minValueRect = valuesRect;
minValueRect.width /= 2.0f;

var maxValueRect = valuesRect;
maxValueRect.width /= 2.0f;
maxValueRect.x += minValueRect.width;
```

Getting the Values

Because the `MinMaxSlider` directly modifies the variables that you pass into it, we need to temporarily store the values of `minProperty` and `maxProperty` into variables. These values will eventually be stored back into the property objects, after being modified by the controls we're about to draw:

```
var minValue = minProperty.floatValue;
var maxValue = maxProperty.floatValue;
```

Creating the Change Check

There's one more bit of setup required before we get into the guts of drawing the controls. We need to ask Unity to tell us if any controls that we're about to draw have had their value changed.

This is an important step, because if we didn't do this, we'd be making changes to the properties every time that we draw the controls, even if the changes weren't applied.

This would usually be fine, except that if *multiple* objects are selected, and they all have a `Range`, then the act of displaying the controls for the `Range` would also change them *all* to a single value, even when the user does nothing at all. Adding a change check prevents this accidental behavior.

```
EditorGUI.BeginChangeCheck();
```

Drawing the Slider

We can finally start drawing controls. We have the data that they need to display, a way to store the results that come back, and the rectangles in which they should appear.

We'll first draw the `MinMaxSlider`:

```
EditorGUI.MinMaxSlider(
  sliderRect,
  ref minValue,
  ref maxValue,
  minLimitProperty.floatValue,
  maxLimitProperty.floatValue
);
```

Drawing the Fields

Next, we'll draw the text fields. Note how we're using the same variables that were passed to the `MinMaxSlider`; doing this means that changing the slider will also update the text fields, and vice versa:

```
minValue = EditorGUI.FloatField(minValueRect, minValue);
maxValue = EditorGUI.FloatField(maxValueRect, maxValue);
```

Checking for Changes

Finally, we can ask Unity if any control was changed since we began the change check. The `EditorGUI.EndChangeCheck` method will return `true` if that's the case:

```
var valueWasChanged = EditorGUI.EndChangeCheck();
```

Storing the Properties

If a control *was* changed, we need to store the new value in the property:

```
if (valueWasChanged) {
  // Store the modified values
  minProperty.floatValue = minValue;
  maxProperty.floatValue = maxValue;
}
```

Testing It Out

With this, we're all done.

Return to Unity, and look at the Inspector. You'll see a custom UI for the Range variable (Figure 16-8).

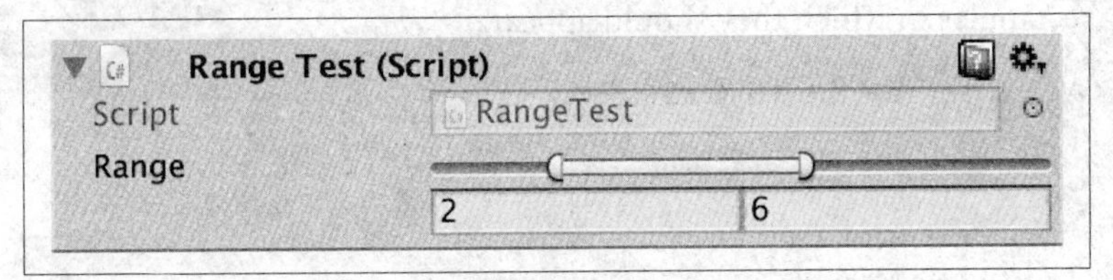

Figure 16-8. The customized property drawer

You may need to de-select and then select the game object for the user interface to update.

With this code written, *any* `Range` property on *any* script will get this custom interface.

Making a Custom Inspector

The last thing we'll discuss in this chapter is creating entirely customized Inspectors. In addition to customizing the appearance of individual properties, you can replace the entire user interface for a component in the Inspector.

We'll look at how you can do this by first creating a simple component, and then creating an entirely new Inspector interface for that component.

Creating a Simple Script

This simple component will change the color of a mesh when the game starts.

1. *Create a new script* called `RuntimeColorChanger`.
2. *Update the* `RuntimeColorChanger` *class to the following code:*

```
public class RuntimeColorChanger : MonoBehaviour {

  public Color color = Color.white;

  void Awake() {
    GetComponent<Renderer>().material.color = color;
  }
}
```

3. *Return to Unity.* Open the GameObject menu, and choose 3D Object → Capsule.
4. *Drag the* `RuntimeColorChanger` *script onto the object.*
5. *Change the* `RuntimeColorChanger`*'s* `Color` *property to red* and hit Play. The capsule will turn red.

Creating a Custom Inspector

So far so good: the script does exactly what we want it to.

We now want to create a custom Inspector that adds a cool feature: we want to have a list of buttons that let us quickly change the color to a predefined color. To do this, we'll create the custom Inspector that adds these buttons.

1. *Create a script* called *RuntimeColorChangerEditor.cs*, and put it in the *Editor* folder.
2. *Replace the contents of RuntimeColorChangerEditor.cs with the following code:*

```
using UnityEngine;
using System.Collections;
using System.Collections.Generic; // needed for Dictionary
using UnityEditor;

// This is an editor for RuntimeColorChangers
[CustomEditor(typeof(RuntimeColorChanger))]
```

```
// It can handle editing multiple things at once
[CanEditMultipleObjects]
class RuntimeColorChangerEditor : Editor {

  // A collection of string-color pairs
  private Dictionary<string, Color> colorPresets;

  // Represents the "color" property on all selected objects
  private SerializedProperty colorProperty;

    // Called when the editor first appears
    public void OnEnable() {

        // Set up the list of color presets
        colorPresets = new Dictionary<string, Color>();

        colorPresets["Red"] = Color.red;
        colorPresets["Green"] = Color.green;
        colorPresets["Blue"] = Color.blue;
        colorPresets["Yellow"] = Color.yellow;
        colorPresets["White"] = Color.white;

        // Get the property from the object(s)
        // that are currently selected
        colorProperty
          = serializedObject.FindProperty("color");
    }

    // Called to draw the GUI in the Inspector
    public override void OnInspectorGUI ()
    {
        // Ensure that the serializedObject is up to date
        serializedObject.Update();

        // Start a vertical list of controls
        using (var area
          = new EditorGUILayout.VerticalScope()) {

          // For each color in the preset list..
          foreach (var preset in colorPresets) {

            // Show a button
            var clicked = GUILayout.Button(preset.Key);

            // If it was clicked, update the property
            if (clicked) {
              colorProperty.colorValue = preset.Value;
            }
          }

          // Finally, show a field that allows for
```

```
            // setting the color directly
            EditorGUILayout.PropertyField(colorProperty);
        }

        // Apply any property that was changed
        serializedObject.ApplyModifiedProperties();
    }
}
```

Once again, we'll go through this large chunk of code in detail.

Setting Up the Class

The first step is to define the class and its role in the Unity system. The `RuntimeColorChangerEditor` class is made to be a subclass of the `Editor` class.

Additionally, we give it the `CustomEditor` attribute, indicating that this class should be used as the editor for any `RuntimeColorChanger` component. Finally, the class is given the `CanEditMultipleObjects` attribute (which, as the name suggests, indicates that multiple objects can be edited at once):

```
// This is editor for RuntimeColorChangers
[CustomEditor(typeof(RuntimeColorChanger))]
// It can handle editing multiple things at once
[CanEditMultipleObjects]
class RuntimeColorChangerEditor : Editor {
```

Defining the Colors and Properties

The class needs to store two main piece of information. First, we need a list of predefined colors that the user can choose from. Additionally, we need an object that represents the `color` property on all of the currently selected objects.

Just like when we were creating the custom property drawer, we represent properties with the `SerializedProperty` object. Doing this means that Unity can provide extra features for us, such as undo:

```
// A collection of string-color pairs
private Dictionary<string, Color> colorPresets;

// Represents the "color" property on all selected objects
private SerializedProperty colorProperty;
```

Setting Up the Variables

When an object containing a RuntimeColorChanger component is selected, the Inspector will create an editor for it. It then calls the OnEnable method, which is your first opportunity to do some setup. In this editor, we prepare the colorPresets dictionary by filling it with predefined colors.

In addition, we need to get the color property to work with. We do this by accessing the serializedObject variable, which is set by Unity; this variable represents all objects that are currently selected.

```
public void OnEnable() {

    // Set up the list of color presets
    colorPresets = new Dictionary<string, Color>();

    colorPresets["Red"] = Color.red;
    colorPresets["Green"] = Color.green;
    colorPresets["Blue"] = Color.blue;
    colorPresets["Yellow"] = Color.yellow;
    colorPresets["White"] = Color.white;

    // Get the property from the object(s)
    // that are currently selected
    colorProperty = serializedObject.FindProperty("color");
}
```

Starting to Draw the GUI

In the OnInspectorGUI, we can implement our own custom Inspector. The first step is to ask the serializedObject to update itself to the current situation in the game scene, which ensures that the controls that we're about to draw will accurately represent the scene:

```
public override void OnInspectorGUI ()
{
    // Ensure that the serializedObject is up to date
    serializedObject.Update();
```

Drawing the Controls

At this point, we can draw the controls for this component. Using a VerticalScope, we draw a button for each of the presents in the colorPresets dictionary. If any of these buttons are clicked, the colorProperty's value is set to the corresponding preset's color value.

After the button is drawn, we display a PropertyField for the color. The PropertyField control displays a control that's appropriate for whatever type the property is—in this case, because colorProperty represents the color variable in RuntimeColorChanger, a color well will appear, allowing the user to choose their own color. In this way, we preserve the ability for the user to make fine-grained choices about the object, as well as providing additional features:

```
using (var area = new EditorGUILayout.VerticalScope()) {

  // For each color in the preset list...
  foreach (var preset in colorPresets) {

    // Show a button
    var clicked = GUILayout.Button(preset.Key);

    // If it was clicked, update the property
    if (clicked) {
      colorProperty.colorValue = preset.Value;
    }
  }

  // Finally, show a field that allows for setting the color
  // directly
  EditorGUILayout.PropertyField(colorProperty);
}
```

Applying Changes

The last thing to do is to ask the selected object (or objects, if multiple objects are selected) to apply the changes that were made. We do this by calling ApplyModifiedProperties on the serialized Object.

```
// Apply any property that was changed
serializedObject.ApplyModifiedProperties();
```

Testing It Out

You can now test out the custom Inspector.

Select the game object, and you'll see your custom Inspector (Figure 16-9). You may need to deselect and then reselect the capsule first.

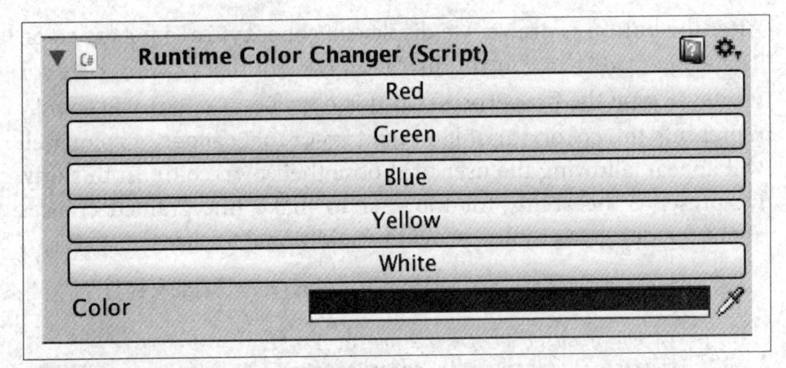

Figure 16-9. The custom inspector

Showing the Default Inspector Contents

Sometimes, you don't need to *replace* the Inspector for a component, but rather just want to add some extra stuff. In these cases, you can use the `DrawDefaultInspector` method to quickly draw everything that the Inspector would normally contain; you can then draw additional controls above or below this:

```
public override void OnInspectorGUI() {

    // Draw the default Inspector controls
    DrawDefaultInspector();

    // Show a motivational message to the
    // developer underneath
    var msg = "You're doing a great job! " +
      "Keep it up!";

    EditorGUILayout.HelpBox(msg, MessageType.Info);
}
```

Wrapping Up

Custom editors can make your life a lot easier. If you have a repetitive task, or if you need a better way to view the data contained inside your objects, then editors can be a real help. <<< It's worth keeping in mind, though, that your players will never see your custom editors. They only exist for you as a developer, so don't get too caught up in making the perfect custom editor—what counts is what those editors help you to create.

CHAPTER 17

Beyond the Editor

Your game is done, the gameplay is polished, and the whole thing looks great. What do you do now?

It's time to look outside the Unity editor itself. Unity provides a number of useful services that you can either use to improve your game, improve the way you make your game, or even develop a revenue stream for your game. In this chapter, we'll look at all three.

We'll also discuss building your game for devices, and making it available to the wider world.

The Unity Services Ecosystem

When people discuss Unity, they're typically referring to the Unity editor—the software that Unity Technologies develops and sells. However, Unity is more than just the editor. In addition to this software, Unity provides a number of services that are designed to improve the quality of life for developers. Three of the most useful ones are the Asset Store, the Unity Cloud Build service, and the Unity Ads platform.

The Asset Store

The Unity Asset Store is an online storefront where programmers, artists, and other content creators for games can sell content designed to be integrated into a game.

The Asset Store is particularly good for people who lack a certain skill; for example, programmers without sufficient skill in (or time for) producing art assets can purchase the 3D models they need, so that they can focus on the programming that they're better at. The same applies for people who need audio, a script in their game, and so on. Content in the Asset Store ranges from small to large; through the store, you can purchase a single 3D model of a car, all the way through to complete asset kits for a certain type of game.

Assets you buy from the Asset Store—especially the *good* assets—are often quite easily noticed by other people as having come from the store. Be aware that relying too heavily on Asset Store assets can make your game look and feel quite samey.

Some of the assets available through the store are of particular interest, because they add features to Unity that it otherwise lacks.

PlayMaker

PlayMaker is a visual scripting tool created by Hutong Games. In a visual scripting system, you define the behavior of your game objects by connecting together predefined modules of code, which are represented as boxes with wires coming out of them.

Visual scripting systems are an alternative to writing code, and are often easier to grasp for newcomers to programming. They're particularly good at representing behaviors that rely heavily on state—for example, an enemy AI that roams around randomly until it sees the player, at which point it jumps into a *seeking* state and charges at the player until either it dies, the player dies, or it loses sight of the player.

PlayMaker is available via the Asset Store (*https://www.assetstore.unity3d.com/#!/content/368*).

Installing PlayMaker. Because PlayMaker provides a completely different approach to defining game behavior, it's worth taking a closer look at it, and set up some simple behavior. To follow these steps, you'll need to purchase PlayMaker from the Asset Store; at the time of writing in mid-2017, it is $65.

We'll be following these steps in a new, empty project configured for 3D graphics.

1. *Download and import the package.* The installation window will appear (Figure 17-1).

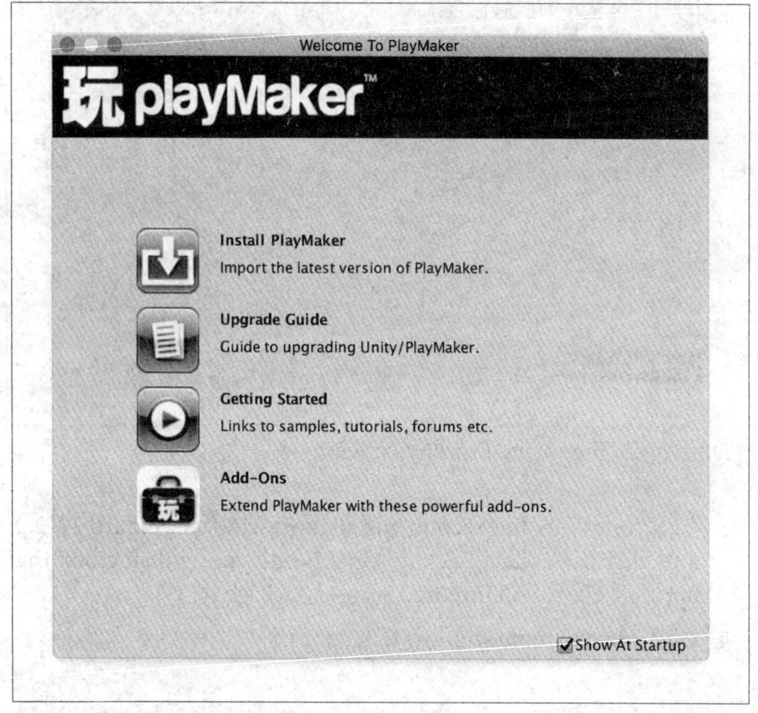

Figure 17-1. The PlayMaker installation window, which appears after you import the package

2. *Click Install.* PlayMaker will check your project to be sure that installation will work, and ensure that you have the most up-to-date version.

PlayMaker will complain if you're not using a version control tool such as Git; you can ignore it, but using version control is a great idea overall.

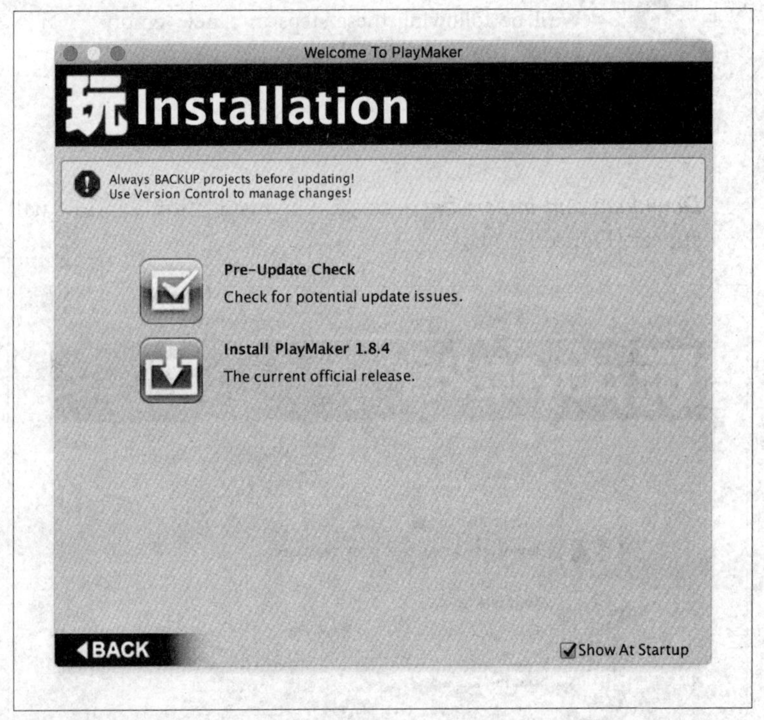

Figure 17-2. The second installation window

3. *Click Install* in the second installation window (Figure 17-2), and click "I Made a Backup, Go Ahead!" in the dialog box that appears. Unity will import a second package.
4. *Click Import* in the window that appears.

Depending on your version of Unity, you may be asked if it's OK if Unity upgrades the code to be compatible with the most recent APIs. You'll need to agree to it in order to proceed.

After the installation process is complete, close any lingering windows. You're now ready to start.

Playing with PlayMaker. PlayMaker's whole idea is based around the concept of *finite state machines*, or *FSM* for short. A finite state machine is a logical system in which an object can be in one of multiple *states*; each state is allowed to change, or *transition*, to a predetermined subset of those states. That is, if you had the states *sitting*, *standing*, and *running*, you could transition from *standing* to either *sitting* or *running*, but you couldn't transition directly from *sitting* to *running*. When the state changes, you have an opportunity to run some behavior.

The behavior we'll add in this brief tutorial will be extremely simple: we'll create a ball that, when it lands on a surface, changes color.

To begin, we'll set up the environment:

1. *Create the sphere* by opening the GameObject menu, and choosing 3D Object → Sphere.

   Select the newly created object, and, using the Transform component in the Inspector, set the position to (0, 15, 0).
2. *Add a Rigidbody component to the sphere.*
3. *Create the ground* by opening the GameObject menu again, and choosing 3D Object → Plane.

   Position this object at (0, 0, 0).
4. Finally, *reposition the Camera* to (0, 9, -16), with a rotation of zero. This will make it view both the ball and the ground.

Your scene should now look like Figure 17-3.

Figure 17-3. The laid-out scene for the tutorial

We'll now start adding the PlayMaker behaviors to the sphere.

1. *Open the PlayMaker Editor* by opening the PlayMaker menu and choosing PlayMaker Editor.

   The PlayMaker Editor tab will appear (Figure 17-4).

You might find it more convenient to attach the tab to the Unity window. To do this, drag and drop the tab at the top of the window to where you'd like it to be.

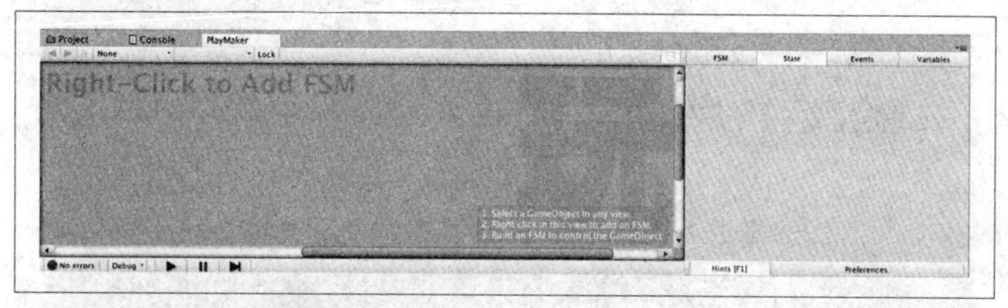

Figure 17-4. The PlayMaker editor

2. *Add an FSM to the sphere.* Select the Sphere, and in the PlayMaker window, right-click and choose Add FSM.

The PlayMaker window displays a number of tips, which can be quite useful, but take up a lot of space. You can disable the hints by pressing the F1 key, or by clicking the Hints button at the lower-right of the PlayMaker window.

By default, the FSM will contain a single state, titled `State1`. In our demo, we'll have two states: `Falling` and `HitGround`. We'll rename the first state, and then add another.

3. *Rename the first state to "Falling"*, by selecting it, going to the State tab at the right of the PlayMaker window, and changing its name to "Falling".
4. *Add the `HitGround` state* by right-clicking in the PlayMaker window, and choose Add State. Rename the new state to "HitGround".

   Your FSM should now look like Figure 17-5.

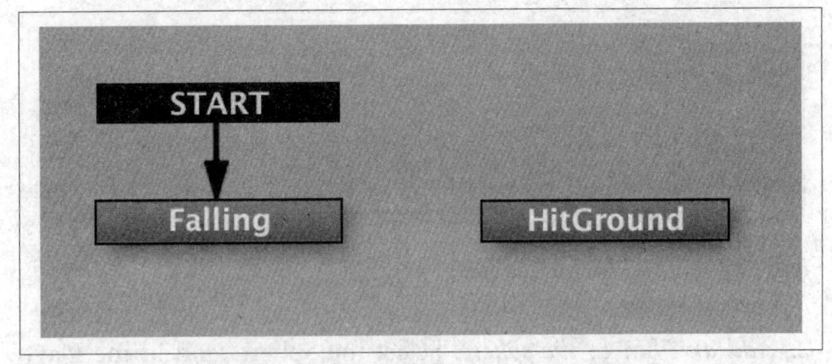

Figure 17-5. The FSM, with states added

We want the state to change when the ball hits the ground. To do this, we'll create a *transition* from the `Falling` state to the `HitGround` state, which fires when the object that the FSM is attached to collides with something.

5. *Add the transition* by right-clicking on the `Falling` state, and choosing Add Transition → System Events → COLLISION ENTER. A new transition will appear, with a warning indicating that the transition isn't connected to a destination state (see Figure 17-6).

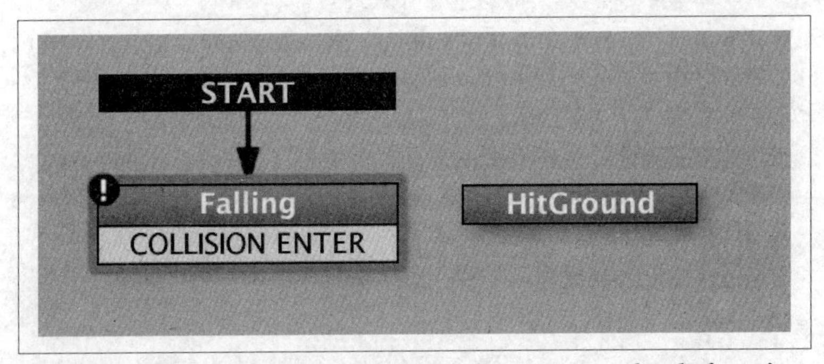

Figure 17-6. The FSM, after you've added a transition, but before it's connected

6. *Connect the transition to the `HitGround` state* by left-clicking on the COLLISION ENTER transition, and dragging to the `Hit Ground` state. An arrow will appear that connects them (Figure 17-7).

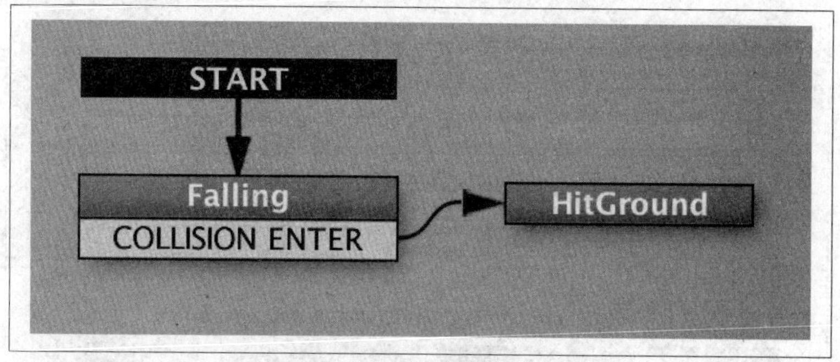

Figure 17-7. The FSM, with the states connected

7. *Test the game* by pressing the Play button. The FSM window will highlight the current state; the Falling action will be highlighted until the moment that the sphere touches the ground.

Next, we need to add an action to run when the object enters the `HitGround` state. Specifically, we want the material to change color.

1. *Add the Set Material Color action to the `HitGround` state* by selecting the state, going to the State tab, and clicking Action Browser. The Action Browser window will appear; scroll down to the Material button, click it, and then select the Set Material Color entry (Figure 17-8). Click Add Action to State; the action will appear in the State tab (Figure 17-9).

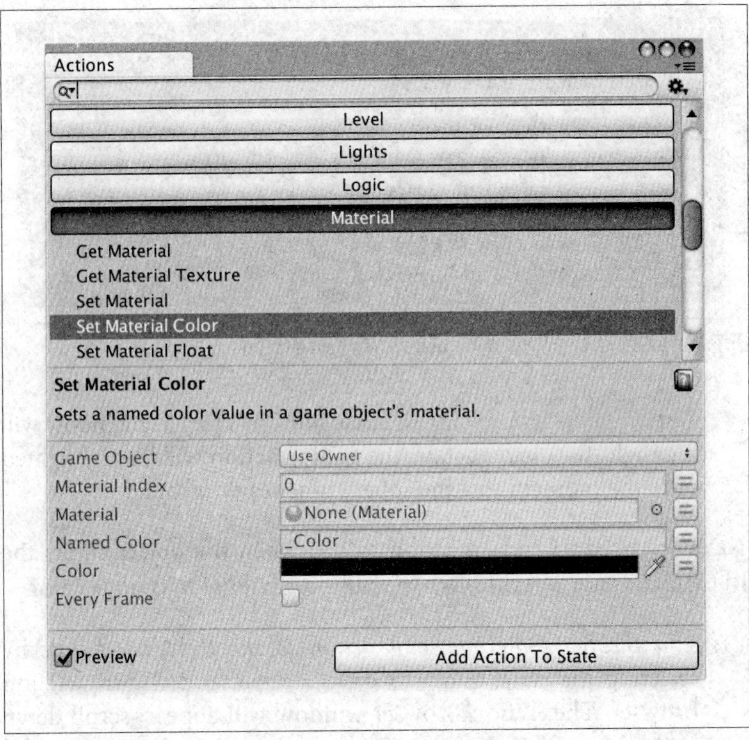

Figure 17-8. The Action Browser

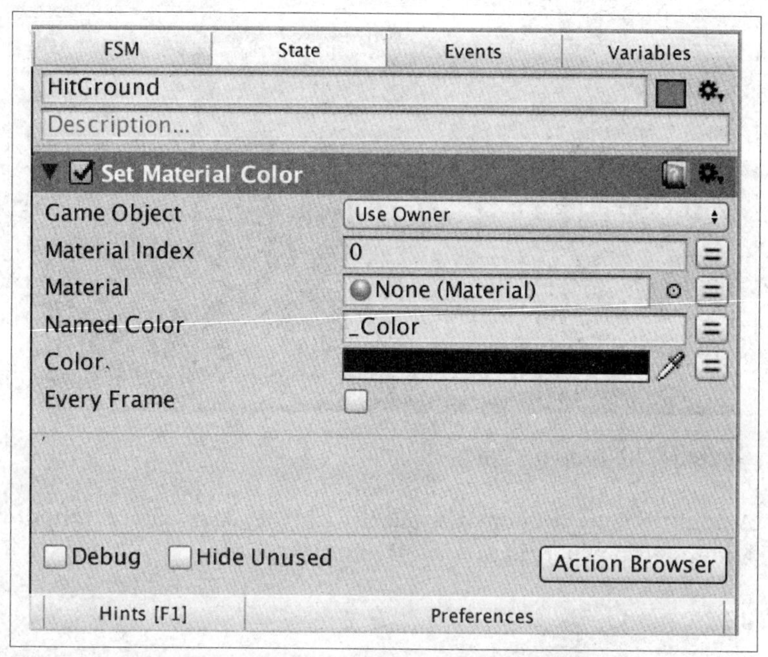

Figure 17-9. The FSM, with the action added and configured

2. *Make the color change to green.* In the State section, change the color to green.
3. *Test the game.* When the ball touches the ground, it will change color to green.

Amplify Shader Editor

Shaders, as you'll recall from Chapter 14, generally involve writing code. However, the visual nature of shaders lends them to visual construction even more than gameplay code; instead of writing code that multiplies two vectors that represent colors together, it would be more intuitive to *see* that happening.

Amplify Shader Editor (Figure 17-10) is one of a couple of such visual shader editors for Unity. By connecting nodes together, Amplify creates and demonstrates your material, and generates the assets for use in your game. This is often faster and easier than writing the shader code yourself, and is particularly useful for people who may find visual creation of visual results to be more intuitive.

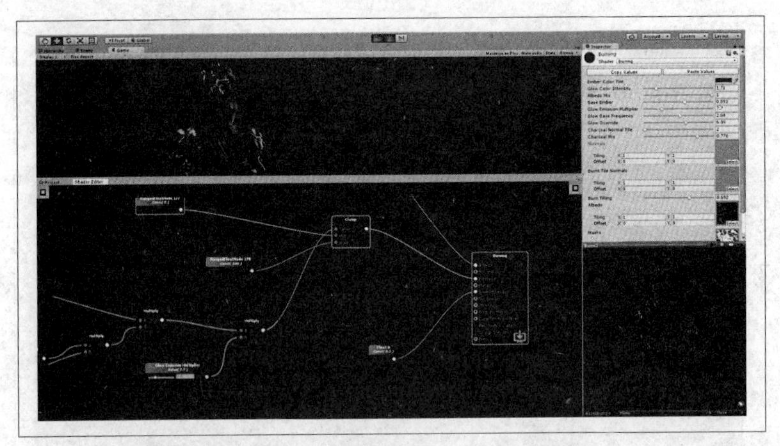

Figure 17-10. Amplify Shader Editor

Amplify Shader Editor is available via the Asset Store (*https://www.assetstore.unity3d.com/en/#!/content/68570*).

UFPS

UFPS, or Ultimate FPS (Figure 17-11), is a simple base for first-person shooter games. While Unity does ship with a first-person controller, it doesn't include other features that are common to first-person games, such as ducking, climbing ladders, or interacting with buttons. UFPS provides implementations of these, as well as features that are specific to shooting-oriented gameplay, such as managing an inventory of weapons, ammunition management, and managing player health.

While UFPS is geared toward building action-oriented shooters, it's equally good at slower-paced games; Fullbright's *Gone Home* (2014), a game whose gameplay revolves around walking around a house and examining the objects, documents, and furniture that are left behind, uses UFPS to handle the first-person presentation.

Figure 17-11. Ultimate FPS

UFPS is available via the Asset Store (*https://www.assetstore.unity3d.com/en/#!/content/2943*).

Unity Cloud Build

Building a project for your destination platform is a complex process that requires a large amount of processing power and time. While you *can* do builds on your own computer, it doesn't always follow that you *should*—especially if you have a large, complex project.

Unity Cloud Build is a service that downloads your source code, builds it, and then makes the build available to you to download (or notifies you if the build failed). When you configure Cloud Build to watch your source code repository, it will be watched for changes; as a result, Unity will automatically build your game as it changes.

It's certainly possible to create your own build server, and not use Cloud Build. However, the process is fiddly, and also consumes one of the two activations that you get with your license. Cloud Build takes some control out of your hands, but replaces it with ease of use.

At the time of writing, Cloud Build is a free service. If you own a Unity Plus subscription, your builds are prioritized, so that they get completed sooner. If you own a Unity Pro subscription, your builds are also able to run in parallel, so that if your game is designed to run on multiple platforms (e.g., both iOS and Android), then both builds will start at the same time.

You can find more information about Cloud Build on Unity's website (*https://unity3d.com/services/cloud-build*).

Unity Ads

Unity Ads is a service that delivers fullscreen video ads for display in your game. When the player views an ad, you get paid a small fee. In this way, you can develop an additional source of revenue for your game.

One particular use case for video ads is *rewarded advertising*, in which your player receives an in-game reward of some kind (such as bonus currency, cosmetic changes, or other content) in exchange for watching ads.

Designing a monetization strategy for games is a huge topic, and can (and does!) take up a whole library of books. To get started with Unity Ads, take a look at the service's page on Unity's website (*https://unity3d.com/services/ads*).

Deployment

When you're ready to take your game out of the editor and put it onto an actual device, Unity needs to *build* your game. This involves three things: bundling up all of your game's assets, compiling its scripts, and installing the built app onto the device. The first two parts are done for you by Unity; the second is something that you'll need to help with yourself.

In this section, we'll talk about how to build your game, for both iOS and Android devices. Before we do that, we need to talk a little bit about some of the setup you'll need to do, and differences between the different versions of Unity.

Setting Up Your Project

You can build your project at any time. However, for best results, it's good to ensure that the player settings for your game are correct. Player settings include your game's name and icon, as well as settings that control which screen orientation your game runs in, what the unique ID string is that identifies your game to the operating system you're installing on, and so on.

To configure them, you'll need access to Player Settings. To do this, open the Edit menu, and choose Project Settings → Player. The Inspector will look like Figure 17-12.

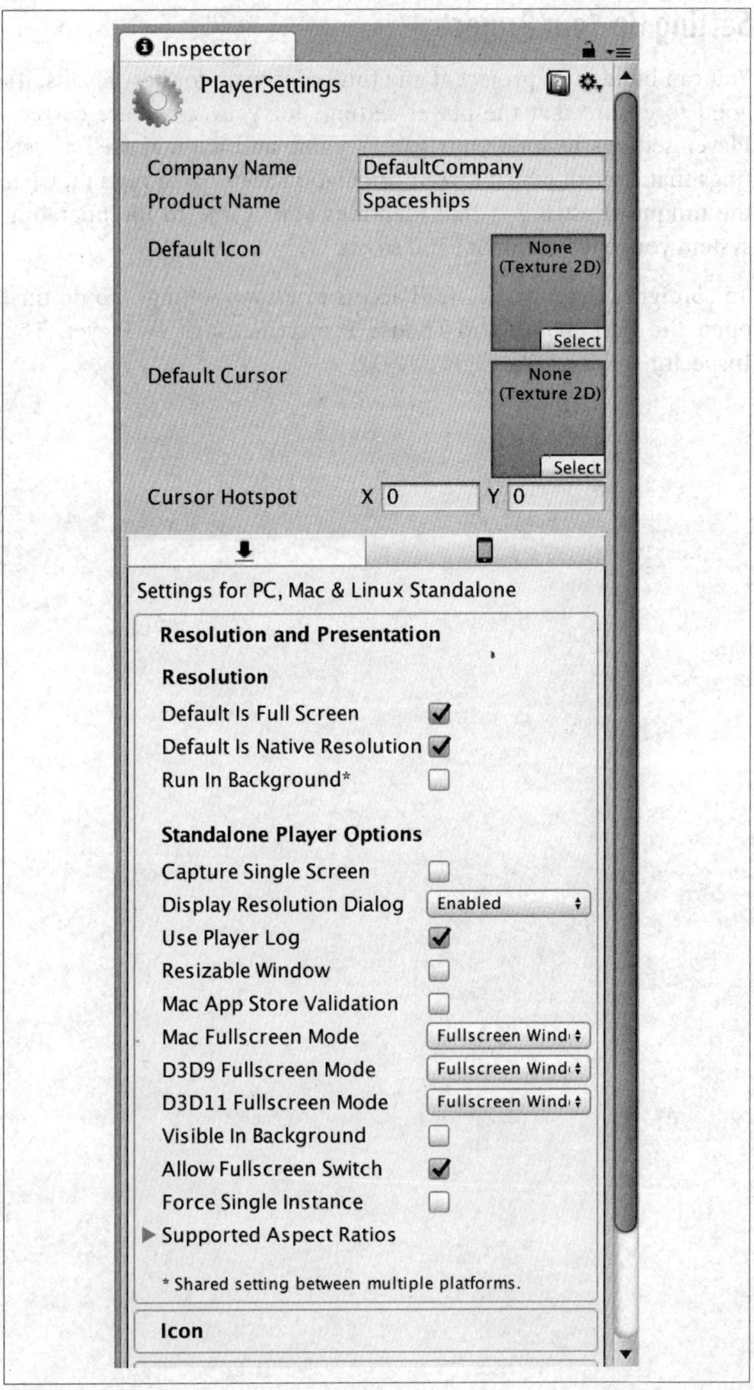

Figure 17-12. The PlayerSettings Inspector

Certain settings are the same across multiple platforms. For example, the name of your game is unlikely to change between platforms, as is the icon. Unity identifies settings that are shared across multiple platforms with an asterisk (*) next to their name.

Every application, on both iOS and Android, needs several things:

- The *product name* of the game, for display on the home screen and the marketplace;
- The *company name* of the game, used on the marketplace;
- The game's *icon*, again for the home screen and marketplace;
- The game's *splash screen*, which is shown while the game is starting up;
- The game's *bundle identifier*, which is a piece of text that uniquely identifies the game on the marketplace and is not shown to the user; this is done by taking a domain name that you own (e.g., *oreilly.com*), reversing it, and adding the name of the game (e.g., *com.oreilly.MyAwesomeGame*).

To test your game, you need to have the name and identifier configured. To release your game to either the iTunes App Store or the Google Play store, you need to have all of these elements.

By default, the product name is set to the name of your project, and the company name is set to "DefaultCompany". If you're happy with that, you can leave the product name (seen at the top of Figure 17-12) unchanged.

To change the bundle identifier, select the platform you want to build for from the menu (underneath Cursor Hotspot), and open the Other Settings section. From there, set the Bundle Identifier to the identifier you'd like to use (see Figure 17-13).

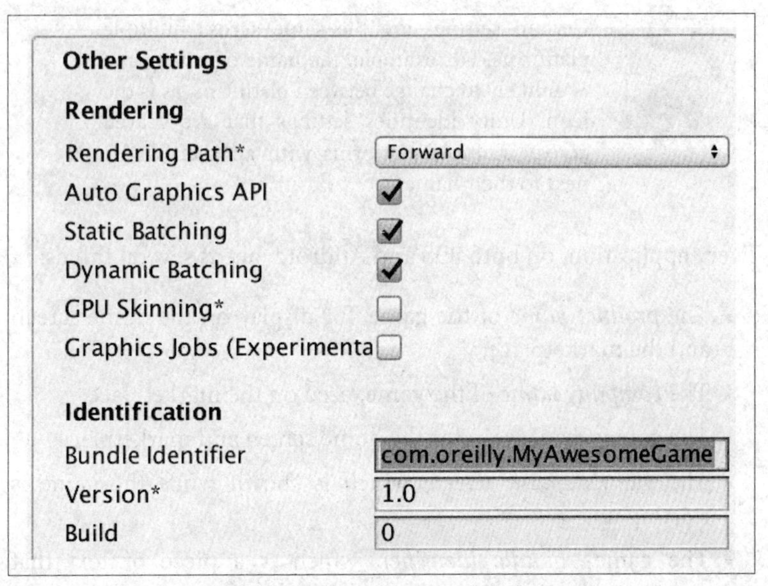

Figure 17-13. Setting the bundle identifier of the project

If you're building a game for both iOS and Android, you don't need to set the bundle identifier twice. The setting is shared between all platforms, as well as the game's version number and several other features.

Setting Your Target

You can only select one target at a time. By default, Unity sets your target to *PC, Mac, and Linux Standalone*, and will default the specific target to the same as the one you're running Unity on—so, for example, if you're using a Mac, the target will default to macOS, and to Windows if you're using a PC.

Downloading Platform Modules

In order to build for a certain platform, Unity first needs to have the right module installed for it. When you first install Unity, the installer asks you which platforms you want to install modules for; if you don't have the module for the platform you want to deploy to, the Build Settings window will look like Figure 17-14. You'll need to click on the button in the window to download and install the appropriate module.

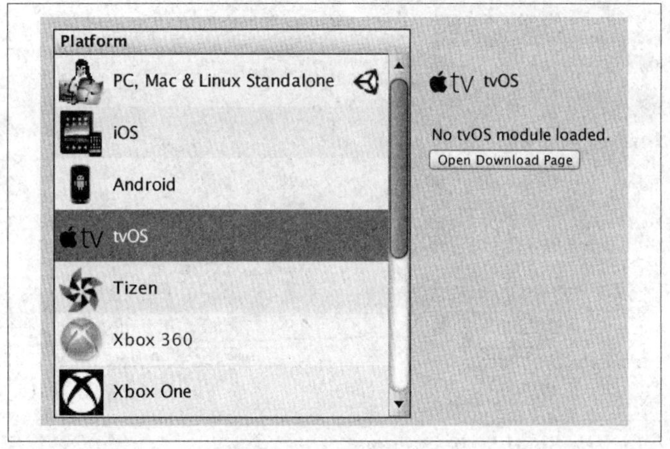

Figure 17-14. The Build Settings window, with a platform selected whose module is not loaded

Given that in Parts II and III we designed and built mobile games, the first thing to do will be to switch to your desired platform. To do this, open the Build Settings window by opening the File menu and choosing Build Settings (see Figure 17-15).

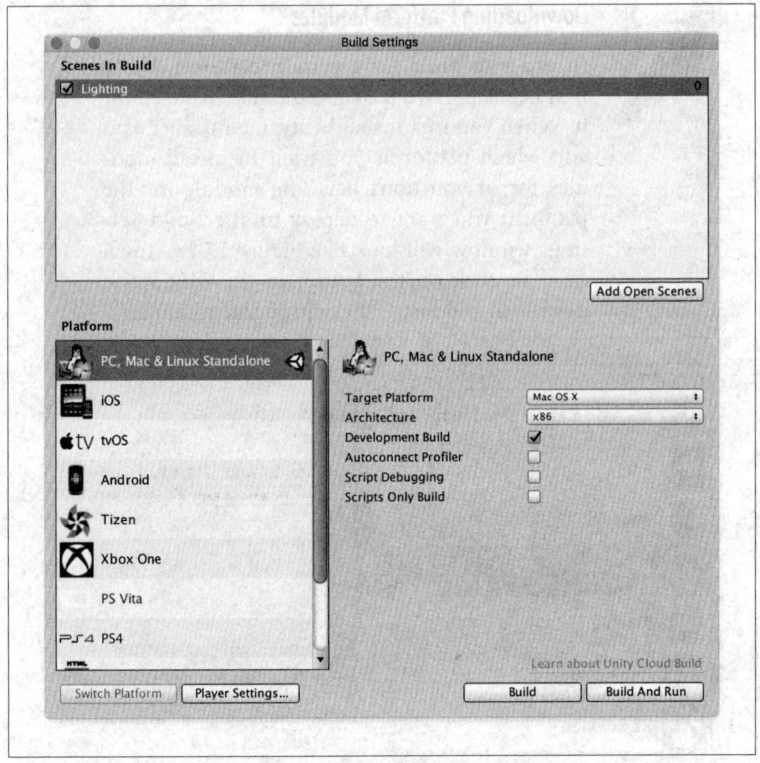

Figure 17-15. The Build Settings menu

From there, it's a simple matter of selecting your target platform, and clicking Switch Platform at the bottom left of the window.

When you switch platforms, Unity will reimport all of your game's assets. If you have a large project, this can take a long time. Be prepared to wait. To assist with this, Unity provides a tool called the Cache Server, which saves a copy of imported assets; for more information, check the documentation (*http://bit.ly/cache-server*).

Splash screens

It's worth pointing out that there are differences between the Free subscription plan and the Plus and Pro plans, as far as built apps go. Users of the Free plan are required to have a splash screen appear at the start of their game, while Plus and Pro subscribers can choose to disable it.

The splash screen is fairly restrained; the Unity logo is displayed, along with the words "Made with Unity," and it lasts for two seconds while your game's initial scene is loaded in the background.

The splash screen can be customized fairly significantly, regardless of which version of Unity you use. In addition to showing the Unity logo, you can also include your own logos, customize the background color, set a background image and opacity, and set the splash screen to either display multiple logos on screen at once or show them sequentially. To customize the splash screen, open the Edit menu, choose Project Settings → Player, and scroll down to Splash Image → Splash Screen. Unity provides lots of documentation on this topic, so for more info, go and take a look (*https://docs.unity3d.com/Manual/class-PlayerSettingsSplashScreen.html*).

Building for Your Platform

The steps for building for iOS and Android are different. In this section, we'll walk you through the steps for each.

Building for iOS

Unity makes it easy to build a version of your game for iOS. In this section, we'll step through the process you need to follow to get your game running on your phone.

Building for iOS is currently only possible on a macOS computer, or via Unity Cloud (who do their builds on Macs).

It's totally free to deploy your games directly to your own personal device. To distribute your games to others, you need to do it via the iTunes App Store. This means enrolling in the Apple Developer Program, which costs $99 per year; you can do so at *https://developer.apple.com/programs/*.

To get started, you'll first need to download Xcode, the iOS development environment:

1. *Download Xcode* from the Mac App Store by launching it, searching for Xcode, and downloading it.
2. Once it's downloaded, *launch Xcode.*

We now need to configure Xcode to use your account. Regardless of whether you're enrolled in the paid Apple Developer Program or not, Xcode needs to use your Apple ID to register you as a developer in order to do the necessary code signing before your game can be installed on your device.

1. *Open the Xcode menu, and choose Preferences.* At the top of the window, click Accounts. In the lower-left corner, click the Add button (+). Choose Add Apple ID from the pop-up menu.
2. *Connect your device to your computer's USB port.*

With Xcode now configured, you're ready to build your Unity game.

1. *Return to Unity, and open the Build Settings window* by opening the File menu, and choosing Build Settings.
2. *Select the iOS platform, and click Switch Platform.* Unity will switch the project over to iOS (Figure 17-16). It might take a few minutes.

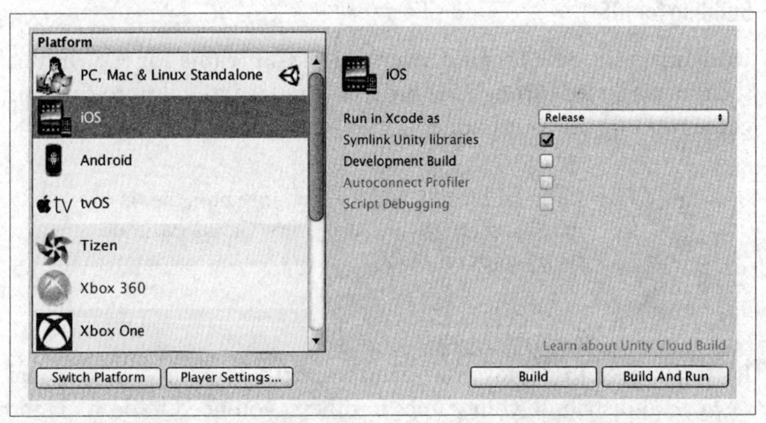

Figure 17-16. The Build Settings window, using the iOS platform

To save space, turn on the Symlink Unity Libraries button. Doing this will make your project not copy the entire Unity library into your project, which can be several hundred megabytes in size.

3. *Click Build and Run.* Unity will ask you where you want to save the project; after you select a folder to put it in, Unity will build the app for iOS, then open the project in Xcode, and instruct Xcode to build and run the app on the connected device.

Code Signing Issues

If you get an error about code signing, select the project at the top-left of the window, select the Unity-iPhone target, select your development team (which may just be your name) from the Team drop-down menu, and click Fix Issue (Figure 17-17). This will sort out your certificates and get things going again; press Command-R to try building again.

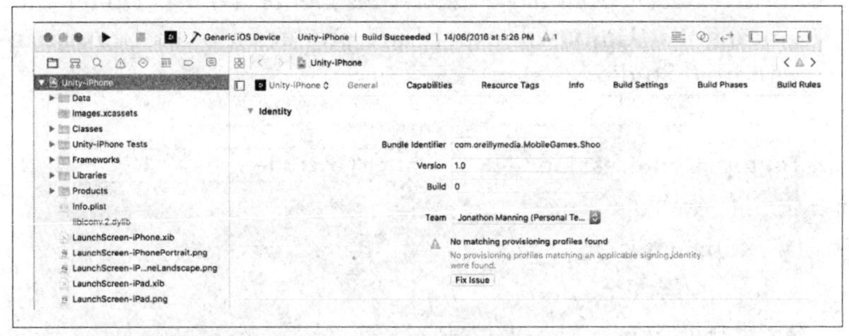

Figure 17-17. The Fix Issue button in Xcode

Building for Android

To build for Android, you'll want to first install the Android SDK. This handles the delivery of the built application to your device:

1. *Download the Android SDK* from the Android Developer site: *http://developer.android.com/sdk*.

2. *Install the SDK* by following the instructions at: *http://developer.android.com/sdk/installing/index.html*.

If you're on Windows, you may need to download an additional USB driver before your computer can communicate with your Android device. You can download the driver at: *http://developer.android.com/sdk/win-usb.html*. You don't need this if you're on macOS or Linux.

You can now tell Unity where the installed Android SDK is:

1. *Open the Unity menu, and choose Preferences → External Tools.* In this window, click on the Browse button next to the SDK field, and browse to the folder where you installed Android Studio.
2. *Open the Build Settings window* by opening the File menu and choosing Build Settings.
3. *Select the Android platform, and click Switch Platform.* Unity will switch the project over to Android.
4. *Select Google Android Project* (see Figure 17-18). Doing this means that Unity will export and produce a project to use in Android Studio.

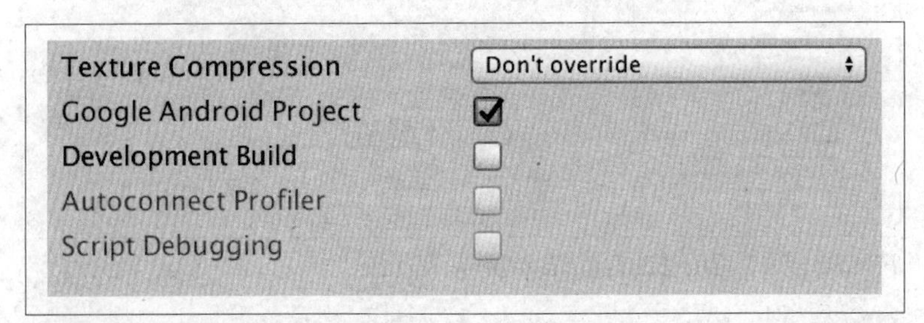

Figure 17-18. Making the Android build generate a Google Android project

5. *Click Export.* Unity will ask you where you want to save the project. After you do, the project will be generated.
6. *Open the project in Android Studio, and click the Play button.* The project will compile, and install on your phone.

While Unity tends to not change significantly between releases, the processes for setting up and building onto mobile devices can change more frequently. To this end, rather than repeat Unity's documentation in a form that might go out of date as quickly as it's printed, we'll instead direct you to Unity's useful step-by-step tutorials on how to get set up for both Android and iOS:

- "Getting started with Android development" (*http://bit.ly/android-gettingstarted*)
- "Getting started with iOS development" (*http://bit.ly/iphone-gettingstarted*)

Both iOS and Android development involve downloading and installing fairly significant amounts of software. We recommend that you set up in a place where you will have a decent internet connection.

Where to Go from Here

Welcome to this: the last section of the book. If you've read this far, you've completed a fairly huge journey, having started from scratch, built two complete games, and then taken the reins of Unity to customize it to your needs.

If you skipped to the end of the book: this is how it ends. Congratulations on spoiling yourself.

Before we part ways, here are some useful resources for future reading:

- Unity's documentation (*http://docs.unity3d.com*) is quite good, and serves as a reference manual for the entire editor. The documentation is split into two sections: the Manual (*http://docs.unity3d.com/Manual/index.html*), which describes the editor, and the Scripting Reference (*http://docs.unity3d.com/ScriptReference/index.html*), which describes every class, method, and function of Unity's scripting API. It's extremely handy to have as a reference.

- Unity's official forum (*http://forum.unity3d.com*) serves as a hub for community discussion, and is a good place to go to get help.
- Unity Answers (*http://answers.unity3d.com*) is an officially supported question and answer forum. If you have a specific question, it's a great place to check first.
- Unity frequently performs live training sessions (*http://unity3d.com/learn/live-training*), in which one of their trainers demonstrates a feature or complete project in a live class session. Even if you can't make it to the live session, they're generally recorded for later viewing.
- Finally, Unity hosts a number of tutorials (*http://unity3d.com/learn/tutorials*), which range from introductory beginner content through to more advanced, highly specific instruction.

We hope that you've enjoyed reading this book. If you make something, no matter how small and no matter if you think it's bad, we'd love to hear about it. *Send us an email* any time.

Index

D

E

F

G

H

S

T

U

V

W

X

Y

About the Authors

Dr. Jon Manning and **Dr. Paris Buttfield-Addison** are cofounders of Secret Lab, where they build games and game development tools. Recently, they've built the ABC Play School iPad games, helped on indie game Night in the Woods, and built the Qantas Joey Playbox.

At Secret Lab, they build the YarnSpinner narrative game framework, and write books for O'Reilly Media.

Jon and Paris formerly worked as mobile developers and product managers for Meebo (acquired by Google), and both have a PhD in Computing.

Jon can be found on Twitter at @desplesda (*https://twitter.com/desplesda*) and online at *http://www.desplesda.net*, and Paris can be found on Twitter at @parisba (*https://twitter.com/parisba*) and online at *http://paris.id.au*.

Secret Lab can be found on Twitter at @thesecretlab (*https://twitter.com/thesecretlab*) and online at *http://www.secretlab.com.au*.

Colophon

The animals on the cover of *Mobile Game Development with Unity* are the thorny devil stick insect (*Eurycantha calcarata*) and longhorn beetle (family *Cerambycidae*).

The thorny devil stick insect is a herbivorous, wingless insect native to Australasia. Males grow to 4–5″ in length, while the larger females tend to be around 6″. While most stick insects tend to live in trees, *Eurycantha calcarata* live on the ground (typically in rainforests), where they forage for food at night, using a combination of camouflage and catalepsy to evade predators. They huddle in groups under shed bark and in tree hollows during the day. These insects are popular pets, and the long thorns on the hind legs of males (for which the species is named) are used as fish hooks in Papua New Guinea.

The longhorn family of beetles possess uniquely long and powerful antennae that often extend to, if not exceeding, the length of the insect's body. Over 26,000 species make up this family, ranging from the titan beetle (the world's lagest insect at 12.6″, excluding legspan) to the tiny genus *Decarthia*, whose three species are only a few millimeters long. Family *Cerambycidae* gets its name from the Greek myth of Cerambus, a shepherd who is transformed into a beetle by a group of nymphs.

Many of the animals on O'Reilly covers are endangered; all of them are important to the world. To learn more about how you can help, go to *animals.oreilly.com*.

The cover image illustrations are by Karen Montgomery, based on engravings from J.G. Wood's *Insects Abroad*. The cover fonts are URW Typewriter and Guardian Sans. The text font is Adobe Minion Pro; the heading font is Adobe Myriad Condensed; and the code font is Dalton Maag's Ubuntu Mono.

O'Reilly Media, Inc.介绍

O'Reilly Media通过图书、杂志、在线服务、调查研究和会议等方式传播创新知识。自1978年开始，O'Reilly一直都是前沿发展的见证者和推动者。超级极客们正在开创着未来，而我们关注真正重要的技术趋势——通过放大那些"细微的信号"来刺激社会对新科技的应用。作为技术社区中活跃的参与者，O'Reilly的发展充满了对创新的倡导、创造和发扬光大。

O'Reilly为软件开发人员带来革命性的"动物书"；创建第一个商业网站（GNN）；组织了影响深远的开放源代码峰会，以至于开源软件运动以此命名；创立了Make杂志，从而成为DIY革命的主要先锋；一如既往地通过多种形式缔结信息与人的纽带。O'Reilly的会议和峰会集聚了众多超级极客和高瞻远瞩的商业领袖，共同描绘出开创新产业的革命性思想。作为技术人士获取信息的选择，O'Reilly现在还将先锋专家的知识传递给普通的计算机用户。无论是书籍出版、在线服务还是面授课程，每一项O'Reilly的产品都反映了公司不可动摇的理念——信息是激发创新的力量。

业界评论

"O'Reilly Radar博客有口皆碑。"

——Wired

"O'Reilly凭借一系列（真希望当初我也想到了）非凡想法建立了数百万美元的业务。"

——Business 2.0

"O'Reilly Conference是聚集关键思想领袖的绝对典范。"

——CRN

"一本O'Reilly的书就代表一个有用、有前途、需要学习的主题。"

——Irish Times

"Tim是位特立独行的商人，他不光放眼于最长远、最广阔的视野，并且切实地按照Yogi Berra的建议去做了：'如果你在路上遇到岔路口，走小路（岔路）。'回顾过去，Tim似乎每一次都选择了小路，而且有几次都是一闪即逝的机会，尽管大路也不错。"

——Linux Journal

出版说明

随着计算机技术的成熟和广泛应用,人类正在步入一个技术迅猛发展的新时期。计算机技术的发展给人们的工业生产、商业活动和日常生活都带来了巨大的影响。然而,计算机领域的技术更新速度之快也是众所周知的,为了帮助国内技术人员在第一时间了解国外最新的技术,东南大学出版社和美国O'Reilly Meida,Inc. 达成协议,将陆续引进该公司的代表前沿技术或者在某专项领域享有盛名的著作,以影印版或者简体中文版的形式呈献给读者。其中,影印版书籍力求与国外图书"同步"出版,并且"原汁原味"展现给读者。

我们真诚地希望,所引进的书籍能对国内相关行业的技术人员、科研机构的研究人员和高校师生的学生和工作有所帮助,对国内计算机技术的发展有所促进。也衷心期望读者提出宝贵的意见和建议。

最新出版的影印版图书,包括:

- 《学习 OpenCV 3(影印版)》
- 《数据科学:R 语言实现(影印版)》
- 《数据驱动设计(影印版)》
- 《设计数据密集型应用(影印版)》
- 《Scikit - Learn 与 TensorFlow 机器学习实用指南(影印版)》
- 《Perl 语言入门 第 7 版(影印版)》
- 《Python 漫游指南(影印版)》
- 《算法技术手册 第 2 版(影印版)》
- 《RxJava 反应式编程(影印版)》
- 《设计人见人爱的产品(影印版)》
- 《你好,创业公司(影印版)》
- 《可持续性设计(影印版)》
- 《基础设施即代码(影印版)》
- 《Cassandra 权威指南 第 2 版(影印版)》
- 《网站运维工程(影印版)》
- 《商业数据科学(影印版)》
- 《深度学习(影印版)》
- 《深度学习基础(影印版)》
- 《面向数据科学家的实用统计学(影印版)》
- 《高性能 Spark(影印版)》
- 《Python 数据分析 第 2 版(影印版)》
- 《Unity 移动游戏开发(影印版)》